ADVANCES IN CATALYSIS

VOLUME 27

Advisory Board

G. K. Boreskov
Novosibirsk, U.S.S.R.

M. Boudart
Stanford, California

M. Calvin
Berkeley, California

P. H. Emmett
Portland Oregon

J. Horiuti
Sapparo, Japan

G.-M. Schwab
Munich, Germany

G. A. Somorjai
Berkeley, California

R. Ugo
Milan, Italy

ADVANCES IN CATALYSIS

VOLUME 27

Edited by

D. D. ELEY
*The University
Nottingham, England*

HERMAN PINES
*Northwestern University
Evanston, Illinois*

PAUL B. WEISZ
*Mobil Research and
Development Corporation
Princeton, New Jersey*

1978

ACADEMIC PRESS

NEW YORK   SAN FRANCISCO   LONDON

A Subsidiary of Harcourt Brace Jovanovich, Publishers

COPYRIGHT © 1978, BY ACADEMIC PRESS, INC.
ALL RIGHTS RESERVED.
NO PART OF THIS PUBLICATION MAY BE REPRODUCED OR
TRANSMITTED IN ANY FORM OR BY ANY MEANS, ELECTRONIC
OR MECHANICAL, INCLUDING PHOTOCOPY, RECORDING, OR ANY
INFORMATION STORAGE AND RETRIEVAL SYSTEM, WITHOUT
PERMISSION IN WRITING FROM THE PUBLISHER.

ACADEMIC PRESS, INC.
111 Fifth Avenue, New York, New York 10003

United Kingdom Edition published by
ACADEMIC PRESS, INC. (LONDON) LTD.
24/28 Oval Road, London NW1 7DX

LIBRARY OF CONGRESS CATALOG CARD NUMBER: 49–7755

ISBN 0–12–007827–9

PRINTED IN THE UNITED STATES OF AMERICA

Contents

CONTRIBUTORS ix
PREFACE xi
CORRIGENDUM xiii

Electronics of Supported Catalysts

GEORG-MARIA SCHWAB

I.	Introduction	1
II.	Normal Supported Catalysts	6
III.	Inverse Supported Catalysts	12
IV.	Two Semiconductors	18
V.	Physical Observations	19
VI.	Conclusions	21
	References	22

The Effect of a Magnetic Field on the Catalyzed Nondissociative *Para*hydrogen Conversion Rate

P. W. SELWOOD

I.	Introduction	23
II.	Conversion Rate Measurements	24
III.	Experimental Results	25
IV.	Correlation of Experimental Results	48
V.	Conversion and Field Effect Theory	50
VI.	Conclusions	51
	References	56

Hysteresis and Periodic Activity Behavior in Catalytic Chemical Reaction Systems

VLADIMÍR HLAVÁČEK AND JAROSLAV VOTRUBA

I.	Introduction	59
II.	Porous Catalyst Particle Problem	60
III.	Catalytic Wire Problem	69
IV.	Continuous Stirred-Tank Reactor	74
V.	Tubular Fixed-Bed Reactor	77
VI.	Monolithic Catalyst	88

VII.	Other Catalytic Systems	93
VIII.	Concluding Remarks	94
	Notation	94
	References	95

Surface Acidity of Solid Catalysts

H. A. BENESI AND B. H. C. WINQUIST

I.	Introduction	98
II.	Critique of Methods for Determination of Surface Acidity	99
III.	Review of Acidity and Catalytic Activity of Specific Solids	120
	References	176

Selective Oxidation of Propylene

GEORGE W. KEULKS, L. DAVID KRENZKE, AND THOMAS N. NOTERMANN

I.	Introduction	183
II.	Activation of Propylene	185
III.	Activation of Oxygen	191
IV.	Active Catalyst Systems	198
V.	Nature of the Active Sites	210
VI.	Conclusions	221
	References	222

σ–π Rearrangements and Their Role in Catalysis

BARRY GOREWIT AND MINORU TSUTSUI

I.	Introduction	227
II.	The σ–π Rearrangement	228
III.	Classification of σ–π Rearrangements	230
IV.	Role of σ–π Rearrangement in Catalysis	235
V.	Conclusion	261
	References	261

Characterization of Molybdena Catalysts

F. E. MASSOTH

I.	Introduction	266
II.	Background	267
III.	Catalyst Preparation	268
IV.	Characterization Techniques	269
V.	State of Molybdena Catalysts	289

VI.	Catalyst Activities	294
VII.	Proposed Models of the Sulfided Catalyst	298
VIII.	Role of Cobalt	302
IX.	Nickel–Molybdena Catalysts	303
X.	Active Sites—Concluding Remarks	304
	References	306

Poisoning of Automotive Catalysts

M. SHELEF, K. OTTO, AND N. C. OTTO

I.	Introduction	311
II.	Catalysts	313
III.	Poisons	314
IV.	Analysis and Examination of Poisons on Catalysts	317
V.	Early Observations of Catalyst Poisoning	318
VI.	Contaminant Retention	321
VII.	Poison Distribution	327
VIII.	Thermal Deactivation versus Poisoning	334
IX.	Mass Transfer Effects in Poisoned Catalysts	337
X.	Poisoning Effects of Individual Elements	341
XI.	Interaction between Poisons and Active Components	352
XII.	Poison-Resistant Automotive Catalysts	357
XIII.	Catalyst Rejuvenation	358
XIV.	Future Developments	361
	References	361

AUTHOR INDEX	367
SUBJECT INDEX	384
CONTENTS OF PREVIOUS VOLUMES	394

Contributors

Numbers in parentheses indicate the pages on which the authors' contributions begin.

H. A. BENESI,* *Shell Development Company, Westhollow Research Center, Houston, Texas 77001* (98)

BARRY GOREWIT,† *Department of Chemistry, Texas A & M University, College Station, Texas 77843* (227)

VLADIMÍR HLAVÁČEK, *Department of Chemical Engineering, Institute of Chemical Technology, Prague, 6, Suchbatárov 1903, Czechoslovakia* (59)

GEORGE W. KEULKS, *Department of Chemistry, Laboratory for Surface Studies, The University of Wisconsin, Milwaukee, Milwaukee, Wisconsin 53201* (183)

L. DAVID KRENZKE, *Department of Chemistry, Laboratory for Surface Studies, The University of Wisconsin, Milwaukee, Milwaukee, Wisconsin 53201* (183)

F. E. MASSOTH, *Department of Mining and Fuels Engineering, University of Utah, Salt Lake City, Utah 84112* (266)

THOMAS M. NOTERMANN,‡ *Department of Chemistry, Laboratory for Surface Studies, The University of Wisconsin, Milwaukee, Milwaukee, Wisconsin 53201* (183)

K. OTTO, *Research Staff, Ford Motor Company, Dearborn, Michigan 48121* (311)

N. C. OTTO, *Research Staff, Ford Motor Company, Dearborn, Michigan 48121* (311)

GEORG-MARIA SCHWAB, *Institute of Physical Chemistry, University of Munich, Munich, Germany* (1)

P. W. SELWOOD, *Department of Chemistry, University of California, Santa Barbara, California 93106* (23)

M. SHELEF, *Research Staff, Ford Motor Company, Dearborn, Michigan 48121* (311)

---

* Present address: Catalytica Associates, Inc., Palo Alto, California 94304.
† Present address: Department of Chemistry, University of Southern California, Los Angeles, California 90007.
‡ Present address: Union Carbide Corporation, Charleston, West Virginia.

MINORU TSUTSUI, *Department of Chemistry, Texas A & M University, College Station, Texas 77843* (227)

JAROSLAV VOTRUBA, *Department of Chemical Engineering, Institute of Chemical Technology, Prague, 6, Suchbatárov 1903, Czechoslovakia* (59)

B. H. C. WINQUIST, *Shell Development Company, Westhollow Research Center, Houston, Texas 77001* (98)

Preface

RAINBOWS AND CATALYSIS

A rainbow is a beautiful and somehow a very systematic and complete phenomenon. It presents colors systematically, according to wavelengths, and it is complete in that it presents to us the colors of the entire visible spectrum. We think that its systematic progression and its completeness are responsible for its beauty.

The science of catalysis covers a large spectrum of phenomena. We observe—with some pride and joy—that this volume presents eight topics which, like the rainbow, form an almost systematic and complete sweep of the major classes of topics in catalysis. It spans from the most classical mechanistic study (P. W. Selwood), to a presentation of a "hard" practical application (M. Shelef *et al*). As we sweep across, we cover characterization studies of catalyst solids in terms of electronic (G. M. Schwab), surface chemical (H. A. Benesi and B. H. C. Winquist), as well as physicochemical and structural (F. E. Massoth) parameters, chemical reaction mechanisms and pathways (G. W. Keulks *et al.,* and B. Gorewit and M. Tsutsui), and a topic on reactor behavior (V. Hlaváček and J. Votruba), which takes us from the single catalyst particle to the macroscopic total reactor operation.

It strikes us that by way of this display of major topics in catalytic science, this volume will serve doubly; as a record of progress as well as a book of general instruction on catalysis.

P. B. WEISZ

Corrigendum

On page xiv of Volume 26 of this publication, the following correction should be noted:

Humphrey Owen Jones Professor of Colloid Science should read John Humphrey Plummer Professor of Colloid Science at Cambridge.

Electronics of Supported Catalysts

GEORG-MARIA SCHWAB

Institute of Physical Chemistry
University of Munich
Munich, Germany

I. Introduction	1
A. Test Reactions	2
B. General Considerations	3
II. Normal Supported Catalysts	6
A. Cu and Ag on MgO	6
B. Ni on Al_2O_3	7
C. Ni on ZnO	8
D. Fe on Al_2O_3	9
E. Ag on SiC	10
III. Inverse Supported Catalysts	12
A. Theory	12
B. NiO on Ag	13
C. NiO on Au	14
D. Fe_2O_3 on Ag	14
E. ZnO on Ag	17
IV. Two Semiconductors	18
V. Physical Observations	19
A. Rectification	19
B. Electronic Conductivity	20
VI. Conclusions	21
References	22

I. Introduction

The term *electronics* normally refers to the theory and function of the electronic devices and circuits used so universally for measurement, control, and computation. Conventional electronics is an important tool also in the study of supported catalysts.

In this paper the term *electronics* is used with a different connotation: it stands for the description of electron transitions to and from, as well as within supported catalysts. It will be shown that in many cases such transitions are responsible for the enhanced or reduced catalytic activities of supported catalysis. In this concept, supported catalysts are described as solid state systems in which a catalytically active component (the catalyst) is in electrical contact with a second solid (the support or carrier) that, by itself, has little or no catalytic activity.

The electronic structure of a solid metal or semiconductor is described by the band theory that considers the possible energy states of delocalized electrons in the crystal lattice. An apparent difficulty for the application of band theory to solid state catalysis is that the theory describes the situation in an infinitely extended lattice whereas the catalytic process is located on an external crystal surface where the lattice ends. In attempting to develop a correlation between catalytic surface processes and the bulk electronic properties of catalysts as described by the band theory, the approach taken in the following pages will be to assume a correlation between bulk and surface electronic properties. For example, it is assumed that lack of electrons in the bulk results in empty orbitals in the surface; conversely, excess electrons in the bulk should result in occupied orbitals in the surface (*1*). This principle gains strong support from the consistency of the description thus achieved. In the following, the principle will be applied to supported catalysts.

New experimental methods such as AES, ESCA, FEM, FIMS, LEED, SEM, and SIMS (*2*) allow for the first time to dispense with the above assumption since they permit direct study of surface processes. Stimulated by these developments, the theoretical treatment tends to turn more toward consideration of single surface atoms and their orbitals or bonds. However, it is clear that this direction of research is still in its beginning and therefore restricted to very special cases. The treatment of the more complex problems presented by supported catalysts by such methods—in experiment and in theory—is still ahead. Thus for the moment the use of band theory is much more promising as it can be based on a large amount of existing data.

A. TEST REACTIONS

During the last two decades it has been found that there is a special group of chemical reactions, essentially redox reactions, for which the catalytic influence of solids can be interpreted in terms of the catalyst's electronic structure and the controlled variations of that structure. The study of single-phase catalysts and the relationship between function and electronic structure of solid state catalysts show that redox reactions may be divided into two classes. *Donor reactions* are reactions in which the rate-determining step involves an electron transition from the reactant molecule to the catalyst; *acceptor reactions* are those where the reactant must accept electrons from the catalyst in order to form the activated state. Broadly speaking, donor reactions mobilize reducing agents like hydrogen or carbon monoxide and thus are involved in reactions such as hydrogenation, dehydrogenation, parahydrogen conversion, deuterium exchange, and so forth. Acceptor reactions mobilize oxygen or other oxidants such as H_2O_2 (3). Following this concept, certain reactions have come to be used as tests for the ability of catalysts to accept or donate electrons. For example, the dehydrogenation of formic acid vapor is a typical donor reaction while the decomposition of hydrogen peroxide is a typical acceptor reaction; the oxidation of carbon monoxide can be of one or the other type, depending on the electronic character of the catalyst.

With metallic catalysts as well as with semiconductors,* it has been found experimentally that donor reactions are best catalyzed by metals with many empty electron states (d metals or univalent B metals) and by p-type semiconductors, whereas acceptor reactions require electron-rich alloys or n-type semiconductors.

These general considerations, which may be summarized as "the electronic factor in catalysis," describe a broad group of catalytic reactions satisfactorily, although not all kinds (4). Thus, it appears hopeful to extend these considerations from single catalysts to mixed catalysts—and especially supported catalysts—with their often striking effects. The technical significance of these catalyst systems provides added incentive for applying the electronic viewpoint, in an attempt to gain a better understanding of their function and effectiveness.

B. General Considerations

Two kinds of influence of a carrier on the activity of a catalyst have been known since many years (5):

1. *Structural promotion:* A highly dispersed support can provide and (or) stabilize a high surface area of the catalyst supported by it. A typical example is ammonia synthesis where the thermal sintering of the iron catalyst is inhibited by alumina (although the phase configuration is different).

2. *Synergetic promotion:* Here the support interacts energetically with the catalyst and produces a new kind of active entity.

Phenomenologically, synergetic promotion is characterized by an increased reaction rate, accompanied by a decreased activation energy, compared with the catalyst alone. Structural promotion, on the other

---

* Insulators are not amenable to treatment of catalytic activity in terms of electronic band theory.

hand, does not alter the activation energy.* Since nearly all industrially important catalytic reactions use supported catalysts of the synergetic type, development of a correlation between these effects and other solid state properties is not only a challenging scientific problem but, if successful, promises to yield rational guidelines for the design and preparation of better industrial catalysts.

The task of tackling this problem from the electronics viewpoint must start from the experience with single catalysts, which indicates that correlations exist between catalytic and bulk electronic properties. In going from single-phase to supported catalysts, the working hypothesis may be stated as follows: is it not possible (or even probable) that the electronic and hence the catalytic properties of a catalyst are influenced by the electronic properties of an inert support with which the catalyst is in intimate, electrically conducting contact? For example, a small amount of a metal, dispersed on the surface of a large amount of a semiconducting carrier, must necessarily adjust its Fermi level to the Fermi level of the semiconductor. The position of the Fermi border of the metal within its Brillouin zone in turn influences not only the conductivity in the well-known manner, but also according to the findings cited earlier, catalytic activity. Cases where metallic catalysts are supported by semiconducting carriers (e.g., Al_2O_3) are very frequently encountered in industrial processes (2, 13). The technical success of such catalyst systems thus appears to become accessible to a new understanding, at least in certain cases.

The hypothesis can be tested if the catalytic activity of a metal can be modified by a controlled shift of the Fermi level of the support. With semiconducting supports such a shift is readily achieved by doping: additions of cations of higher charge than that of the matrix cations produces quasi-free electrons and/or removes defect electrons and raises the Fermi level; addition of lower charged cations has the opposite effect. This calls for investigation of metal catalysts on doped semiconductors as supports.

The inverse case, a semiconducting catalyst supported by a metal, termed *inverse supported catalyst,* has been studied systematically only in the last few years. Here, even more drastic effects can be expected because normally the number of free electrons in a metal is several orders of magnitude higher than in semiconductors. The effects are indeed considerably larger as will be shown below. However, the principles and the theory involved are more complex (6–8).

* It should be mentioned here that there are also some cases of "anomalous promotion" in which both reaction rate and activation energy are increased. In these special cases, the frequency factor is increased sufficiently to more than compensate for the increased activation energy.

In a metal, the Fermi level is located within the conduction band. In a semiconductor, this level usually is found in the forbidden gap between the valence band and the conductivity band; by doping it can be shifted up or down relative to the band edges. The activation energy of a catalyzed reaction depends on the distance of the Fermi level from the band edges: for acceptor reactions it is related to the distance from the conduction band, for donor reactions to the distance from the valence band. The exact theory will not be presented here; it has been given by Hauffe (6) and by Steinbach (9).

An important consideration for the electronics of semiconductor/metal supported catalysts is that the work function of metals as a rule is smaller than that of semiconductors. As a consequence, before contact the Fermi level in the metal is higher than that in the semiconductor. After contact electrons pass from the metal to the semiconductor, and the semiconductor's bands are bent downward in a thin boundary layer, the space charge region. In this region the conduction band approaches the Fermi level; this situation tends to favor acceptor reactions and slow down donor reactions. This concept can be tested by two methods. One is the variation of the thickness of a catalyst layer. Since the bands are bent only within a boundary layer of perhaps 10^{-5} to 10^{-6} cm in width, a variation of the catalyst layer thickness or particle size should result in variations of the activation energy and the rate of the catalyzed reaction. A second test consists in a variation of the work function of the metallic support, which is easily possible by preparing homogeneous alloys with additive metals that are either electron-rich or electron-poor relative to the main support metal.

In a few cases (10) catalysts have been studied that consist of mixtures of two semiconductors of the same chemical composition but with different levels of doping. The considerations here are analogous to those presented above. Of course, the ultimate proof as to whether electronic factors are indeed responsible for the catalytic effects discussed above will have to come through physical measurements of the electronic properties of catalyst/support systems.

In this section, we have alternatively mentioned catalytic activity and catalytic activation energy as the characteristics influenced by support effects. In the following examples, we will concentrate on changes of the true or apparent activation energy* as prime indicator. This simplifying procedure is justified by the empirical fact that for a group of closely comparable reactions (here, identical reactions on identical catalysts having variable supports or promotors) there is generally a linear relationship

* The term *true activation energy* refers to the activation energy for the rate-determining step, free of adsorption energy terms.

between the logarithm of the frequency factor and the activation energy. This means that below a certain "isocatalytic" temperature the catalyst with the lower activation energy is indeed the more active one. On the other hand, the frequency factor depends not only on the activation entropy but also on the active surface. Since this surface may vary as a result of different levels of support doping, it seems safer to rely on a parameter such as the activation energy that is characteristic of the internal nature of the reaction.

II. Normal Supported Catalysts

A. Cu and Ag on MgO

The first indication of a synergetic support action was provided by a fortuitous observation. During some studies of the catalytic action of alloys it was found (*11*) that the Laves-phase alloys $CuMg_2$ and Cu_2Mg showed a lower activation energy for the dehydrogenation of formic acid vapor than does pure copper, although both alloys have higher resistivities than copper. This was in contradiction to a general rule holding for series of alloys, namely that the better catalyst is also the better electric conductor. The explanation of these unexpected results was that the catalyst in contact with the formic acid vapor was decomposed according to

$$Cu_2Mg + HCOOH \rightarrow 2Cu + MgO + H_2 + CO$$

In support of that explanation, X-ray analysis of the catalyst after use indicated the presence of MgO. Hence, the catalytically active phase was finely divided copper in intimate contact with magnesia, quasi as carrier. The same phenomenon was observed with the Zintl-phase alloys of silver and magnesium. Such catalysts were then deliberately prepared by coprecipitation of copper and silver oxides with magnesium hydroxide, followed by dehydration and reduction. Table I shows that these supported catalysts had the same activation energies as those formed by *in situ* decomposition of copper and silver alloys with magnesium.

Since insulating magnesium oxide can be doped to form n-type or p-type semiconductors at moderate temperatures (*12*), we have here a first example of the electronic effect of a semiconducting support. However, it is unknown what type of conduction was present during the catalytic reaction in the originally undoped specimens used for the preceding experiments.

TABLE I
Activation Energies of Model Catalysis and Alloys for Formic Acid Dehydrogenation

Catalyst	Activation energy (kcal/mole)
Cu	20
Cu_2Mg	15
2Cu, 1MgO	16
$CuMg_2$	19.5
1Cu, 2MgO	18
Ag	19
AgMg	16.5
1Ag, 1MgO	15

B. Ni on Al_2O_3

The results with magnesia led us to a planned series of experiments with doped aluminas.* Nickel was evaporated *in vacuo* onto the surface of grains of undoped or doped alumina or, alternatively, onto compact nickel. These preparations were then used as catalysts for the donor model reaction of formic acid dehydrogenation as above. Table II shows the results.

It is seen again that the activation energy typical of the pure metal is lowered by contact of the metal with an oxidic support. Even more important, p-type doping with bivalent oxides lowers, n-type doping with tetravalent oxides increases the activation energy relative to that observed with undoped alumina. This is consistent with the concept

TABLE II
Activation Energies of Nickel on Different Supports

Catalyst	Support	Activation energy
Ni	Ni	26.5
	Al_2O_3	20.5
	Al_2O_3 + 2% BeO	19.0
	Al_2O_3 + 5% NiO	7.0
	Al_2O_3 + 5% TiO_2	24.0
	Al_2O_3 + 2% GeO_2	23.0

* The proof that α- and γ-Al_2O_3 can be doped to form normal p-type or n-type semiconductors was established subsequently (*13*).

discussed above, namely that the Fermi level of the thin metal layer is influenced by that of the mole abundant support. Very similar results were also found with cobalt and with silver as catalytic layers. With nickel it was also found that the optical reflexion edge and the zero-field magnetic susceptibility were changed—direct proofs for existence of an electronic effect (14).

C. Ni on ZnO

Because of the potential importance for industrial-scale catalysis, we decided to check (i) whether an influence of a semiconductor support on a metal catalyst was present also if the metal is not spread as a thin layer on the semiconductor surface but rather exists in form of small particles mixed intimately with a powder of the semiconductor, and (ii) whether a doping effect was present even then. To this end the nitrates of nickel, zinc (zinc oxide is a well-characterized n-type semiconductor) and of the doping element gallium (for increased n-type doping) or lithium (for decreased n-type character) were dissolved in water, mixed, heated to dryness, and decomposed at 250°–300°C. The oxide mixtures were then pelleted and sintered 4 hr at 800° in order to establish the disorder equilibrium of the doped zinc oxide. The ratio Ni/ZnO was 1:8 and the eventual doping amounted to 0.2 at % (15).

Ethylene hydrogenation was used as test reaction. It is a donor reaction, but has a peculiar kinetics. On nickel and copper it follows an Eley–Rideal mechanism, with hydrogen from the gas phase reacting with chemisorbed ethylene. Up to around 100°C, the ethylene adsorption is saturated, and the apparent activation energy does not contain a term for ethylene adsorption. At higher temperatures ethylene is less and less adsorbed. Since the ethylene adsorption energy is subtracted from the true activation energy of hydrogenation, formally negative values of the resulting apparent activation energy are observed; this results in the well-known temperature optimum of reaction rate. If the support ZnO has an electronic influence as discussed above, the expectation is that doping with lithium will decrease the activation energy, increase the reaction rate and, at the same time, increase the adsorption energy of the donor ethylene. Thus, the temperature optimum should still appear. On the other hand, doping with gallium should increase the activation energy of the reaction and decrease the ethylene adsorption energy, thus removing the conditions for a formally negative activation energy and the associated temperature optimum. This expectation is borne out by the experiments. Table III shows the energy parameters calculated from the measurements.

The full lines in Fig. 1 show the observed interval, the dotted curves

TABLE III
Activation Energies of Ethylene Hydrogenation over Nickel Supported by Doped Zinc Oxide (kcal/mole)

Doping level	Activation energy at low temperature	Adsorption energy of ethylene	Activation energy at high temperature
0.2 atom % Li	13.4	13.8	−0.4
0.2 atom % Ga	16.5	11.6	+4.9

denote extrapolations. Both table and figure confirm the expectations. Figure 1 shows that the kinetic behavior of the reaction is changed as a consequence of the energy shifts: on the gallium-doped nickel the reaction rate maximum has disappeared since the difference in the last column of Table III is now positive. At the same time, the absolute reaction rate is higher for lithium-doped zinc oxide support.

D. Fe on Al$_2$O$_3$

Encouraged by these results, we used the donor test (or model) reaction of formic acid dehydrogenation for the study of the well-known ammonia

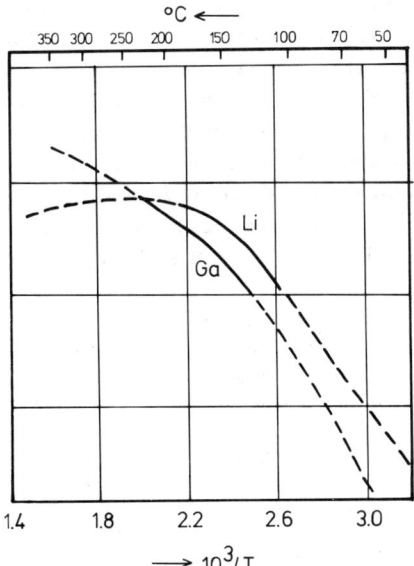

Fig. 1. Arrhenius plots for hydrogenation rate of ethylene over nickel on doped zinc oxide.

synthesis catalyst Fe–Al$_2$O$_3$–K$_2$O. As mentioned before, this catalyst represents an example of purely structural promotion, at least as far as ammonia synthesis is concerned. However, we felt that the same catalyst might show synergetic behavior in the formic acid reaction, and our experiments (*16–18*) confirmed this expectation. Table IV shows some results which broadly parallel those obtained with the nickel/alumina system (Table II).

It should be noted that the results for the formic acid decomposition donor reaction have no bearing for ammonia synthesis. On the contrary, if that synthesis is indeed governed by nitrogen chemisorption forming a nitride anion, it should behave like an acceptor reaction. Consistent with this view, the apparent activation energy is increased from 10 kcal/mole for the simply promoted catalyst (iron on alumina) to 13–15 kcal/mole by addition of K$_2$O. Despite the fact that it retards the reaction, potassium is added to stabilize industrial synthesis catalysts. It has been shown that potassium addition stabilizes the disorder equilibrium of alumina and thus retards its self-diffusion. This, in turn, increases the resistance of the iron/alumina catalyst system to sintering and loss of active surface during use.

E. AG ON SIC

A very clear-cut example for the influence of the electronic factor in supported catalysts, again involving a thin-layer metal type, is represented in Fig. 2. Here the carriers are commercially available samples of doped carborundum (SiC) which by itself is catalytically entirely inactive. In the abscissa of Fig. 2 we have arranged these samples in the order of their conductivity as stated by their manufacturers. The concentration of positive holes increases towards the right and that of the quasi-free electrons towards the left. Grains of these supports approximately 1 mm in size were covered with a thin layer of silver by the usual mirror produc-

TABLE IV
Apparent Activation Energy of Ammonia Synthesis Catalysts for Formic Acid Dehydrogenation

Catalyst	Support	Activation energy (kcal/mole)
Fe	None	20
Fe	Al$_2$O$_3$ (commercial n-type)	18
Fe	Al$_2$O$_3$ + K$_2$O (p-type)	7–13
Fe	Al$_2$O$_3$ + GeO$_2$ (n-type)	22

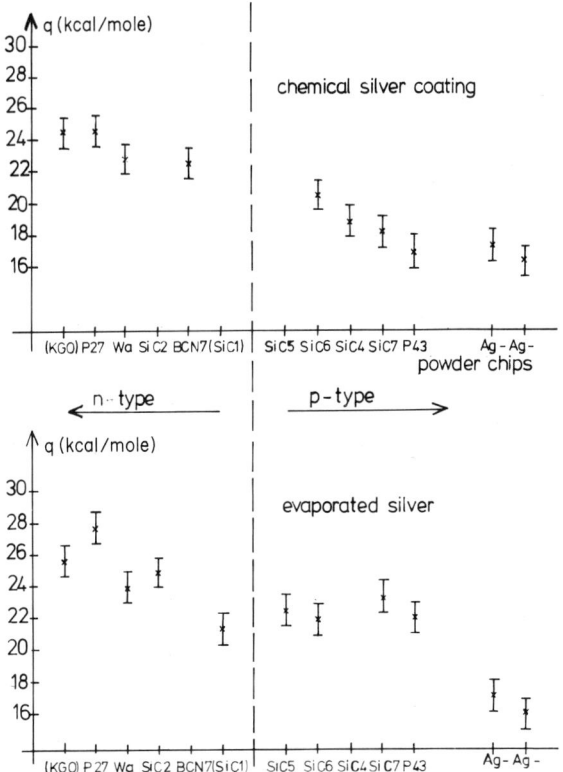

FIG. 2. Activation energy of formic acid dehydrogenation over silver on doped silicon carbide (19). (Copyright by the Université de Liège. Reprinted with permission.)

tion method (reduction of a silverammine solution; upper part of the diagram) or by vacuum evaporation of silver (lower part). The ordinate in the figure is the activation energy of the formic acid dehydrogenation, our familiar test reaction. To the far right, two points are given for metallic silver. The general trend is evident: the activation energy increases with increasing n-character (decreasing p-character) of the support, as in the examples discussed previously (19).

Similar cases of synergetic promotion of metals by semiconducting supports have been found in many other laboratories. Solymosi (7) observed the electronic effect in formic acid dehydrogenation catalyzed by nickel supported on doped carriers such as TiO_2, Cr_2O_3, MgO, ZnO, Al_2O_3, and NiO. In all cases the activation energy was lowered by addition of p-type dopants and increased by n-type dopants to the support; the same was true with germanium as support. On the other hand the opposite effect was observed for hydrazine decomposition, as expected for an acceptor reaction. Batta et al. (20) also found that the activation energy for the formic acid test reaction showed the same trend for mixed oxide catalysts and for nickel when supported by these oxides. This observation tends to confirm the electronic interaction of the catalyst and support phases. Finally, a paper by Pfeil (21) may be mentioned which describes the dehydrogenation of cyclohexane over platinum-on-alumina catalysts. While activation energies were not reported in that paper, a decrease of activity with increasing level of n-type doping by ZrO_2, Ta_2O_5, and WO_3 was found, consistent with the expectation for a donor reaction. Somewhat surprisingly, p-type doping was without influence, at least on the yield.

III. Inverse Supported Catalysts

A. Theory

In the initial part of this paper, we mentioned already that the effects described in the preceding section on "normal" supported catalysts are understandable only if the amount of semiconductor in the mixed catalysts is very large compared to the amount of metal: only then will the number of electrons a semiconductor is able to extract from the metal be sufficient to affect the electron concentration and the Fermi level of a metal. This condition is fulfilled in all of the preceding examples.

These considerations suggest that in the study of catalyst–support interaction it would be promising to reverse the situation and investigate semiconducting catalysts supported by metallic carriers: the very large number of electrons in the metallic support can easily modify the spatial distribution and the average free energy of the relatively few electrons in a semiconducting catalyst, at least in a thin layer near the interphase contact. In this case, it is no longer necessary to make up for a mismatch in electron concentrations by a large support-to-catalyst mass ratio.

To explore this concept in more detail, let us consider Fig. 3 which was drawn originally by the physicist P. Hilsch (22) to explain the influence of a semiconducting coating on the superconductivity of a metal. In Fig. 3(a), the phases of the metal and the dielectric (semiconductor) are separated. Normally, metals have a smaller work function ϕ_m than semiconductors (ϕ_s) and therefore a higher Fermi level E_F. E_C is the lower edge of the conduction band in the semiconductor, E_V the upper edge of its valence band. In Fig. 3(b), the two phases are in electric contact. Electrons have diffused from the metal into the semiconductor until their Fermi levels have equalized. This has resulted in buildup of a space charge in a bound-

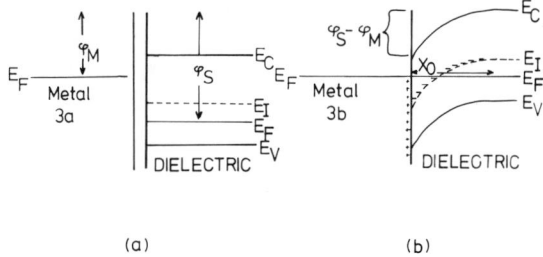

FIG. 3. Energy diagram of the combination metal/dielectric (a) before and (b) after contact (after R. Hilsch) (*19*). (Copyright by the Université de Liège. Reprinted with permission.)

ary layer extending to a depth X_0 in the semiconductor. E_C and E_V have been bent downwards, and the distance of the Fermi level from E_V has increased. Within the boundary layer, this effect simulates n-type doping of the catalyst which deactivates the boundary layer as a catalyst for donor reactions. On the other hand, the distance from E_F to E_C has decreased, and this favors acceptor-type catalytic reactions. From this explanation, it is clear that electronic interaction effects can change the activity and activation energy only if the semiconductor catalyst surface lies within a distance X_0 (the width of the boundary layer) from the metal support. Beyond that distance the semiconducting catalyst remains unaffected by the metal. There are two possibilities to test this hypothesis experimentally: we can vary the thickness of the semiconducting catalyst layer (or, in a powder mixture, the grain size), or we can vary ϕ_m by alloying or entirely changing the metal.

B. NiO on Ag

The first of these crucial experiments was carried out in a planned manner only in 1966 (*23*); surprisingly, no use of this interesting possibility seems to have been made before then. Nickel layers of different thickness were deposited electrolytically on silver foils, and the layers were then oxidized thermally by oxygen, forming NiO layers of between 75 Å and 15,000 Å thickness. With these compound catalysts, the activation energy of the CO combustion was then measured. The catalytic oxidation of CO is a donor reaction which in the temperature range used in these experiments is nearly uncatalyzed by silver, but well catalyzed by NiO. The experimental results are shown in Fig. 4. On the abscissa, the logarithm of the NiO layer thickness is plotted, on the ordinate the apparent activation energy of the reaction. Layers thicker than 500 Å exhibit the activation energy of compact NiO (16 kcal/mole). Below that value (that is, within the boundary layer) the activation energy rises steeply up

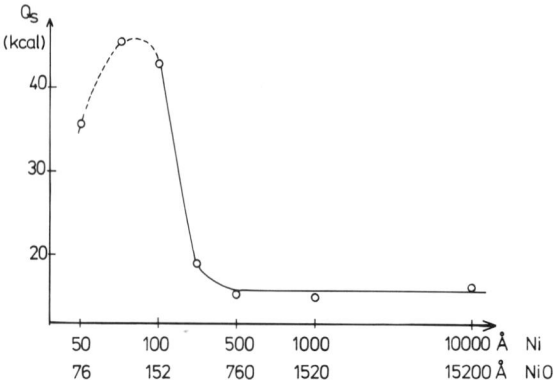

FIG. 4. Activation energy of carbon monoxide oxidation over nickel oxide of different layer thickness on silver (26). (Copyright by Akademische Verlags-Gessellschaft. Reprinted with permission.)

to 46 kcal/mole. The dotted part of the curve probably means the transition of catalyst action to that of silver because of holes in the very thin nickel oxide layer.

C. NiO on Au

The second type of critical experiment involves varying the support. This possibility was also investigated in more detail in our laboratory (24). Figure 5 shows the results when silver, gold, and palladium were used as supports for nickel oxide. The work function of these metals increases in the above order while the effect on the activation energy of nickel oxide decreases in the same order. This shows directly that a low work function and higher Fermi level of the metal favors the electron transfer and inhibits a donor-type reaction. It may be noted that the opposite trend in the activation energy of the pure "naked" metals (NiO = 0) is in agreement with previous work (25). In another series of experiments (24), the maximum activation energy was determined for gold-palladium alloys of different compositions. The curve shown in Fig. 6 indicates a decrease of the activation energy with increasing palladium content, that is, with decreasing occupation of the s band and, hence, increasing work function of the support. (The black square represents an alloy of 15% platinum in gold.)

D. Fe_2O_3 on Ag

In the preceding example, the effect of the metallic support consisted of an increase of the activation energy and consequently an inhibition of the

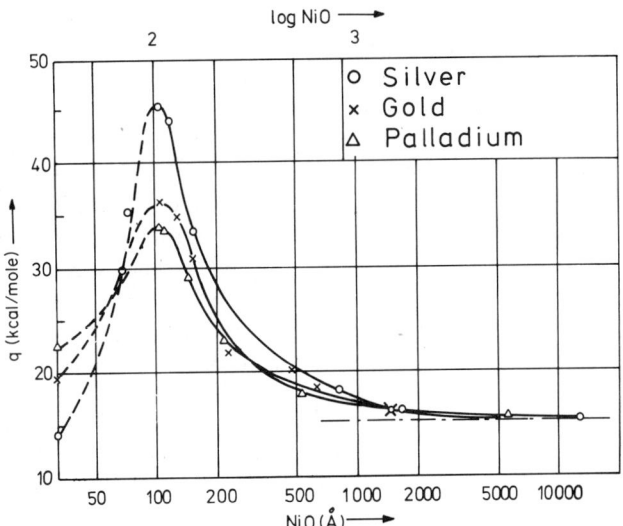

FIG. 5. Activation energy of carbon monoxide oxidation over nickel oxide of different layer thickness on three metals (24). (Copyright by Akademische Verlags-Gessellschaft. Reprinted with permission.)

reaction. From a practical point of view, it would be much more interesting if a decrease of the activation energy and thus a real promotion could be achieved by the same general approach. This possibility was in fact realized soon afterwards (26). The test reaction involved was the

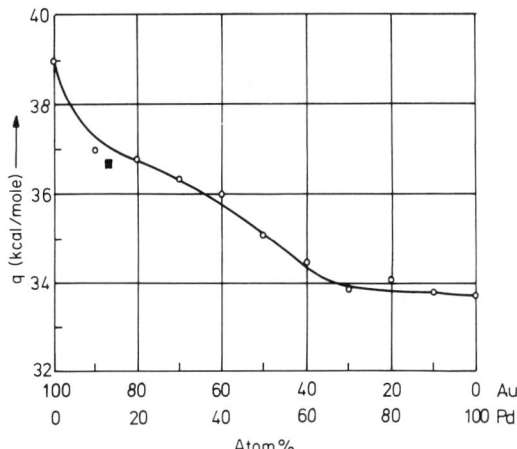

FIG. 6. Activation energy of carbon monoxide oxidation over nickel oxide on gold–palladium alloys (24). (Copyright by Akademische Verlags-Gessellschaft. Reprinted with permission.)

oxidation of sulfur dioxide to sulfur trioxide, a reaction of considerable technical interest. The catalyst was α-Fe$_2$O$_3$ although it is probable that during the reaction iron is present as Fe$_3$O$_4$. The support again was silver powder which was covered with a thin layer of the iron oxide. The silver support was alloyed (doped) either with 5.2 atom% mercury (increasing the relative electron concentration from 1.0 to 1.052) or with 20 atom% palladium (decreasing the relative electron concentration from 1.0 to 0.8) by coprecipitation of metal powders from solution with hydrazine. All powders had specific surfaces around 200 cm^2/g. They were covered with a 64 Å thick layer of α-Fe$_2$O$_3$ by evaporation of an organosol. The alloying percentages above apply to the catalyst composition after annealing at 600°C. Conversions were measured for a mixture of 7% SO$_2$ and 93% oxygen. The results are given in Fig. 7 and in Table V.

For the same catalyst, the activation energy is drastically decreased and the reaction rate enormously increased with increasing electron concentration and decreasing work function of the support. This behavior

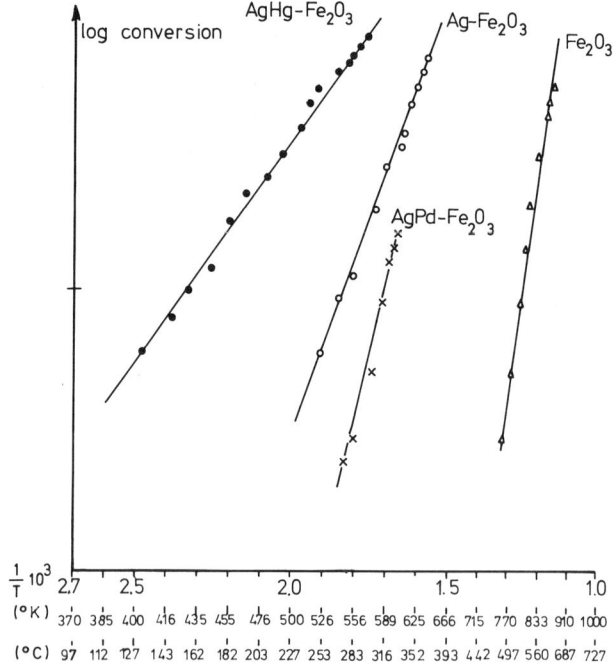

FIG. 7. Arrhenius diagram for oxidation rate of sulfur dioxide over ferric oxide on doped silver (19). (Copyright by the Université de Liège. Reprinted with permission.)

TABLE V
Apparent Activation Energies of SO_2 Oxidation over Iron Oxide Supported by Silver and its Alloys

Support	q(kcal/mole)	Log k_0 (%)	Isocatalytic temperature	Temperature range (°C)	Electron concentration
—	31	9.1	—	480–600	—
Ag + Pd	24	9.5	427	250–340	0.8
Ag	13	5.84	427	200–360	1.0
Ag + Hg	7	4.2	427	100–300	1.05

indicates that the reaction over Fe_2O_3 is an acceptor reaction; most probably oxygen acts as electron acceptor and becomes chemisorbed with a negative charge. Admittedly, for this particular reaction an increase of activity (800-fold in the case of Ag/Hg) is of no economic significance since the yield already approaches 100% with currently used catalysts. However, this type of promotion could be of considerable importance for other technical reactions.

E. ZnO on Ag

As in the case of normal supported catalysts, we tried with this inverse supported catalyst system to switch over from the thin-layer catalyst structure to the more conventional powder mixture with a grain size smaller than the boundary layer thickness. The reactant in these studies (27) was methanol and the reaction its decomposition or oxidation; the catalyst was zinc oxide and the "support" silver. The particle size of the catalyst was 3×10^{-3} cm; hence, not the entire particle in contact with silver can be considered as part of the boundary layer. However, a part of the catalyst particle surface will be close to the zone of contact with the metal. Table VI gives the activation energies and the start temperatures for both methanol reactions, irrespective of the exact composition of the products.

The reaction is of the acceptor type (28). Consistent with this and the views presented above, the data of Table VI suggest that contact with silver has bent the bands in zinc oxide and thus lowered the activation energy and the start temperature. However, a bifunctional catalytic action of the silver–zinc oxide system canot be excluded with certainty.

On the basis of the results cited above and in view of the fact that metals generally show a lower work function than their own oxides, a

TABLE VI

Activation Energies and Starting Temperatures of Methanol Reactions over Silver, Zinc Oxide, and their Mixture (1:1 by wt)

Reaction	Catalyst	Start temperature	Activation energy
Decomposition	ZnO	250°C	41
	Ag	(380°C)	Only wall reaction
	ZnO + Ag	260°C	32
Oxidation	ZnO	250°C	47
	Ag	130°C	37.5
	ZnO + Ag	70°C	24.5

metal oxide should show a much changed activity and activation energy when it is a covering oxidation layer on its basic metal. Unfortunately, the choice of reactions to test this broad hypothesis is rather restricted because the reactants must neither reduce the oxide nor oxidize the metal. However, in several cases—for the dehydration of formic acid and of ethanol over zinc oxide on zinc, and for formic acid dehydration over Cu_2O on copper—a lowered activation energy was indeed observed (29).

IV. Two Semiconductors

Only two studies have been reported where both the active catalyst and the carrier were semiconductors. In one of these studies (20), a mixture of magnesia and chromia was used. The activation energy here depends strongly on the preheating treatment because magnesium chromite spinel is formed. Although the observed changes in activation are electronic effects in the sense of our definition, they do not directly contribute to the understanding of electronic interactions in catalyst-support systems. The second study (10) is more suitable in this respect. Here, two different samples of the same oxide (zinc oxide as an n-type conductor in one case, nickel oxide as a p-type conductor in the other case) were mixed to produce a semiconductor–semiconductor catalyst system. One of these samples was doped in the direction of higher, and the other in the direction of lower conductivity (electron or hole concentration). Especially in the case of doped zinc oxides, a considerable lowering of the activation energy for carbon monoxide oxidation was observed. The results for the p-type mixed nickel oxides system were ambiguous.

V. Physical Observations

A. Rectification

The observed changes of true or apparent activation energies of catalyzed reactions show that the electronic state of a support can influence the energy needed for the transfer of electrons from the reactant through the catalyst into the support (donor reactions) or from the support through the catalyst into reactant (acceptor reactions). This energy, in turn, should be related to the barrier for electron transport across the metal–semiconductor interface—that is, the degree of rectification exerted by that interface. Thus, the electronic mechanisms assumed for the explanation of solid state catalytic phenomena might be confirmed directly by checking this rectification concept for representative catalyst-support structures.

It is, of course, well known that metal–semiconductor interfaces frequently have rectifier characteristics. It is significant, however, that this characteristic has been confirmed specifically for systems that have been used as inverse supported catalysts, including the system NiO on Ag described above as catalyst for CO-oxidation. In the experimental approach taken, nickel was evaporated onto a silver electrode and then oxidized in oxygen. A space charge-free counter-electrode was then evaporated onto the nickel oxide layer, and the resulting sandwich structure was annealed. The electrical characteristic of this structure is represented in Fig. 8. The abscissa (U) is the applied potential; the ordi-

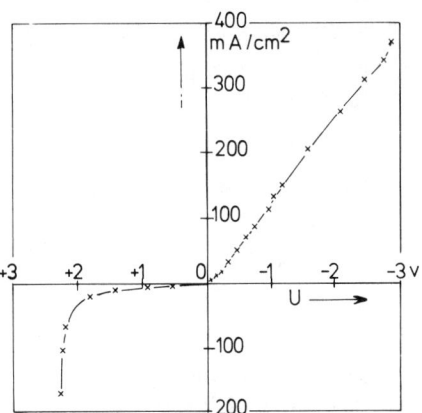

FIG. 8. Current characteristic of a silver–nickel oxide sandwich (*19*). (Copyright by the Université de Liège. Reprinted with permission.)

nate (i) shows the current density. When silver is negative (right side of the figure), current flows freely (resistance 0.1 ohm · cm^{-2}) whereas in the opposite direction a resistance of 200 ohm · cm^{-2} is measured up to a breakthrough potential of +2.1 V. Hence, electrons indeed move freely from the metal to the semiconductor, consistent with the electronic interpretation of support effects in solid state catalysis (30).

B. Electronic Conductivity

The influence of metal–semiconductor interaction on electron transfer was proved in a rather convincing way through conductivity measurements on compound pellets (31–33). From a purely geometrical point of view one should expect that, because of a partial short circuiting by the metal, the conductivity of a semiconducting pellet will increase by approximately 10% if 10% of silver powder is added before pelleting. In reality, the increase observed was much higher: with n-type conductors such as ZnO, Fe$_2$O$_3$, and TiO$_2$, it amounted to 100% and more and this effect decreased with increasing temperature, proving an electronic interaction. Even more convincing was the behavior observed for the

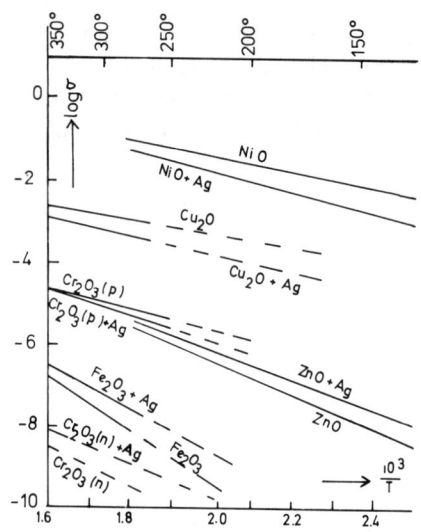

Fig. 9. Temperature dependence of conductivity of semiconductor pellets with and without added 10% wt of silver powder. The logarithm of structure-independent conductivity is plotted as a function of the reciprocal absolute temperature (degrees Celsius on top of the figure). Full lines are measured, dotted parts extrapolated (19). (Copyright by the Université de Liège. Reprinted with permission.)

p-type conductors NiO and Cu_2O where the presence of 10% silver *decreased* electronic conductivity by a comparable amount. The explanation is that, with p-type conductors the injected electrons neutralize a fraction of the positive holes and thus decrease the concentration of free carriers. Chromia shows both effects, depending on whether pretreatment resulted in a p-type conductor or, through exposure to hydrogen, in an n-type conductor. The different effects on conductivity of mixing various metals and semiconductors are summarized in Fig. 9.

ThO_2 shows a very interesting behavior. Although it is an n-type semiconductor, it reacts like the p-conductors mentioned in the silver-conductivity test. This is consistent with the fact that ThO_2 is the only oxide with a work function lower than that of silver: it will tend to emit electrons into the metal rather than accept them.

These observations again confirm the basic hypothesis used for the electronic interpretation of catalytic efects in catalyst-support systems, namely that in boundary layers (or in grains of correspondingly small size) electrons are exchanged between catalysts and support.

VI. Conclusions

The experimental results described in this review support the concept that, in certain reactions of the redox type, the interaction between catalysts and supports and its effect on catalytic activity are determined by the electronic properties of metals and semiconductors, taking into account the electronic effects in the boundary layer. In particular, it has been shown that electronic effects on the activity of the catalysts, as expressed by changes of activation energies, are much larger for inverse mixed catalysts (semiconductors supported and/or promoted by metals) than for the more conventional and widely used "normal" mixed catalysts (metals promoted by semiconductors). The effects are in the order of a few electron volts with inverse systems as opposed to a few tenths of an electron volt with normal systems. This difference is readily understandable in terms of the different magnitude of, and impacts on electron concentrations in metals versus semiconductors.

It is hoped that these facts and concepts are amenable to practical applications in the *planned synthesis of active catalysts* for certain suitable reactions. Implementation of such a systematic approach to catalyst design will require that the character of the reaction (donor or acceptor) and the work functions of catalysts and supports are known or determined.

Acknowledgments

Most of the investigations quoted in this article have been supported very effectively by the Deutsche Forschungs-Gemeinschaft (DFG), to whom the author is much indebted. He also thanks Dr. F. Kalhammer (EPRI, Palo Alto, California) for revising the manuscript.

References

1. G. M. Schwab, *Umsch. Wiss. Tech.* **65**, 766 (1965).
2. *Ger. Fed Minist. Res. Technol.* "*Chem. Technik,*" Vol. 2: "*Catalysis,*" Bonn, 1976.
3. e.g., G.-M. Schwab, *Angew. Chem.* **67**, 433 (1955).
4. e.g., G.-M. Schwab, *Mem. Soc. Roy. Sci. Liège* (6) **1**, 11 (1971).
5. e.g., G.-M. Schwab and H. Schultes, *Z. Phys. Chem., Abt. B* **9**, 265 (1930).
6. K. Hauffe and Th. Wolkenshein, *Symp. Electron. Phenom. Chemisorp. Catal., Moscow, 1968, Berlin 1969, p. 1.*
7. F. Solymosi, *Catal. Rev.* **1**, 233 (1967).
8. G.-M. Schwab, *Chem.-In.-Tech.* **39**, 1191 (1967).
9. F. Steinbach, *Fortschr. Chem. Forsch.* **25**, 117 (1972).
10. W. Komatsu, H. Ooki, I. Naka, and A. Kobayashi, *J. Catal.* **15**, 43 (1969).
11. J. Block, W. Müller, and D. Schultze, *Naturwissenschaften* **44**, 582 (1957).
12. *Gmelin's Handb. Inorg. Chem. Syst.* No. 27 (Mg), p. 44f (1937).
13. G.-M. Schwab, A. Beer, and J. Foitzik, *Z. Angew. Phys.* **14**, 763 (1962).
14. G.-M. Schwab, *Angew. Chem.* **73**, 399 (1961); G.-M. Schwab, D. Schultze, and J. Block, *Angew. Chem.* **71**, 101 (1958).
15. G.-M. Schwab and G. Mutzbauer, *Z. Phys. Chem.* **32**, 367 (1962).
16. G.-M. Schwab, *Dechema Monogr.* **49**, 99 (1964).
17. G.-M. Schwab and R. Putzar, *Chem. Ber.* **92**, 2132 (1959).
18. G.-M. Schwab and R. Putzar, *Z. Phys. Chem.* **31**, 342 (1962).
19. G. Pfahler, Dissertation, Univ. of Munich, 1968; see G.-M. Schwab, *Mem. Soc. R. Sci. Liège* (6) **1**, 39 (1971).
20. I. Batta, T. Bánsági, F. Solymosi, and Z. G. Szabó, *Acta Chem. Acad. Sci. Hung.* **41**, 219 (1964).
21. W. Pfeil, *Z. Phys. Chem. (Leipzig)* **243**, 52 (1970).
22. P. Hilsch and D. G. Naugle, *Z. Phys.* **201**, 1 (1967).
23. G.-M. Schwab and R. Siegert, *Z. Phys. Chem.* **50**, 91 (1966).
24. G.-M. Schwab and B. Matthes, *Z. Phys. Chem.* **94**, 243 (1975).
25. G.-M. Schwab and K. Gossner, *Z. Phys. Chem.* **16**, 41 (1958).
26. G.-M. Schwab and H. Derleth, *Z. Phys. Chem.* **53**, 1 (1967).
27. G.-M. Schwab and K. Koller, *J. Am. Chem. Soc.* **90**, 3078 (1968); G.-M. Schwab and H. Seemüller, *Mem. R. Soc. Sci. Liège* (6) **1**, 55 (1970).
28. F. Steinbach, *Fortschr. Chem. Forsch.* **25**, 117 (1972).
29. G.-M. Schwab and H. Zettler, *Chimia* **23**, 489 (1969).
30. G.-M. Schwab and F. Bruncke, *Z. Naturforsch., Teil A* **24**, 1266 (1969).
31. G.-M. Schwab, *Sitzungsber. Math.-Naturwiss. Kl. Bayer. Akad. Wiss. Muenchen* 7* (1968).
32. G.-M. Schwab and A. Kritikos, *Helv. Phys. Acta* **41**, 1166 (1968).
33. G.-M. Schwab and A. Kritikos, *Naturwissenschaften* **55**, 228 (1968).

The Effect of a Magnetic Field on the Catalyzed Nondissociative *Para*hydrogen Conversion Rate

P. W. SELWOOD

Department of Chemistry
University of California
Santa Barbara, California

I. Introduction 23
II. Conversion Rate Measurements 24
III. Experimental Results 25
    A. Alumina, Lanthana, Lutetia, and Yttria 26
    B. The Paramagnetic Rare Earths 32
    C. Chromia 38
    D. Cobalt Monoxide 44
    E. Manganese Monoxide 45
    F. Chromium Dioxide, Europium Monoxide, and Nickel (Metal) . . . 46
IV. Correlation of Experimental Results 48
V. Conversion and Field Effect Theory 50
VI. Conclusions 51
    References 56

I. Introduction

The purpose of this paper is to describe all available results on the extrinsic field magnetocatalytic effect discovered several years ago (*1*). It is well known that the *ortho*–*para*hydrogen conversion may proceed by either of two mechanisms. One, the nondissociative, involves a nuclear spin singlet–triplet transition caused by interaction with magnetic sites present in the catalyst. The other, the dissociative mechanism, involves separation and recombination of the atoms. No effect of a magnetic field on the dissociative conversion has ever been confirmed. All of the following remarks have reference to the nondissociative process. By magnetocatalytic effect is meant a change of a catalyzed reaction rate caused by a field produced externally or internally. An extrinsic field is one coming from the outside. The experimental results are classified according to the catalyst, or class of catalyst. There appear to be no results on a possible field effect for the homogeneous nondissociative conversion. [Note added in proof: A review of magnetic field effects on reaction rates by P. W. Atkins and T. P. Lambert appeared in *Annu. Rep. Chem. Soc.* **72** (A), 67 (1975), but the effects described in the present paper are not mentioned.]

II. Conversion Rate Measurements

The catalyzed *ortho–para* hydrogen conversion rate may be measured in either a flow, or a static, reactor. The former is the more convenient, the latter is generally used for obtaining absolute rates. Both methods have been used to study the extrinsic field effect, but most of the data have been obtained by the flow method.

A typical flow reactor (2) used for this purpose is shown in Fig. 1. Cylinder hydrogen is passed through a commercial palladium–silver purifier, converted to the desired ortho–para ratio, passed through a trap to minimize possible catalyst poisoning, then through the reaction cell. This consists of a Vycor U-tube containing a Vycor fritted disk to support the catalyst sample. Gas flow is up through the disk. The hydrogen then goes to a 4-wire thermal conductivity cell connected to a 1-mV recorder and to exit. For measurements at low temperatures the reactor conversion is, of course, from ortho toward para, otherwise it is from para toward ortho. Magnetic fields in the $1–10^2$ Oe range (a field of 1 Oe is the same as one of 79.6 A m^{-1}) are obtained by placing the reactor in a solenoid of about 5000 turns. The core of the solenoid should be large enough to permit conve-

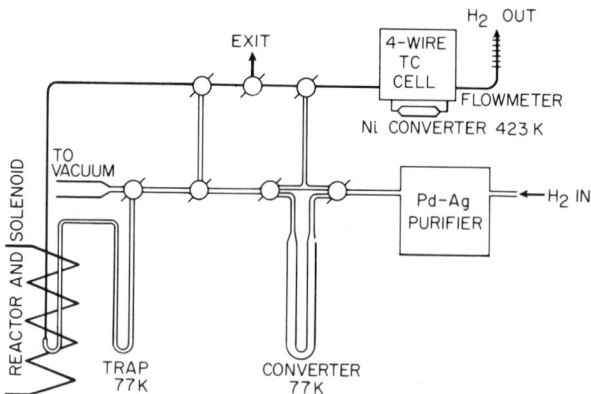

FIG. 1. Flow reactor assembly for measuring extrinsic field effects.

nient temperature control of the sample. For fields in the kOe range the reactor is placed in the gap of a large electromagnet. It must be remembered that an iron-core magnet may have a residual field of 10 Oe or more at zero current. All fields are monitored with a commercial instrument.

For reasonable accuracy by the flow method it is necessary to maintain the degree of conversion at between 30 and 50% of complete conversion from the initial to the equilibrium ortho–para ratio at the reactor temperature. This may be done by varying the amount of catalyst or the flow rate of the hydrogen. Catalyst masses of from 10 mg to 10 g and monitored flow rates of 0.1 to 1 cm^3 s^{-1} (STP) have often been used. Atmospheric pressure has been used for many of the reported results.

Specific conversion rates are calculated in the usual way for a flow reactor: $k = (F/S) \ln[(C_{eq} - C_0)/(C_{eq} - C_x)]$, where F is the flow rate (mol s^{-1}), S the total catalyst surface (m^2), C_{eq} the ortho–para equilibrium ratio at the reactor temperature, C_0 the ratio for hydrogen entering the reactor and C_x the ratio for hydrogen leaving the reactor. For different samples of the same catalyst the zero field conversion reproducibility is seldom better than by a factor of 5, but the fractional change $\Delta k_H = (k_H - k_0)/k_0$ may often be reproduced to $\pm 5\%$. In some cases a change of $\pm 0.5\%$ is measurable. (k_H is the specific rate in a field H, k_0 that in zero or negligible field).

The static reactor method used for absolute rate determinations, and almost always for *ortho–para*deuterium studies, is generally that based on the micro-Pirani gauge analysis chamber as adapted by Ashmead *et al.* (*3*). The time necessary for a single determination of the extrinsic field effect by this method is unfortunately likely to be measured in hours or days rather than in seconds as for the flow reactor. To date the only application of this method to the extrinsic field effect appears to be that of Eley *et al.* (*4*). Van Cauwelaert and Hall (*5*) have described a recirculating adaptation of the static reactor that would seem to be useful for studying the field effect.

Details of catalyst preparation and pretreatment will be described later.

III. Experimental Results

It has been known since the early work of Farkas and Sachsse (*6*) that the ortho–para conversion may be catalyzed by paramagnetic species. That the mechanism for this kind of conversion is nondissociative is shown by the absence of hydrogen–deuterium equilibration at a comparable rate under similar conditions. But proof that the nondissociative conversion always proceeds by a magnetic process is not quite so convincing. Some diamagnetic substances catalyze the conversion and it has been thought that nuclear magnetic moments may be responsible or, alternatively, that interaction of the catalyst with molecular hydrogen may generate a paramagnetic species. But there are some cases for which it is very difficult to identify the catalyst site.

Extrinsic field effects have been studied on a large number of solid catalysts of quite diverse magnetic properties. Presentation of results on these substances will be started with four oxides that are normally considered to be diamagnetic.

A. Alumina, Lanthana, Lutetia, and Yttria

Alumina has not been reported to shown an extrinsic field effect. It is mentioned here because of its frequent use in alumina-supported catalyst systems and because it has some structural similarities to lutetia and yttria, both of which show field effects that can only be described as bizarre. The conversion activity over crystalline α-Al_2O_3 in the form of powdered sapphire is virtually negligible (7). But Acres *et al.* (8) have shown that high purity α-Al_2O_3 evacuated for many hours at 823 K, and having a surface of 9.2 m^2 g^{-1}, had a readily measurable activity at 77 K. The activity was mainly, but not all, nondissociative. This observation has been confirmed by others (5). It shows that the use of Al_2O_3, to obtain greater dispersion, is not generally a very satisfactory support for studies of the extrinsic field effect on paramagnetic solids although in some cases, to be described, the writer has used it without encountering serious difficulties.

Lanthana was reported by Taylor and Diamond (9) to show some low temperature conversion activity. This has been confirmed by the writer (2) who also determined that a large part of the activity at room temperature is nondissociative. The specific activity of La_2O_3 is small compared with those of all the paramagnetic rare earths, but it shows an extrinsic field effect that may be observed at low fields. The sign of the effect is negative. A sample of surface 2 m^2 g^{-1} pretreated in hydrogen at 823 K showed k_0 at 298 K to be about 10^{-7} mol m^{-2} s^{-1} with a fractional change of about -4% in a field of 40 Oe. Higher activation temperatures would probably have increased k_0. It appears, therefore, that La_2O_3 is not an entirely satisfactory support for other oxides, in work of this kind, unless the specific activity and the surface concentration of the supported oxide are large. Nevertheless, La_2O_3 is better than Al_2O_3 for this purpose and, with minor corrections, it may be used.

In the preparation of La_2O_3, and of all other rare earths, a problem may arise from contamination with carbon which is itself a conversion catalyst. This carbon may generally be detected by a slight discoloration when the sample is heated in hydrogen. It may be eliminated by repeated heating to about 800 K in oxygen and then in hydrogen until the surface is quite free from discoloration. This test is satisfactory for the colorless rare earths, but less so for neodymia with its robin's egg blue. Alternatively, the rare earth may be prepared by ignition to the oxide from the recrystallized nitrate followed by transfer to the reactor with minimum exposure to the atmosphere. The product obtained in this way generally has a rather low specific surface.

Lutetia of high purity preheated above 773 K in flowing hydrogen shows conversion activity. At room temperature and up to about 523 K this activity is nondissociative (10). Nearly identical results are obtained by heating the Lu_2O_3 *in vacuo*. The activity is not poisoned by hydrogen but it is instantly destroyed by oxygen. Figure 2 shows specific conversion rates at room temperature as a function of pretreatment temperature in hydrogen. For a 0.5 g sample of Lu_2O_3 of surface 52 m^{-2} g^{-1} and preheated in hydrogen at 973 K the specific conversion rate at 298 K, atmospheric pressure and zero field, was 1×10^{-6} mol m^{-2} s^{-1}. Any dissociative contribution to the activity at this temperature was less than 1% of the total. The activity was much less at 77 K. The absolute zero field conversion rate of Lu_2O_3 measured by Madhusudhan and Selwood (11) at 300 K, a pressure of 2.4×10^3 N m^{-2} and activated in hydrogen at 823 K was about 8×10^{-10} mol m^{-2} s^{-1}.

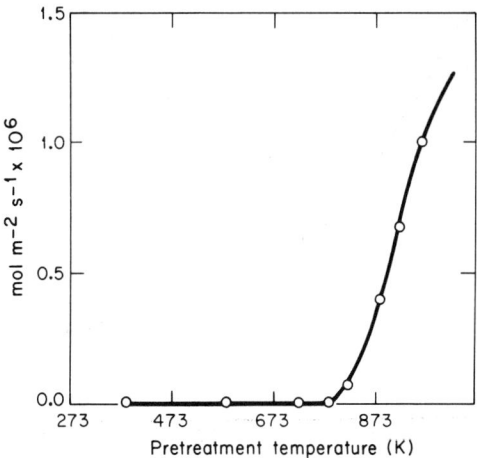

FIG. 2. Specific conversion rates at 298 K, over Lu_2O_3, as a function of pretreatment temperature in H_2.

Figure 3 shows Δk_H, the fractional rate change for a sample of Lu_2O_3 activated in hydrogen at 873 K, measured at 298 K, at atmospheric pressure, and over an extrinsic field range of 0 to 40 Oe. It will be noted that Δk_H is zero up to about 3 Oe, then there is a rapid decrease in rate for a total change of about -42% above 35 Oe. There is no further rate change up to 18 kOe. Absolute rate measurements (*11*) over a similar sample at 40 Oe gave a change of about -50%. For all these measurements the field-induced rate change appears to be instantaneous (a slow approach to the new steady state previously observed is now believed to be due to an artifact in the analysis system). There is no change of hydrogen flow rate and no change in the ortho–para equilibrium constant. Over Lu_2O_3 there is no very obvious change of Δk_H with changing activation temperature. That is to say, Δk_H is approximately independent of k_0.

From Fig. 3 it is clear that the conversion rate is very sensitive to the applied field in the 4 Oe region. If the current through the solenoid is adjusted to produce a field of 4 Oe in the same sense as the Earth's field, then the resultant field at the reactor is about 4.6 Oe. If now the solenoid current is reversed the resultant field becomes 3.4 Oe. This change is easily measurable as shown in Fig. 4. The result shown was actually observed on a dilute lanthana-supported erbia catalyst sample but similar results may be obtained on activated Lu_2O_3 with more convenience. These results show that the Earth's field could be measured by making

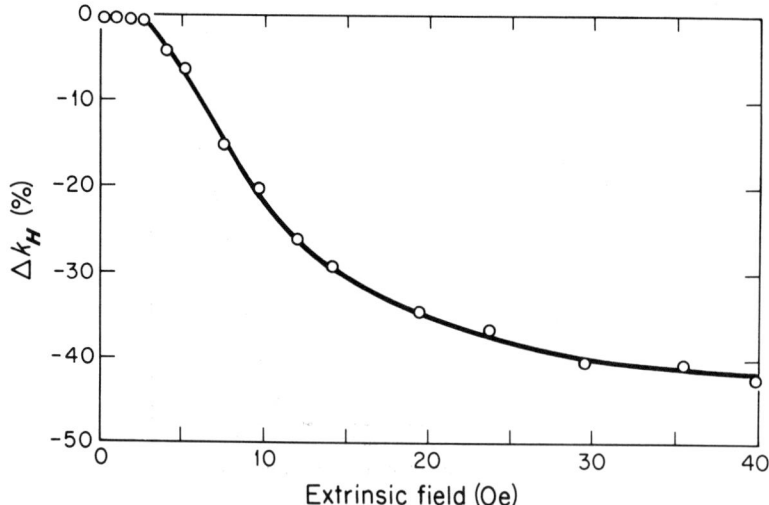

FIG. 3. Fractional decrease of conversion rate, Δk_H, as a function of applied field to 40 Oe over Lu_2O_3 pretreated in H_2 to 873 K. Measurements at 298 K.

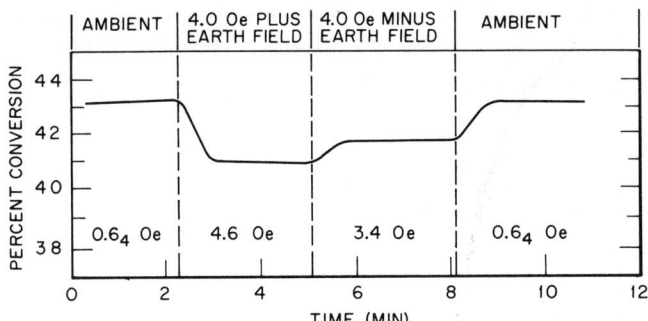

FIG. 4. Influence of the Earth's magnetic field on the conversion rate. These data were obtained on La_2O_3-supported Er_2O_3 although activated Lu_2O_3 and Y_2O_3 give the same, or even better, results.

ortho–*para*hydrogen conversion measurements. If the solenoid is mounted so that it may be rotated perpendicular to its axis, then the direction as well as the intensity of the Earth's field could be measured.

Yttria, after activation, shows magnetocatalytic effects much like those of Lu_2O_3. For instance, Rossington and Capozzi (12) found that evacuation of high purity Y_2O_3 at 773 K produced, in the 298 K region, an absolute nondissociative conversion rate of about 4.8×10^{-8} mol m^{-2} s^{-1}. Baron and Selwood (10) reported that activation in hydrogen at 823 K generated a rate of 1.5×10^{-6} mol m^{-2} s^{-1} at 298 K and atmospheric pressure. The rate at 77 K was much lower. Next to Lu_2O_3, Y_2O_3 has the largest fractional negative field effect for any catalyst thus far studied. Figure 5 shows the fractional rate change, Δk_H, as a function of field (log scale) up to 10 kOe. It was found that Δk_H was independent of k_0 in different preparations. All activity was instantly destroyed on exposure to oxygen.

This section will be concluded with some remarks on possible sources of nondissociative activity on the four diamagnetic solids described in the preceding paragraphs. (It is obvious that little progress can be made in finding the cause of the extrinsic field effect if we do not know the cause of the zero field activity.) There is an abundance of evidence that an intrinsic magnetic field will catalyze this kind of conversion and that the field generally comes from some obvious source of unpaired electrons in the catalysts. Thus all paramagnetic species catalyze the reaction. The possibility that the nondissociative conversion could proceed by a nonmagnetic mechanism has been the subject of speculation, but no actual evidence for such a process has ever been presented. The earlier literature related to this problem, as it applied to Al_2O_3, has been reviewed by Van Cauwelaert and Hall (13).

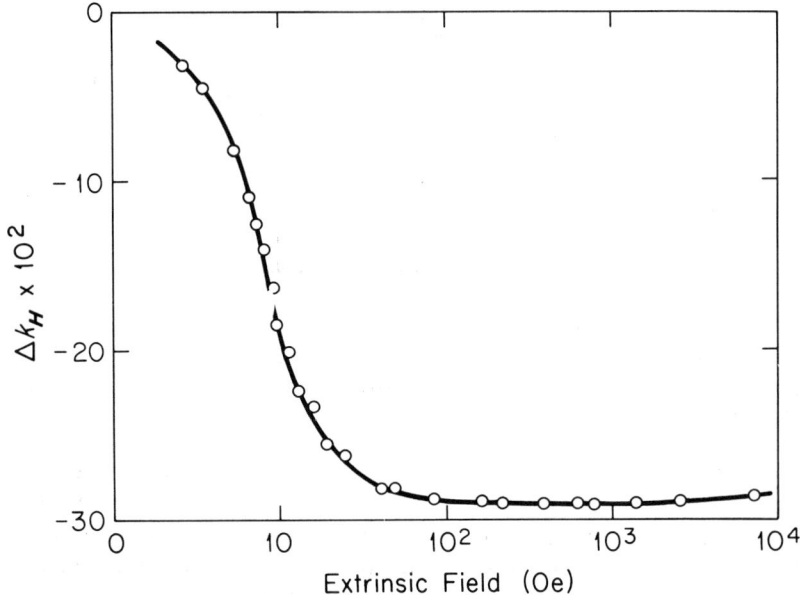

FIG. 5. Fractional (%) rate decrease vs. applied field to 10 kOe (log scale) over Y_2O_3 heated in H_2 at 823 K. Measurements at 298 K.

It is well known that hydrogen, and deuterium, especially in liquid form, undergo self-conversion. This is attributed to mutual interaction of the nuclear magnetic moments. The process is slow because the nuclear moments are small compared with electron moments. The abundant natural isotopes ^{27}Al, ^{139}La, and ^{175}Lu all have fairly large nuclear magnetic moments and it has been thought that activation of the oxides at high temperature, especially in a reducing atmosphere, might remove oxide ions from the surface. Molecular hydrogen would then have better access to the field of the nucleus and thus the conversion would be accelerated. But Y_2O_3 shows almost the same catalytic activity pattern as Lu_2O_3 and the moment of the abundant yttrium nucleus is small. It is, therefore, difficult to attribute the high activity of Lu_2O_3 and Y_2O_3 to nuclear magnetism unless no other source of conversion activity can be found.

A second possible mechanism for conversion on Al_2O_3, discussed by Van Cauwelaert and Hall (5, 13), involves the generation of a strongly adsorbed paramagnetic species such as H_2^+ formed through interaction of the solid with molecular hydrogen. Although no ESR signal attributable to such a species was observed we cannot ignore this possibility over Al_2O_3

at low temperature. But at room temperature H_2^+ would certainly be expected to react with molecular hydrogen. This mechanism seems improbable for the nondissociative activity shown by Lu_2O_3 and Y_2O_3.

The final mechanism that has received extensive investigation involves reduction of surface atoms from diamagnetic to paramagnetic states. This explanation was successfully invoked by Sandler (14) to explain the nondissociative conversion over titanium dioxide after activation *in vacuo* up to 793 K for several days. In this case reduction to Ti^{3+} was indicated not only by catalytic activity but also by the blue color characteristic of Ti_2O_3. Weller and Montagna (15) have shown that treatment of Al_2O_3 with hydrogen up to 823 K produces results that are consistent with, although not conclusive, for surface reduction of Al^{3+} to the paramagnetic Al^{2+}.

Baron and Selwood (10) showed that the development of conversion activity over Y_2O_3 was accompanied by the appearance of an ESR line at $g = 2.03$. This could be attributed to Y^{2+}. The intensity of the line was approximately linear with the specific conversion rate developed. Similar experiments on Lu_2O_3 were somewhat less convincing because of complexities caused by the larger nuclear moment of ^{175}Lu. Further information is available from the quantity of water evolved during the activation process. Arias and Selwood (16) heated high purity samples of Lu_2O_3 and of Y_2O_3 to 1123 K in air, then in hydrogen with continuous monitoring of the water evolved. A small amount of water was given off in the 623–773 K region but this was not associated with the development of appreciable catalytic activity (at 300 K). In the 823–973 K region water was evolved, the amount corresponding to less than 0.1% of the total oxide ion content of the whole sample. At higher temperatures very little more water was generated and little more catalytic activity was developed. Conversion activities were measured without removal of samples from the reactor. Typical results are shown in Table I.

TABLE I
Maximum Activation and Conversion Activities for Lu_2O_3 and Y_2O_3

	Lu_2O_3	Y_2O_3
Surface ($m^2\ g^{-1}$)	7.1	4.0
Water evolved ($\mu mol\ m^{-2}$)	2.4	1.5
k_0 ($\mu mol\ s^{-1}\ m^{-2}$)	1.0	1.6
Δk_H at 40 Oe	−0.42	−0.30

If the activation reaction may be written

$$2Y^{3+} + O^{2-} + H_2 \rightarrow 2Y^{2+} + H_2O$$

then two paramagnetic sites are generated for every water molecule evolved. For the rare earth C structure the unit subcell face length shown in Fig. 6 is 1.06 nm for Y_2O_3 and 1.04 for Lu_2O_3. If two R^{3+} ions must be reduced for each O^{2-} removed and if no more than one O^{2-} may be removed from those normally coordinated to R^{3+}, then it seems probable that only one O^{2-} could be removed from any one cell face such as the 010 shown in Fig. 6. The maximum number of Y^{2+} ions that could then be generated is $2/(1.06 \times 10^{-9} \text{ m})^2 = 1.8 \times 10^{18} \text{ m}^{-2}$. The result for Lu^{2+} is about the same. This corresponds to 1.5 μmol m^{-2} of water from the Y_2O_3 and nearly the same for Lu_2O_3. There is, as shown in Table I, a good correlation between the mechanism suggested, namely reduction to R^{2+}, and the experimental results. These considerations may have applications to the paramagnetic rare earths.

B. The Paramagnetic Rare Earths

In some respects the rare earths are almost like oxides of isotopes of the same metal. In other respects they differ in ways that are so subtle as almost to defy understanding. Data will be presented for extrinsic field effects at relatively high and low fields over most of the rare earths in normal (referred to here as "self-supported") form and also in "lanthana-supported" form. One example of a rare earth in "alumina-supported" form will be given.

Table II shows the zero field conversion rates k_0 and also the fractional

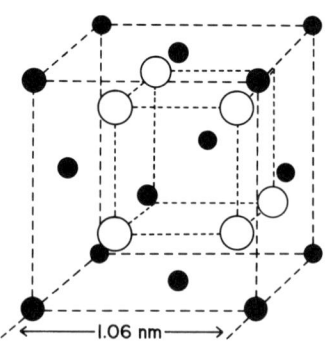

Fig. 6. The rare earth "C" subcell, ● = R^{3+}, ○ = O^{2-} (The true unit cell actually contains 16 R_2O_3 groups rather than the two such groups implied here).

TABLE II
Conversion Rates for Self-Supported Rare Earths and Fractional Rate Changes at 298 K in a Field of 18 kOe

R_2O_3	$k_0 \times 10^5$ (mol m^{-2} s^{-1})	Δk_{18}	$(k_0/\beta^2)^a \times 10^6$ (mol m^{-2} s^{-1})
Pr_2O_3	2.0	1.0	1.5
Nd_2O_3(C)	2.2	1.6	1.7
Sm_2O_3	0.26	0.5	1.2
Eu_2O_3	1.6	0.8	1.2
Gd_2O_3	3.5	0.4	0.58
Tb_2O_3	11.2	1.2	1.2
Dy_2O_3	11.1	0.7	0.99
Ho_2O_3	10.9	0.7	1.0
Er_2O_3	7.1	0.6	0.8
Tm_2O_3	7.0	1.5	1.2
Yb_2O_3	0.94	2.1	0.47

[a] β is the Bohr magneton number for R^{3+}.

change, Δk_{18}, in a field of 18 kOe. These data were obtained (2, 17) at room temperature and atmospheric pressure. At 77 K the conversion rate was much lower. Sample preparation consisted of ignition in air at 1073 K followed by *in situ* treatment in purified hydrogen at 823–873 K. Praseodymia and terbia were treated at higher temperature to insure reduction to the sesquioxide. Values of Δk_H at 40 Oe for the lanthana-supported samples are shown in Table III, all at atmospheric pressure and 298 K.

TABLE III
Fractional Rate Changes for Self-Supported and Lanthana-Supported Rare Earths at 298 K in a Field of 40 Oe

R_2O_3	Self-supported Δk	La_2O_3-supported Δk_H
Pr_2O_3	−0.1	—
Nd_2O_3	−0.02	—
Sm_2O_3	−0.07	—
Eu_2O_3	0	−0.24
Gd_2O_3	0	−0.03
Tb_2O_3	0	−0.31
Dy_2O_3	0	−0.23
Ho_2O_3	0	—
Er_2O_3	0	−0.28
Tm_2O_3	0	—
Yb_2O_3	−0.11	−0.26

These samples were prepared by impregnation of La_2O_3, of surface 2.0 m² g⁻¹, with rare earth nitrate solution of concentration sufficient to give approximately the same nominal surface concentration of the paramagnetic species as occurs in the self-supported oxide. Moderate differences in preparative treatment caused changes of 2 to 3 fold in k_0, but no significant change in Δk_H except as described later.

Figure 7 shows Δk_H for self-supported Pr_2O_3 up to 40 Oe. This is typical of all those self-supported rare earths showing any effect at low field, namely Pr_2O_3, Nd_2O_3, Sm_2O_3, and Yb_2O_3. Figure 8 shows Δk_H up to 18 kOe for Er_2O_3. This is typical of all self-supported samples at high field. Figure 9 shows Δk_H for Gd_2O_3 and for Gd_2O_3/La_2O_3. This figure also shows Δk_H for Er_2O_3/La_2O_3 in a pattern that is typical for all lanthana-supported samples except Gd_2O_3/La_2O_3.

To the data given in Tables II and III and in Figs. 7 to 9 some other information will be added. Terbia supported on alumina, Tb_2O_3/Al_2O_3 (*17*), shows a negative field effect resembling that for Er_2O_3/La_2O_3 up to 100 Oe but no measurable effect at 18 kOe. The Al_2O_3 surface was 160 m² g⁻¹ and the sample contained 0.14 atom % Tb_2O_3. The zero field conversion rate k_0 was about 1×10^{-4} mol m⁻² s⁻¹. At 40 Oe Δk_H was -22%. As was the case for Lu_2O_3 and Y_2O_3, the extrinsic field caused no change of hydrogen flow rate, no change of equilibrium constant, and the same effect is observed for para to ortho conversion as for the reverse. Some samples of the paramagnetic rare earths show an increase of k_0 with rising pretreatment temperature as is also the case for lutetia and yttria. In

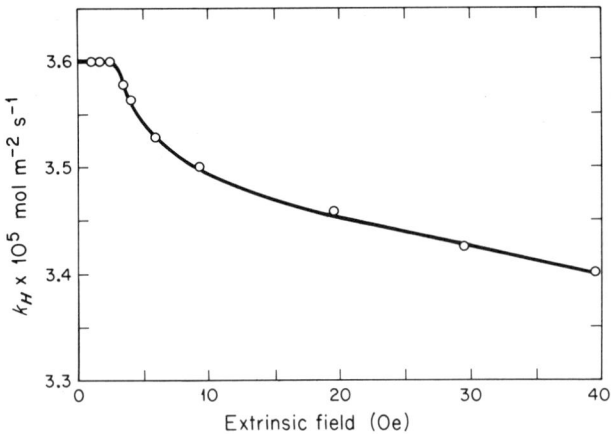

FIG. 7. Conversion rate, k_H, vs. field for Pr_2O_3 from 0 to 40 Oe, at 298 K.

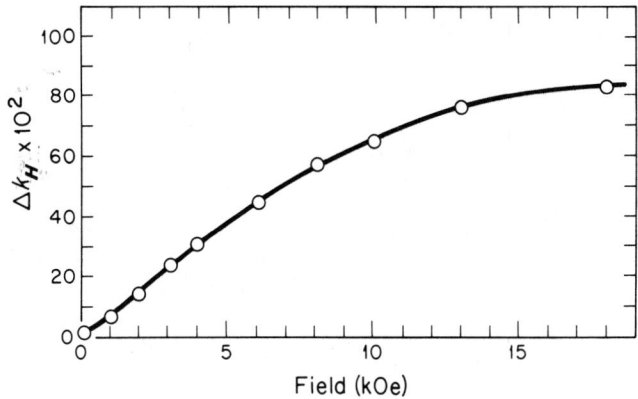

FIG. 8. Fractional rate change, Δk_H, for Er_2O_3 from 0 to 18 kOe, at 298 K.

general, as already pointed out, Δk_H is independent of k_0 but, for self-supported ytterbia, there is a peculiar dependence of Δk_H on k_0 at low fields (16). A sample of Yb_2O_3 of surface 5.6 m^2 g^{-1} had a decreasing Δk_H as the pretreatment temperature was raised from 823 to 973 K. As shown in Fig. 10, Δk_H at 40 Oe became almost zero at the highest activation temperature.

Additions to the above data have been provided by Eley et al. (4) who made absolute conversion rate measurements over Nd_2O_3 for both

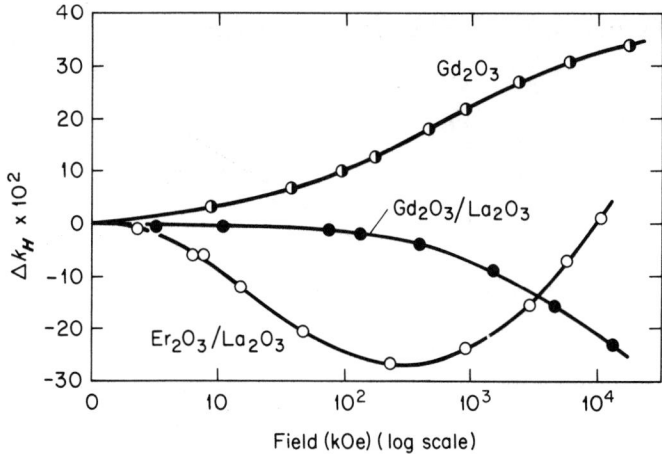

FIG. 9. Fractional rate changes, Δk_H, for Gd_2O_3, Gd_2O_3/La_2O_3 and Er_2O_3/La_2O_3 from 0 to 18 kOe (log scale), at 298 K.

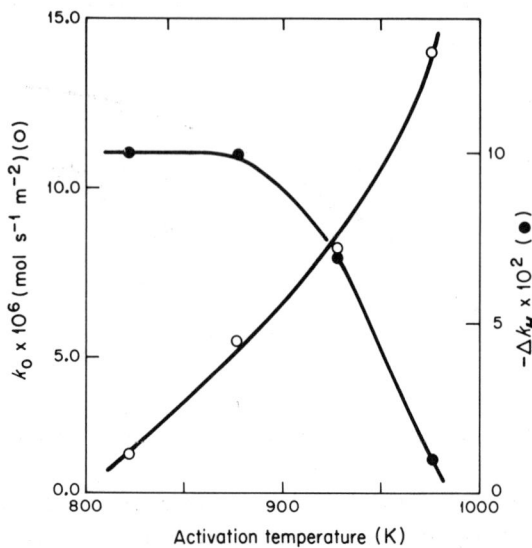

FIG. 10. Conversion rates, k_0, and fractional field effect (%) at 40 Oe as a function of activation temperature in H_2 on Yb_2O_3 at 298 K.

o–p-H_2 and o–p-D_2. Some of the results are shown in Fig. 11. The field effect shown in Table II for Nd_2O_3 was confirmed in fields up to 2.1 kOe and it was found that Δk_H for o–p-D_2 at 2.1 kOe is opposite in sign from that for o–p-H_2. Eley et al. also noted that Δk_H may be sensitive to the extent of surface dehydration during pretreatment. For instance, a sample

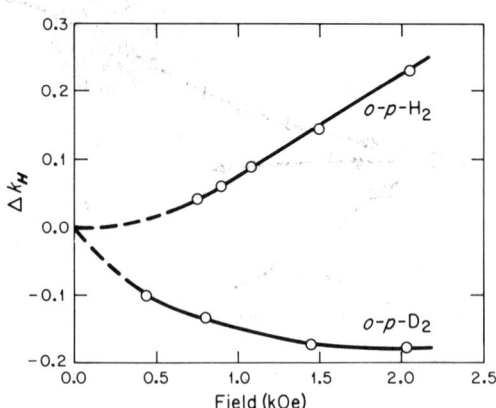

FIG. 11. Fractional field effects and absolute rates at 273 K for o–p-H_2 and o–p-D_2 on Nd_2O_3. [After Eley et al. (4).]

outgassed below 673 K gave no field effect although k_0 was readily measurable.

This section will be concluded with some further comments on the sources of nondissociative conversion activity over the rare earths and also on an area of possible disagreement between the results of Eley *et al.* (*4*) and the writer (*2, 16, 17*) for the field effect over neodymia. The classical paper of Wigner (*18*) showed a dependence of conversion rate on β^2 where β is the Bohr magneton number of the paramagnetic species. This has been confirmed experimentally in many reports of which only two will be mentioned. The first is the absolute rate measurements of Ashmead *et al.* (*19*) over Nd_2O_3, Sm_2O_3, Gd_2O_3, Dy_2O_3, and Er_2O_3. Excellent constancy of k_0/β^2 was found for these catalysts. The second report is that shown in Table II where, although β^2 varies by a factor of 50, k_0/β^2 varies by a factor of only 2 from the average. There can be no doubt that β^2 is rate controlling although it is by no means the only variable in the rate expression. In view of this our first problem is to find a source for the change of k_0 in Lu_2O_3 and Y_2O_3 with increasing pretreatment temperature. Eley *et al.* (*4*) relate pretreatment temperature to progressive removal of OH^- and this view is supported by tracer estimation of residual hydrogen at each step. This removal would, presumably, allow closer access of molecular hydrogen to the paramagnetic ion or, possibly, increased efficiency of conversion through the generation of electrostatic fields (*5*). There seems little doubt that dehydroxylation does, indeed, increase catalyst activity but it does not necessarily generate a catalyst site. If the nondissociative conversion reaction requires a paramagnetic catalyst, then activated yttria and lutetia must contain some paramagnetic species such as Y^{2+} and Lu^{2+}. If, as shown earlier (*16*), there is evidence that the maximum site densities are the same for Lu_2O_3 and Yb_2O_3, then it is possible to estimate the magnetic moment developed on the Lu_2O_3 because k_0/β^2 should be the same for both catalysts. From the data given in Figs. 2 and 10 it is found that the paramagnetic species on Lu_2O_3 activated at 923 K has a Bohr magneton number of about 1.6 and on Y_2O_3 it is about 2.0. These differ only slightly from the "spin-only" number for one electron. The results taken together with the ESR evidence virtually confirm the existence of a paramagnetic species on activated Lu_2O_3 and Y_2O_3. But they do not prove that the species is a d electron. Lutetia is very similar to ytterbia. The change $Yb^{3+}(4f^{13}) \rightarrow Yb^{2+}(4f^{14})$ creates a diamagnetic species. On the other hand, if the change in ytterbia should be $Yb^{3+}(4f^{13}5d^0) \rightarrow Yb^{2+}(4f^{13}5d^1)$ or $Yb^{2+}(4f^{13}5d^06s^1)$ then may it not occur for all the other rare earth ions or, at least, for those in the C crystal structure? This, in turn, suggests that activation of the rare earths may actually consist of two processes and that the catalytic sites formed do not necessarily have identical properties.

We now attempt to classify all the rare earths according to the several kinds of field effects they exhibit.

1. All paramagnetic self-supported rare earths show a large positive rate change in fields of over 1 kOe. The effect reaches saturation at about 10 kOe.
2. Excepting Eu_2O_3 those paramagnetic rare earths of relatively low magnetic susceptibility show a moderate negative field effect at about 40 Oe. This becomes positive at higher fields.
3. Excepting Gd_2O_3/La_2O_3 all lanthana-supported rare earths show a negative field effect at about 40 Oe, but this becomes more nearly positive at 10 kOe. The same is true of Tb_2O_3/Al_2O_3.
4. The diamagnetic earths Lu_2O_3 and Y_2O_3 show negative field effects above about 3 Oe. This approaches saturation at about 40 Oe.

It might be thought from its anomalous behavior that in self-supported Eu_2O_3 the Eu^{3+} is actually reduced to Eu^{2+}, but this change is accompanied by a marked increase in β which is not reflected in an abnormal k_0. The anomalous behavior of Gd_2O_3/La_2O_3 may be related to the fact that Gd^{3+} is the only trivalent rare earth ion in an S spectroscopic state.

C. Chromia

All solids having a relatively high electron spin density exhibit Heisenberg exchange interactions leading, at low enough temperature, to magnetic phase transitions. If the exchange interaction is negative (antiparallel) the sample becomes antiferromagnetic below the transition temperature called the Néel point, T_N. If the interaction is positive (parallel) the sample becomes ferromagnetic below the transition temperature called the Curie point, T_C. Numerous complexities related to these effects are described by Schieber [20]. Transition temperatures for the rare earth sesquioxides described in earlier sections are all too low to be of consequence for the *ortho–para*hydrogen conversion. This section and the next will be devoted to three solids having Néel points not too far from room temperature. These will be followed by three solids having Curie points in an accessible, if less convenient, temperature range.

For pure α-chromia, α-Cr_2O_3, $T_N = 308$ K. At this temperature and up to about 350 K the conversion mechanism is less than 1% dissociative. But this is true only if the pretreatment consists of flowing hydrogen at

773 K followed by rapid cooling to room temperature or by cooling in vacuum or pure helium. Details of preparative procedures, the kinds of activity generated and the possible reasons thereof have been described by the writer (21). It is also to be noted that the catalyst surface prepared in this way is slowly poisoned by molecular hydrogen at room temperature, more rapidly as the temperature is raised (21). These considerations sharply limit the maximum temperature at which the extrinsic field effect may be studied on this catalyst.

Figure 12, derived from the results of Misono and Selwood (22) and of Ng and Selwood (23), shows Δk_H for α-Cr_2O_3 as a function of extrinsic field up to about 18 kOe at 311.2 K. For a sample of surface 6.0 m² g^{-1} and at atmospheric pressure, k_0 at 311.2 K was about 9.8 μmol m^{-2} s^{-1}. This rate varied somewhat, as expected, with small differences in pretreatment conditions. The rate was also sufficiently sensitive to the temperature of measurement, especially near T_N, as to require control to ±0.1 K.

Figure 13 shows Δk_H vs. H for the same sample of α-Cr_2O_3 under the same conditions except that the reactor temperature was 306.7 K. The value of k_0 at 306.7 K was about 16.0 μmol m^{-2} s^{-1}. At a reactor temperature of 77 K, k_0 was about 5 μmol m^{-2} s^{-1} and Δk_H at 17 kOe was negative but quite small.

As a partial summary of these effects over α-Cr_2O_3, Fig. 14 shows conversion rates at zero field and at 17 kOe over the temperature range 300 to 318 K. Some further details related to these effects are to be found in Ref. 23.

It was mentioned above that α-Cr_2O_3, in various forms, is poisoned by

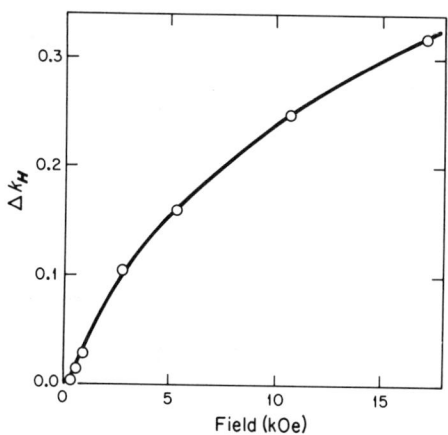

FIG. 12. Fractional rate change, Δk_H, vs. extrinsic field for α-Cr_2O_3 at 311.2 K.

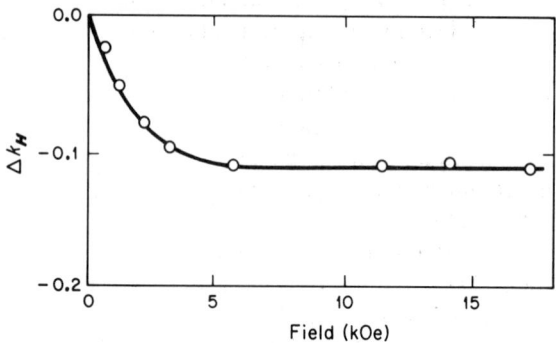

FIG. 13. Fractional rate change, Δk_H, vs. extrinsic field for α-Cr$_2$O$_3$ at 306.7 K.

hydrogen for the nondissociative conversion. The extrinsic field effect has been measured (21) at 18 kOe, and at various stages of poisoning, both above and below T_N. An example of the results is shown in Fig. 15 which gives Δk_{18} as a function of diminishing k_0. It will be seen that Δk_{18} is independent of k_0 at 318 K and only very slightly dependent at 290 K.

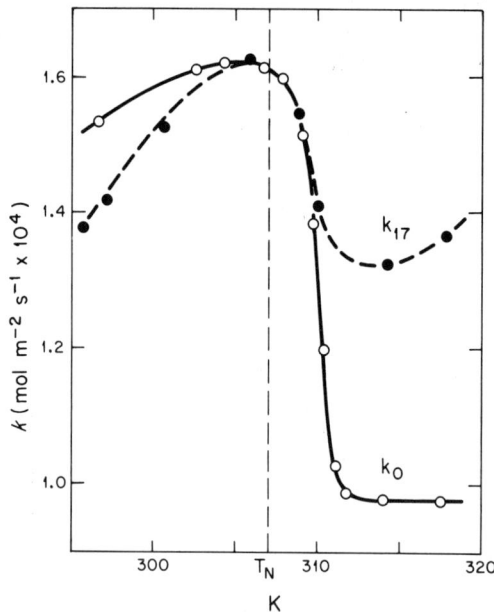

FIG. 14. Conversion rates, in zero field, k_0 (○), and in 17 kOe, k_{17} (●) on α-Cr$_2$O$_3$ from 300 K to 318 K.

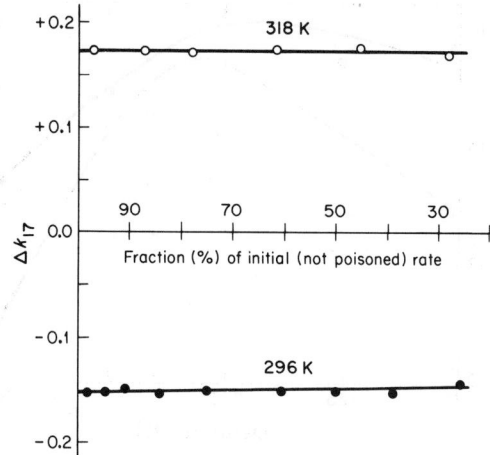

FIG. 15. Fractional field effect at 17 kOe, Δk_{17}, for α-Cr_2O_3 subject to progressive H_2 poisoning. Poisoning and measurements made at 318 K (O) and at 296 K (●). The time involved, from left to right, was about 2 hr.

Solid solutions of α-Cr_2O_3 in α-Al_2O_3 have received investigation because, in part, of the wealth of other information available on this system. In such solutions the Néel point is lowered and, with decreasing chromia concentration, it disappears. Extrinsic field measurements have been made (22) on powdered ruby of surface 2 m g^{-1} at several Cr_2O_3 concentrations. Samples were pretreated in hydrogen at 773 K, then cooled quickly to minimize poisoning. For a 1.0 atom % Cr_2O_3 sample k_0 and $k_{7.5}$, from 173 to 373 K, are shown in Fig. 16. Figure 17 shows Δk_H as a function of field, at 173 K. There was no significant difference between ruby of 1.0 and 0.10 atom % Cr_2O_3. At the lowest field used, 200 Oe, there was no indication of a negative Δk_H.

Chromia supported on powdered sapphire (α-Al_2O_3) of surface 2 m^2 g^{-1} has also been studied (22). Results for k_0 and $k_{7.5}$ on a sample containing 0.17 atom% Cr_2O_3 are shown in Fig. 18. Figure 19 shows Δk_H as a function of field at 173 and 298 K.

There appear to be no extrinsic field studies on the familiar amorphous chromia gel catalyst although it is known (24) to have only a slight indication of the zero field anomaly shown by α-Cr_2O_3 at T_N. A gel catalyst aged in hydrogen has a small positive field effect at 323 K and a small negative effect at 273 K.

This section will be concluded with some comments on the various patterns of activity shown by α-Cr_2O_3 in the several forms described. Prior to activation in hydrogen the nondissociative conversion activity is

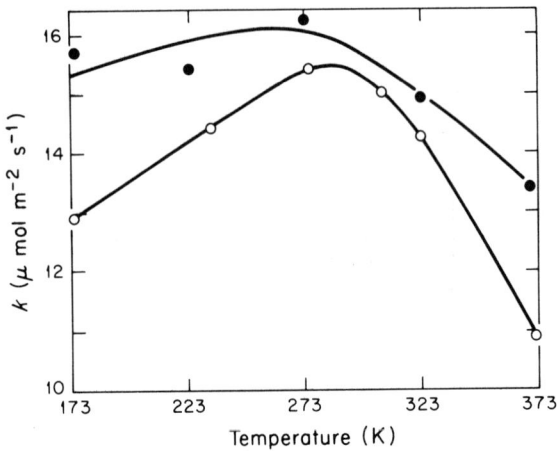

FIG. 16. Conversion rates in zero field (○) and in 7.5 kOe (●) over 1 atom% ruby (Cr_2O_3–Al_2O_3).

quite low but activation makes it surprisingly high. This change is probably due to removal of OH^- and perhaps of O^{2-} with simultaneous reduction of the Cr^{3+} to Cr^{2+}. The latter has the larger magnetic moment. Thus any attempt to related activity to structure is complicated by the likelihood that the catalyst is actually Cr^{2+} supported on Cr_2O_3. The

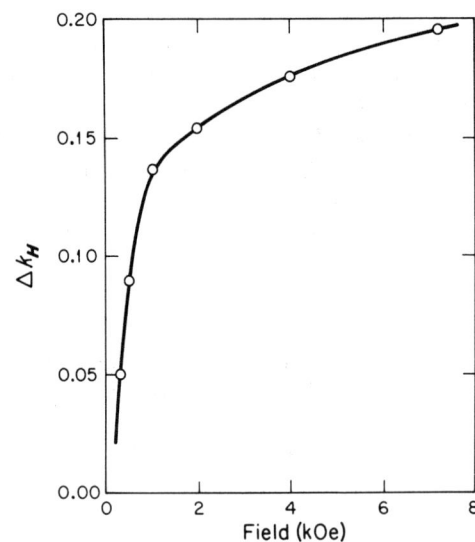

FIG. 17. Fractional field effect, Δk_H, for 1 atom% ruby (Cr_2O_3–Al_2O_3) at 173 K.

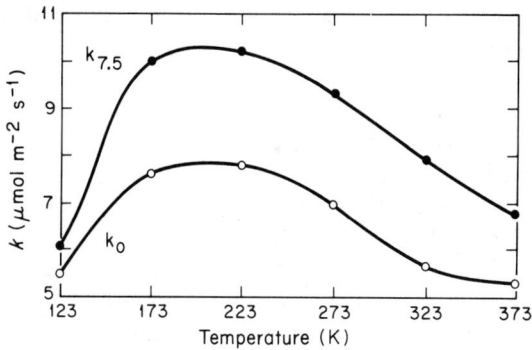

FIG. 18. Conversion rates in zero field (○) and in 7.5 kOe (●) over 0.17% Cr_2O_3 supported on α-Al_2O_3.

poisoning effect caused by hydrogen is probably due to steric blocking by hydride ion (21), but this is normally of concern only if the field measurements are performed above room temperature and cannot be concluded promptly.

The abrupt changes of reaction rate that occur at the Néel point over α-Cr_2O_3 are, of course, examples of the Hedvall effect (type I). If the α-Cr_2O_3 is dispersed on a diamagnetic support (25), or by being prepared in a high specific surface form as in chromia gel (26), or dissolved in α-Al_2O_3 (7), then the normal coordination with respect, especially, to Cr–Cr is diminished. It may then be expected that the various anomalies observed near the phase transition will progressively disappear and this is actually observed.

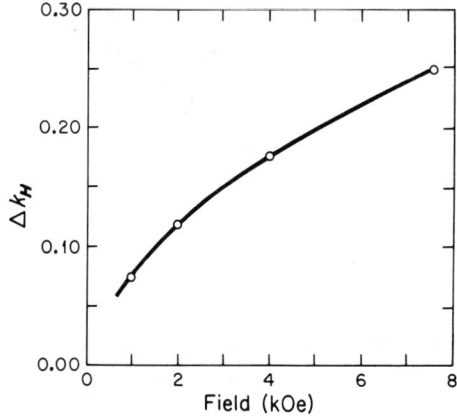

FIG. 19. Fractional field effect, Δk_H, for 0.17% Cr_2O_3 supported on α-Al_2O_3 at 173 K.

The results available on chromia may be summarized, in qualitative terms, as follows:

1. Both k_0 and Δk_H are closely related to T_N.
2. For preparations in which the chromium ion is dispersed the relation to T_N is progressively lost.
3. In the paramagnetic phase, above T_N, Δk_H is positive tending to become independent of field at 10 kOe and higher.
4. In the antiferromagnetic phase below T_N, Δk_H is negative and tends to become independent of field at 1 kOe and higher.
5. There is no indication of any effect at fields of a few oersteds.
6. The effect on Δk_H (at least at 18 kOe) of progressive hydrogen poisoning is negligible.

D. Cobalt Monoxide (23)

The Néel point of cobalt monoxide is 291 K. For obvious reasons this oxide cannot be heated in hydrogen without reduction to the metal. A sample of surface 5.8 m² g⁻¹ was pretreated *in situ* by heating it in purified helium at 823 K. The catalyst was then cooled to near room temperature before the admission of hydrogen. Figure 20 shows Δk_H for CoO as a function of extrinsic field up to 18 kOe at 301 K. At this temperature k_0 was about 7.2 μmol m⁻² s⁻¹. Figure 21 shows results for the same sample at 275 K. Figure 22 shows k_0 and Δk_H at 17.3 kOe over the temperature range 275–300 K.

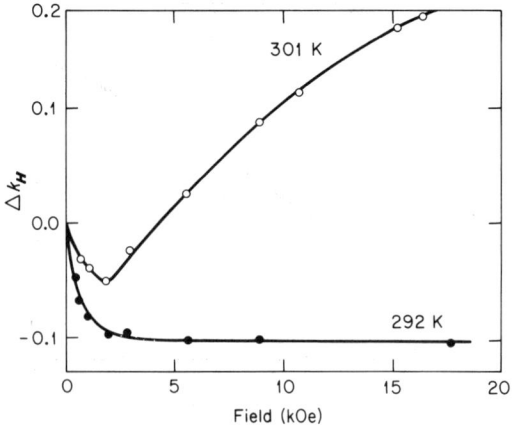

FIG. 20. Fractional field effect, Δk_H, for CoO at 301 K (○) and at 292 K (●).

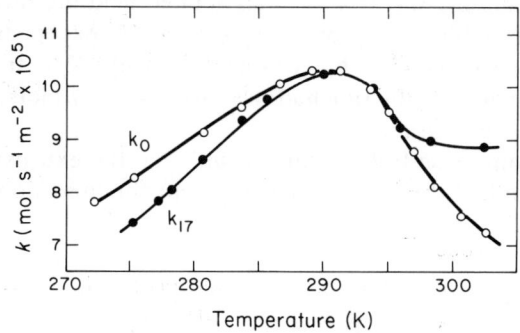

FIG. 21. Conversion rates in zero field (○) and in 17 kOe (●) for CoO from 272 K to 303 K.

Like α-Cr_2O_3 the CoO shows a change of zero field conversion rate (a Hedvall effect) near T_N although the plots of k_0 vs. temperature differ considerably for the two catalysts. Like α-Cr_2O_3 the CoO shows a positive field effect above T_N and a negative effect below. But the changes that occur for CoO are not so abrupt as are those for α-Cr_2O_3. Below T_N it is found that Δk_H for CoO changes appreciably with temperature. For a short distance above T_N a negative field effect persists at moderate fields. Further details are given in Ref. 23.

E. Manganese Monoxide (23)

A few results are available on MnO (T_N = 122 K). A sample preheated in hydrogen at 823 K and of surface 6.0 m² g⁻¹ showed a conversion rate at

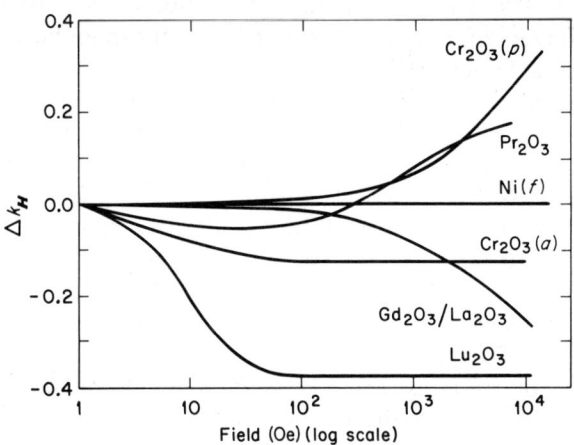

FIG. 22. Major types of Δk_H vs. H (log scale).

296 K of $k_0 = 6.0$ μmol m^{-2} s^{-1}. In an applied field the rate rose rapidly in a pattern resembling that for α-Cr$_2$O$_3$ (Fig. 12) except that at 17 kOe the total fractional change, Δk_{17}, was slightly over 200%. In the antiferromagnetic phase at 77 K the rate had risen to $k_0 = 8.3$ μmol m^{-2} s^{-1}, but no field effect was measurable.

It is clear from these fragmentary results that the extrinsic field behavior of MnO is much like that of CoO except for the astonishingly large value of Δk_H.

F. Chromium Dioxide, Europium Monoxide, and Nickel (Metal)

Few of the many known ferromagnetic solids are suitable as catalysts for the nondissociative *ortho*–*para*hydrogen conversion. This is especially true if measurements are needed in the neighborhood of the magnetic phase transition, T_C. The reasons for this are threefold: the solid may decompose at the temperature necessary to free the surface from contaminants, the Curie point may be so low that the experimental difficulties are formidable, and many such solids show strong dissociative conversion activity near T_C. Of the three solids named above none is very satisfactory.

Chromium dioxide, CrO$_2$, of surface 28.5 m^2 g^{-1} and $T_C = 394 \pm 0.5$ K has been reported by Arias and Selwood (27) to have negligible activity as normally prepared. But pretreatment with pure flowing helium at 623 K produces a surface with conveniently measurable activity at room temperature. All activity was quickly lost in hydrogen near T_C. It is necessary, therefore, to restrict measurements to the ferromagnetic phase. At 298 K and atmospheric pressure $k_0 = 1.2$ μmol m^{-2} s^{-1}. At 17.5 kOe the change in conversion rate was less than 0.5%. It cannot be said that there is any measurable field effect.

Europium monoxide, EuO, is another ferromagnetic solid that may be used for studying the extrinsic field effect although the low Curie temperature restricts the usefulness of the results. Data are available (27) on two samples, one from the insulative part of the phase diagram with $T_C = 69.3 \pm 1.0$ K and surface 1.4 m^2 g^{-1}. (All samples of EuO are very sensitive to moist air and must be handled accordingly.) This sample is referred to below as EuO (*ins*). Another sample of EuO, from the conductive part of the diagram, had $T_C = 129 \pm 3$ K and surface 0.8 m^2 g^{-1}. This is referred to as EuO (*con*).

The sample of EuO (*ins*) had a constant average conversion activity of $k_0 = 20$ μ mol m^{-2} s^{-1} in the temperature range 64.5 to 113 K, with no indication of any abrupt change near T_C. It did not prove possible to measure Δk_H for this sample below T_C. But at 78 K, Δk_H was +0.24 at 17.5 kOe. at 300 K, k_0 was 60 μmol m^{-2} s^{-1} and $\Delta k_{17.5}$ was +0.46.

For EuO (*con*), k_0 averaged 14 μmol m^{-2} s^{-1} over the temperature range 78 to 142 K with no trend or abrupt change near T_C. For this sample $\Delta k_{17.5}$ at 78 K was less than ±0.1 which, under the difficult experimental conditions, was within the limit of reproducibility. At 300 K, $k_0 = 39$ μmol m^{-2} s^{-1} and $\Delta k_{17.5} = +0.40$.

To summarize these results, it may be said that both samples of europium monoxide showed appreciable zero field activity below and above T_C. Both showed a fairly strong positive field effect in the paramagnetic region. If there was any field effect below T_C for the one sample measured it was quite small.

Nickel metal has a Curie point at 631 K. At this temperature, and down to about 150 K, it shows extraordinarily strong catalytic activity for the hydrogen–deuterium equilibration, for the *dissociative ortho–para*hydrogen conversion and for many other reactions. At 77 K, however, the dissociative activity is small compared with the nondissociative and the latter is easily measured. In 1953, it was reported by Justi and Vieth (*28*) that an extrinsic field strongly increases the conversion rate and also that of certain other reactions. In view of the possible significance of this claim it will be examined in detail. The catalyst used was a nickel capillary tube of 1-mm internal diameter and a total inner surface of 31 cm². The pretreatment was heating in air at 733 K followed by hydrogen at 543 K. Hydrogen was forced through the capillary at from 30 to 360 cm³ min^{-1}. Reaction temperatures were in the range 346 to 378 K. From this information it may be concluded that the conversion rates were in the order of 20 mmol m^{-2} s^{-1}. The rates rose very rapidly with increasing applied magnetic field and, in some cases, the rates were doubled or tripled at 10 kOe. The rates also showed a strong increase with rising temperature. No magnetocatalytic effects were observed over nickel powder.

In view of the nature of the catalyst, the pretreatment, the reaction temperature range, the very large specific conversion rate, and the positive temperature coefficient it may be concluded that the reaction studied by Justi and Vieth was not the nondissociative conversion. Furthermore, Schwab and Kaiser (*29*) reported that they had been unable to confirm the Justi and Vieth results (over nickel wire). The writer (*30*) found no field effect for the conversion at 298 K greater than ±0.5% over pretreated nickel wire up to 4 kOe. It is necessary to conclude that the Justi and Vieth work is of no significance insofar as the present paper is concerned, and also that there is doubt concerning the reality of their results.

The writer (*30*) has also attempted to measure the extrinsic field effect over pretreated nickel wire, and over nickel supported on silica gel, at 77

K. The nondissociative conversion is readily observed at this temperature but no field effect in excess of ±0.5% was found up to 18 kOe. In conclusion, it appears that, while nickel metal shows considerable nondissociative conversion activity, this is obscured by the dissociative activity above about 150 K and this is far below T_C. Thus, in the ferromagnetic phase no field effect has been found.

Summarizing the results on ferromagnetic catalysts it may be said that, where measurable, the nondissociative conversion activity shows a strong positive extrinsic field effect above T_C, but none in the ferromagnetic phase. There are no anomalies near T_C comparable with those observed near T_N in antiferromagnetic catalysts. But this may be due to the failure to find any suitable ferromagnetic substances with T_C above about 130 K. It is quite possible that the conversion rate controlling mechanism is different at this temperature as compared with that in the 300 K region (27).

IV. Correlation of Experimental Results

Certain correlations are possible but it must be kept in mind that comparisons of activity in heterogeneous catalysis are likely to be treacherous. Minor differences in surface pretreatment may make major differences in properties. Nevertheless, it is clear that there are recognizable patterns of extrinsic field effects. The several kinds are shown in Fig. 22 wherein Δk_H is plotted against H (log scale). The curves drawn are meant to be representative, but except for relatively minor details, all of the samples studied to date fall into one of the six patterns shown. This covers over 30 different catalyst preparations, a variety of temperature-sensitive magnetic phases, and various pretreatments. Figure 22 shows one example of catalysts found to have the indicated kind of field effect. Pretreatment conditions are stated in Table IV which includes all samples on which measurements have been made. Pretreatments are abbreviated as follows: ($H_2$773q298) means that the sample had been heated in hydrogen for an hour or more at 773 K and then cooled rapidly to 298 K. One or two temperatures at which each Δk_H pattern has been observed are also shown in Table IV.

A few additional comments will be made on some experimental anomalies. The plot of Δk_H vs. H for α-Cr_2O_3 above T_N resembles those for several of the paramagnetic rare earths such as Er_2O_3 but it rises more abruptly. Both α-Cr_2O_3 and Er_2O_3 have a high electron moment density. But if the α-Cr_2O_3 is diluted by being dissolved in, or supported on, α-Al_2O_3 it still shows approximately the same extrinsic field pattern. The

TABLE IV
Catalysts and Conditions for Extrinsic Field Patterns shown in Fig. 22

Type (Fig. 22)[a]	Catalysts[a]	Pretreatment	Reactor temperature (K)
$Cr_2O_3(p)$	R_2O_3 (R = Eu, Gd, Tb, Dy, Ho, Er, Tm)	($H_2$823q298)	298
	α-$Cr_2O_3(p)$	($H_2$773q298)	311
	Cr_2O_3–Al_2O_3[b]	($H_2$773q298)	173–373
	Cr_2O_3/Al_2O_3[c]	($H_2$773q298)	173–373
	$CoO(p)$	(He823q298H_2)	301
	$MnO(p)$	($H_2$823q298)	296
	$EuO(ins)(p)$	($H_2$773q298)	78–300
	$EuO(con)(p)$	($H_2$733q298)	300
Pr_2O_3	R_2O_3 (R = Pr, Nd, Sm, Yb)	($H_2$823q298)	298
	R_2O_3/La_2O_3 (R = Eu, Tb, Dy, Er, Yb)[c]	($H_2$823q298)	298
	Tb_2O_3/Al_2O_3[c]	($H_2$823q298)	298
$Ni(f)$	CrO_2 (f)	(He623q298H_2)	298
	$EuO(con)(f)$	($H_2$773q298)	78
	$Ni(f)$	($H_2$523evac77)	77
$Cr_2O_3(a)$	$CoO(a)$	(He823q298H_2)	275
	$Cr_2O_3(a)$	($H_2$773q298)	307
Gd_2O_3/La_2O_3[c]	Gd_2O_3/La_2O_3[c]	($H_2$823q298)	298
Lu_2O_3	La_2O_3	($H_2$823q298)	298
	Lu_2O_3	($H_2$823q298)	298, 77
	Y_2O_3	($H_2$823q298)	298, 77

[a] (a) Antiferromagnetic phase below T_N; (f) ferromagnetic phase below T_C; (p) paramagnetic phase above T_N or T_C.
[b] R_2O_3–Al_2O_3: a solid solution.
[c] R_2O_3/Al_2O_3: a supported oxide.

only important difference is that the effects at the Néel point are eliminated. Such samples do not show the large negative weak field effects found for the La_2O_3-supported rare earths or for Tb_2O_3/Al_2O_3. It is clear that the differences between Cr_2O_3/Al_2O_3 and, say, Tb_2O_3/Al_2O_3 cannot be attributed solely to magnetic dilution of a paramagnetic species.

Another anomaly is to be found in the complicated changes that occur very close to T_N. For instance, the maximum negative values of Δk_H are found for α-Cr_2O_3 and CoO slightly above the Néel temperature, but this occurs only at fairly high fields.

It was mentioned earlier that Eley et al. (4) observed no field effect for Nd_2O_3 pretreated at less than 673 K. It was thought that this indicated some relationship between Δk_H and the degree of dehydroxylation of the surface. Progressive dehydroxylation of catalyst surfaces often increases k_0. Eley et al. (31) have presented strong evidence that this is related to improved exposure of the paramagnetic ion to molecular hydrogen. This is almost certainly true, even if the ion suffers some change of oxidation state simultaneously. A possible explanation for these effects is to be found by examining Δk_H vs. H for the Pr_2O_3 type in Fig. 22. This curve, in slightly modified form, is that also observed for Nd_2O_3. The measurements of Δk_H on Nd_2O_3 carried out by Eley et al. (4) were at a maximum of 2.1 kOe which, in Fig. 22, is about the field where Δk_H changes sign. A possible reason for the effect reported by Eley et al. is that as k_0 increases with dehydroxylation the field at which Δk_H changes sign becomes lower. Something of this kind was observed by Arias and Selwood (Fig. 10) who found that the negative Δk_H over Yb_2O_3 at 40 Oe disappeared when the pretreatment temperature was raised to 973 K. The progressive poisoning of α-Cr_2O_3 by molecular hydrogen (16) may be considered as approximately the reverse of the dehydroxylation process, although the mechanism may be different. Obviously, access to the paramagnetic ions by the hydrogen is diminished as k_0 becomes low, but no change of Δk_H was observed. A reason for this may well be that because the measurements were made at 18 kOe the possible influence of a negative Δk_H was not observed. But the dehydroxylation results of Eley, the effect of dehydroxylation on Yb_2O_3, and the effect of progressive hydrogen poisoning of α-Cr_2O_3 all point in the direction of a change in the relative proportions of sites responsible for a positive or a negative field effect. The changes observed appear to be less a matter of formation or destruction of sites but rather of increasing or decreasing accessibility to the sites by molecular hydrogen. Although this evidence is far from being conclusive it supports the view that two different kinds of sites for the conversion reaction may be generated in many catalysts.

Not shown in Fig. 22 or Table IV are the results of Eley et al. (4) on the ortho-paradeuterium conversion over Nd_2O_3 pretreated in vacuum for 150 hr at 673 K. The pattern of activity found most closely resembles that of Lu_2O_3.

V. Conversion and Field Effect Theory

It is obvious that no theory of the extrinsic field effect can be developed without a good understanding of conversion theory for the nondissociative process in the absence of an applied magnetic field. The status of k_0 calculations for various systems has been reviewed many times and will be summarized very briefly here. The original development by Wigner

(18) showed that the conversion rate is proportional to $m_c^2 m_p^2$ where m_c is the magnetic moment of the paramagnetic catalyst species and m_p that of the proton. Refinements were made by many authors including Kalckar and Teller (32), Harrison and McDowell (33), Leffler (34), Nielsen and Dahler (35), Golovin and Buyanov (36, 37), and by Scalapino and Petzinger (38). In all this work there has been no serious challenge to the significance of the paramagnetic species in the catalyst. The work of Scalapino and Petzinger has made it possible, in favorable cases, to calculate conversion rates of a realistic order, and to provide some insight to the peculiar and abrupt changes in conversion rates observed over some catalysts in the temperature region of the magnetic phase transition, T_N.

The situation with respect to the extrinsic field effect for the nondissociative ortho–parahydrogen conversion is quite different. It is implied, though not specifically so stated, in the work of Leffler (34) that a change of electron spin relaxation time caused by the extrinsic field could affect the conversion rate. If this is true, it might be expected that those rare earth ions with an even number of unpaired electrons would show a sharply different field effect from those with an odd number. There is little or no evidence (17) that this occurs, although the anomalous field effect for lanthana-supported gadolinia (2) might be considered a possible example. An attempt has been made by Ilisca and Gallais (39) to relate the field effect to spin wave interactions on ferromagnetic catalysts. As described in Section III.F there is no good experimental evidence that such an effect occurs with the catalyst in the ferromagnetic phase. Golovin and Buyanov (40) have tried to relate the field effect to their earlier proposal that the zero field conversion may occur without uncoupling of the nuclear spins in the hydrogen molecule. It is suggested by them that the transition probability is enhanced by the addition, to the normal intrinsic field, of a time-averaged projection of the paramagnetic moment in the direction of the extrinsic field. It is difficult to understand why such an induced increment would not be completely destroyed by thermal effects. Furthermore, an assumption of long relaxation times appears to be without confirmation. It will also be noted that the proposed mechanism offers no explanation for the negative field effects in the 10 Oe region such as are shown by yttria and lutetia. Further theoretical work on this elusive problem is much to be desired.

VI. Conclusions

Understanding the several extrinsic field effects has proved to be difficult because of our failure, in more than a few cases, to identify the catalyst species involved. This concluding section will be devoted to a brief summary of the more important experimental facts, to some further discussion of catalyst sites, and to a few speculative remarks concerning the field effects on these sites.

The experimental facts that appear to be the most significant are as follows. Over many paramagnetic solids including α-Cr_2O_3, CoO, EuO (above T_N or T_C) and the rare earth sesquioxides, the catalyzed nondissociative *ortho*–*para*hydrogen conversion rate may be increased (up to two- or threefold) by applying a magnetic field to the catalyst. The rate rises rapidly in the 1 kOe region and approaches a constant value in the 10 kOe region. Below the Néel point the effect abruptly becomes negative. Below the Curie point there is no effect. Several of the rare earths show a moderate decrease of conversion rate in the 10 Oe range. If the rare earths are supported at a low concentration on Al_2O_3 or La_2O_3 all except Gd_2O_3 show a large negative weak field effect. The diamagnetic oxides yttria and lutetia show a strong negative effect if they are first heated above 823 K in vacuum or hydrogen. These negative effects are measurable at about 3 Oe and the rate becomes constant above about 40 Oe. In the 4 Oe region the rates are so sensitive that the influence of the earth's field is detectable. All the effects mentioned above are instantaneous and reversible. An observation of the o–p-D_2 conversion rate over Nd_2O_3 shows a negative rate change at 2.1 kOe. Under identical conditions the o–p-H_2 rate change is positive.

We turn now to consideration of possible sites for catalytic activity in these various oxides. For a solid such as Cr_2O_3 the obvious source of an intrinsic magnetic moment is the Cr^{3+} ion which, with its three unpaired d electrons has a (spin-only) Bohr magneton number of about 3.8, the orbital component being "quenched." This is, of course, the case above the Néel point at ~307 K although many catalyst preparations such as chromiagel, dilute Cr_2O_3/Al_2O_3 and dilute solid solutions in diamagnetic solids are paramagnetic far below the normal T_N. Surface catalytic activity depends not only on bulk purity and surface purity, but it depends also on surface coordination and surface stoichiometry. These in turn depend on the temperature and time of pretreatment, the nature of the gas, if any, present during pretreatment and even on the rate of cooling to reaction temperature. Many chromia preparations initially contain surface hydroxyl groups. These may be removed by heat treatment in vacuum or in hydrogen. There is thus a decrease of coordination and, if the dehydroxylation proceeds by a reaction such as $2OH^- \rightarrow O^{2-} + H_2O$, the reacting hydrogen molecule will have closer access to the paramagnetic chromium ion. In such a case there is no change of surface stoichiometry. But if dehydroxylation proceeds by $2Cr^{3+} + 2OH^- + H^2 \rightarrow$

$2Cr^{2+} + 2H_2O$, there will be not only a change of coordination and of stoichiometry but also an increase of magneton number to about 4.9. The result will be a large increase of conversion activity as actually occurs. But it is still not possible to say that the surface-active species is Cr^{2+} rather than Cr^{3+}. Furthermore, the active surface at this stage is highly sensitive to poisoning by molecular hydrogen at slightly elevated temperatures. This problem has been discussed in detail by many authors (21). Our point here is that an increase of accessibility of a chromia preparation by dehydroxylation, or a decrease by hydrogen poisoning, has no important effect on Δk_H. Furthermore, Cr_2O_3, CoO, and MnO (above T_N) show no evidence of a negative effect at low fields. This is true of Cr_2O_3 even at high dilution by solution in, or supported on, the diamagnetic Al_2O_3.

The paramagnetic rare earths, because of the 4f-electron shielding effect, might be thought to offer fewer complexities than the 3d catalysts such as chromia. The zero field conversion activities are as nearly proportional to β^2 as can be expected for measurements of this kind. But the negative field effects shown by all except Gd_2O_3/La_2O_3 are very difficult to explain. It has been shown by Eley et al. (31) that the increase in zero field activity (for Nd_2O_3 and Dy_2O_3) that occurs with rising temperature of pretreatment cannot be due to a change of stoichiometry. There is no reason to believe that this does not apply to all the paramagnetic rare earths but some further comments will be made about Eu_2O_3 and Yb_2O_3. The Eu^{3+} ion is fairly easily reduced to Eu^{2+} and it might be surmised that at elevated temperatures hydrogen could reduce the sesquioxide in the surface layer (16). The ion Eu^{2+} is isoelectronic with Gd^{3+} which has a much higher magneton number than Eu^{3+}. If such a reaction actually occurs it would be expected that raising the pretreatment temperature would cause an exceptionally large increase in zero field conversion activity. But this is not the case. Our conclusion is that the Eu^{2+} oxidation state cannot be ruled out but that it is not very probable in this situation. The same problem arises to an aggravated degree in the case of ytterbia. The Yb^{3+} is also capable of being reduced to Yb^{2+} although not quite so readily as for Eu^{3+}. In the case of Yb^{2+} the usual product is diamagnetic. Such a reaction would, therefore, be expected to cause an abrupt loss of conversion activity, but this does not occur. It might be surmised that the extra electron goes into the 5d or 6s level, but there does not appear to be any good evidence that this is true. Our conclusion with respect to the paramagnetic rare earths as conversion catalysts is that the paramagnetic ion in each case is almost certainly R^{3+}, with perhaps possible exceptions for Eu_2O_3 and Yb_2O_3. The increase in activity with increasing pretreatment temperature is almost certainly a matter of improved accessibility. One final point to be mentioned in connection with the rare earths is that the specific zero field conversion activity is considerably less than that of α-Cr_2O_3, after comparable pretreatment (17).

The observed conversion activity on diamagnetic solids is generally more difficult to explain. The case of titanium dioxide has already been mentioned (14). Possible sources of the nondissociative conversion activity on alumina have been examined by Hall (41). It has often been suggested that reduction to the paramagnetic Al^{2+} ion could provide this activity, but there is no EPR evidence to support this view. It is not possible to exclude nuclear magnetism and another possible source, already mentioned (13), is that the intense electric fields generated during pretreatment could ionize H_2 with the formation of the paramagnetic species H_2^+. Finally, we mention the possible formation of the O^- ion. But this reacts rapidly with hydrogen at room temperature.

The various possible sources of conversion activity mentioned above must all be considered as sources in the two diamagnetic oxides lutetia and yttria both of which show strong activity and strong negative field effects after pretreatment *in vacuo* or in hydrogen above 823 K, as previously described. But here again it is not possible to make a definite assignment except that the evidence presented indicates a mechanism involving paramagnetic sites each carrying one unpaired electron. [Note added in proof: This idea has been developed more fully by the author in *J. Catal.* **50,** 15 (1977).]

This review will be concluded with some speculative remarks on possible sources of the several extrinsic field effects. There is one kind of behavior that appears to have a simple explanation. The ferromagnetic catalysts all show zero field effect below T_C. The reason for this must be that in all such substances the Weiss "molecular" field is of the order of 10^5 to 10^7 Oe. An extrinsic field of a few kOe could produce little additional effect.

It appears from the experimental evidence that there may be two independent kinds of catalytic conversion sites. Ignoring, for the present, the effects that occur near and below magnetic phase transitions we designate those sites giving a positive change of conversion rate as k^+ sites. Those yielding a negative rate change will be designated k^- sites. The observed conversion rate is then always $k = k^+ + k^-$ and the observed fractional rate change in an extrinsic field is $\Delta k_H = [(k_H^+ + k_H^-) - (k_0^+ + k_0^-)]/(k_0^+ + k_0^-)$. Hence the magnitude of Δk_H will depend on the ratio k^+/k^-. A relatively large Δk_H^- will be obscured if k_0^+ is more than two orders larger than k_0^-.

Evidence pointing to the existence of k^+ and k^- sites will be reviewed:

1. On some catalysts, of which α-$Cr_2O_3(p)$, $CoO(p)$ and $MnO(p)$ are

examples, there is no indication of any k^- contribution to the activity. On some catalysts, such as Lu_2O_3, there is no indication of any k^+ contribution. What this means in experimental terms is that the ratio k^+/k^- must be over 10^2 for Cr_2O_3 and under 10^{-2} for Lu_2O_3. On most of the paramagnetic rare earths both k^+ and k^- may be observed.

2. For k^+ sites Δk_H tends to become independent of field above 10^4 Oe. On k^- sites saturation occurs at below 10^2 Oe.

3. The relative proportions of k^+ and k^- sites may sometimes, as for Yb_2O_3, be changed by more vigorous pretreatment of the catalyst.

4. The k^+/k^- ratio may be lowered by diluting the paramagnetic rare earths as by supporting Er_2O_3 on La_2O_3 or Al_2O_3. But dilution of Cr_2O_3 with Al_2O_3 does not cause any appearance of k^- sites.

5. In the absence of a magnetic phase transition the k^+ sites, as in Cr_2O_3–Al_2O_3, retain considerable activity at 77 K. But the k^- sites lose about 2 orders of activity.

6. A fairly crude estimate of the magneton number for activated Lu_2O_3 and Y_2O_3 suggests that only one electron spin per site is active. But for nearly all the other catalysts studied the Bohr magneton number is much larger and it varies widely.

The hypothesis of two independent kinds of sites provides a qualitative explanation for several obscure observations. For instance, the typical plot of Δk_H vs. H for Pr_2O_3 shown in Fig. 22 could be constructed by choosing appropriate values for k_0^+ and k_0^-. Dilution of a paramagnetic rare earth by supporting it on La_2O_3 or Al_2O_3 results for all except Gd_2O_3, in a large decrease of the k^+/k^- ratio. But whether this is primarily a decrease in k^+ or an increase in k^- it is not yet possible to say. The failure of $Cr_2O_3(p)$ to develop a measurable k^- activity through diamagnetic dilution must mean that the structure, or mechanism, responsible for k^- is not possible for Cr_2O_3. It may, or may not, be a coincidence that Cr_2O_3 and Gd_2O_3 are both in S spectroscopic states, the former effectively through orbital quenching and the latter with its $4f^7$ electrons. Similarly, the progressive disappearance of k^- activity with increasing activation temperature on Yb_2O_3 may simply mean that k^+ is increasing more rapidly than k^-. Also the failure of $Cr_2O_3(p)$ to show any evidence of k^- sites as hydrogen poisoning progresses tends to confirm that in Cr_2O_3 there are no k^- sites. The other extreme is for Lu_2O_3 and Y_2O_3 in which there appear to be no active k^+ sites. This in turn suggests that the maximum negative value of Δk_H observed (about -40%) represents the maximum change that can be achieved, by a field, for this kind of site. It will also have been noted that the catalysts of relatively high bulk magnetic susceptibility are those in which little or no evidence of k^- sites is found.

This suggests that in at least some such samples the intrinsic field may be large enough to saturate the k^- sites so that the observed k^- is always the same as k_H, regardless of H. But this must, of course, be regarded as pure speculation.

There are some experimental results for which it is difficult to offer any explanation. These include the negative sign of Δk_H for o–p-D_2 over Nd_2O_3 and the curious fact that, for this system, although a maximum effect of -17% is not reached until about 1.5 kOe the form of the plot is otherwise much like that of o–p-H_2 over Lu_2O_3. Another anomaly is the reversal of sign of Δk_H at the Néel point of Cr_2O_3 and CoO, and a third is the unique behavior of Gd_2O_3 supported at low concentration on La_2O_3. But the most surprising aspect of this work is the finding that a magnetic field only slightly larger than that of the Earth could affect any catalyzed rate process. The only similar observation that has come to the writer's attention is the finding, as described in a review by Avakian (42), that fields in the range of 10 Oe can alter the fluorescence decay rate in substances such as anthracene. These various observations suggest a reexamination of the abundant data in the literature on hydrogen–metal oxide systems by methods that use magnetic fields. These include NMR, ESR, and magnetic susceptibility determinations. It is obvious that even for a process as relatively simple as the physical adsorption of hydrogen on a refractory oxide there are still unsuspected complexities. (Note added in proof: The author understands that a theoretical treatment of the complicated effects observed on Cr_2O_3, CoO and MnO is in preparation by C. F. Ng.)

Acknowledgment

This work was done under grant from the Army Research Office.

References

1. M. Misono and P. W. Selwood, *J. Am. Chem. Soc.* **90**, 2977 (1968).
2. P. W. Selwood, *J. Catal.* **22**, 123 (1971).
3. D. R. Ashmead, D. D. Eley, and R. Rudham, *Trans. Faraday Soc.* **59**, 207 (1963).
4. D. D. Eley, H. Forrest, D. R. Pearce, and R. Rudham, *J. Chem. Soc., Chem. Comm.* p. 1176 (1972).
5. F. H. Van Cauwelaert and W. Hall, *Trans. Faraday Soc.* **66**, 454 (1970).
6. L. Farkas and H. Sachsse, *Sitzungsber. Preuss. Akad. Wiss.*, 268 (1933). *Phys.-Math. Kl.*
7. P. W. Selwood, *J. Am. Chem. Soc.* **88**, 2676 (1966).
8. G. J. K. Acres, D. D. Eley, and J. N. Trillo, *J. Catal.* **4**, 12 (1965).
9. H. S. Taylor and H. Diamond, *J. Am. Chem. Soc.* **55**, 2613 (1933); **57**, 1251 (1935).
10. K. Baron and P. W. Selwood, *J. Catal.* **28**, 422 (1973).
11. C. P. Madhusudhan and P. W. Selwood, *J. Catal.* **45**, 106 (1976).

12. D. R. Rossington and V. F. Capozzi, *J. Am. Ceram. Soc.* **57,** 474 (1974).
13. F. H. Van Cauwelaert and W. K. Hall, *J. Colloid Interface Sci.* **38,** 138 (1972).
14. Y. L. Sandler, *J. Phys. Chem.* **58,** 54 (1954).
15. S. W. Weller and A. A. Montagna, *J. Catal.* **21,** 303 (1971).
16. J. A. Arias and P. W. Selwood, *J. Catal.* **33,** 284 (1974).
17. P. W. Selwood, *J. Catal.* **19,** 353 (1970).
18. E. Wigner, *Z. Phys. Chem., Abt. B* **23,** 28 (1933).
19. D. R. Ashmead, D. D. Eley, and R. Rudham, *J. Catal.* **3,** 280 (1964).
20. M. M. Schieber, "Experimental Magnetochemistry." North-Holland Publ., Amsterdam and Wiley (Interscience), New York, 1967.
21. P. W. Selwood, *J. Am. Chem. Soc.* **92,** 39 (1970).
22. M. Misono and P. W. Selwood, *J. Am. Chem. Soc.* **91,** 1300 (1969).
23. C. F. Ng and P. W. Selwood, *J. Catal.* **43,** 252 (1976).
24. J. A. Arias and P. W. Selwood, *J. Catal.* **30,** 255 (1973).
25. R. P. Eischens and P. W. Selwood, *J. Am. Chem. Soc.* **69,** 1590, 2698 (1947).
26. P. W. Selwood, M. Ellis, and C. F. Davis, Jr., *J. Am. Chem. Soc.* **72,** 3549 (1950).
27. J. A. Arias and P. W. Selwood, *J. Catal.* **35,** 273 (1974).
28. E. Justi and G. Vieth, *Z. Naturforsch., Teil A* **8,** 538 (1953).
29. G.-M. Schwab and A. Kaiser, *Z. Phys. Chem.* **22,** 220 (1959).
30. P. W. Selwood, unpublished observations (1973).
31. D. D. Eley, H. Forrest, and R. Rudham, *J. Catal.* **34,** 35 (1974).
32. F. Kalckar and E. Teller, *Proc. R. Soc., Ser. A* **150,** 520 (1935).
33. L. G. Harrison and C. A. McDowell, *Proc. R. Soc., A* **220,** 77 (1953).
34. A. J. Leffler, *J. Chem. Phys.* **43,** 4410 (1965).
35. S. N. Nielsen and J. S. Dahler, *J. Chem. Phys.* **46,** 732 (1967).
36. A. V. Golovin and R. A. Buyanov, *Kinet. Katal.* **10,** 113 (1969).
37. A. V. Golovin and R. A. Buyanov, *Kinet. Katal.* **11,** 957 (1970).
38. D. J. Scalapino and K. G. Petzinger, *Phys. Rev. B* **8,** 266 (1973).
39. E. Ilisca and E. Gallais, *Phys. Rev. B* **6,** 2858 (1972).
40. A. V. Golovin and R. A. Buyanov, *Kinet. Katal.* **16,** 121 (1975).
41. W. K. Hall, *Acc. Chem. Res.* **8,** 257 (1975).
42. P. Avakian, *Pure Appl. Chem.* **37,** 1 (1974).

Hysteresis and Periodic Activity Behavior in Catalytic Chemical Reaction Systems

VLADIMÍR HLAVÁČEK AND JAROSLAV VOTRUBA

Department of Chemical Engineering
Institute of Chemical Technology
Prague, Czechoslovakia

I. Introduction	59
II. Porous Catalyst Particle Problem	60
A. Experimental Observation of Multiple Steady-State Phenomena and Periodic Activity	64
III. Catalytic Wire Problem	69
A. Experimental Observation of Multiplicity and Periodic Activity	70
IV. Continuous Stirred-Tank Reactor	74
A. Experimental Observation of Multiplicity and Periodic Activity	75
V. Tubular Fixed-Bed Reactor	77
A. Experimental Observations	81
VI. Monolithic Catalyst	88
A. Experimental Observation of Multiple Steady-State Phenomena	89
VII. Other Catalytic Systems	93
A. Catalytic Systems with Heat Recirculation	93
B. Trickle-Bed Reactors	93
VIII. Concluding Remarks	94
Notation	94
References	95

I. Introduction

The essential topics of this review article are the experimental aspects of multiplicity of steady states as well as periodic activity of open chemical reacting systems "catalyst-gas." In the last two decades a great number of theoretical papers were published on this subject which indicated a number of "pathological" phenomena to be expected in chemically reacting systems. The next step toward a deeper understanding of all these phenomena is possible only through research programs which try to prove these effects on an experimental way; we shall, therefore, try to focus our attention somewhat more closely on experimental aspects of multiplicity and periodic activity. A knowledge of multiplicity of steady states and their stability to small and large perturbations has an important application in the steady state and transient operation of catalytic reactors encountered in the chemical and petroleum processes. The authors shall not give a detailed survey of their own work, nevertheless, examples from our own experimental work will be used to illustrate some points.

The paper is not equation oriented since after the period of theoretical investigation, only a small percentage of experimental papers published is completely supported with theory and very often only a qualitative explanation is presented. Hence in this paper we shall review the experimental information published in the literature concerning multiplicity of steady states and periodic activity in the systems catalyst-gas, making an attempt to explain qualitatively these phenomena on the basis of the theory developed.[1] The number of experimental observations surveyed here which are not supported by a theory will surely indicate that there are many roads open for fundamental research in this area.

II. Porous Catalyst Particle Problem

The starting point of a number of theoretical studies of packed catalytic reactors, where an exothermic reaction is carried out, is an analysis of heat and mass transfer in a single porous catalyst since such system is obviously more conductive to reasonable analytical or numerical treatment. As can be expected the mutual interaction of transport effects and chemical kinetics may give rise to multiple steady states and oscillatory behavior as well. Research on multiplicity in catalysis has been strongly influenced by the classic paper by Weisz and Hicks (5) predicting occurrence of multiple steady states caused by intrapellet heat and mass intrusions alone. The literature abounds with theoretical analysis of various aspects of this phenomenon; however, there is a dearth of reported experiments in this area. Later the possiblity of oscillatory activity has been reported (6).

The significance of the key parameters which govern concentration and temperature distribution in a single porous catalyst may be illustrated most effectively by presentation and manipulation of the differential equations which are supposed to describe the system.

---

[1] For those who are interested in recent developments of the theory the excellent book by Aris (*1*) and papers by Aris (*2*), Schmitz (*3*), and Ray (*4*) are recommended.

The following assumptions are inherent in the model:

a. Mass and heat transfer in a porous structure may be phrased in terms of diffusion model making use of diffusion coefficient and thermal conductivity, respectively.
b. First-order Arrhenius kinetics is considered.
c. Volume change caused by the reaction is unimportant.
d. An average value of the fluid-film coefficients is assumed to govern the gas-to-particle heat and mass transfer.

The mass and enthalpy balances are then

$$\text{Lw} \frac{\partial y}{\partial t} = \frac{\partial^2 y}{\partial x^2} + \frac{a}{x} \frac{\partial y}{\partial x} - \Phi^2 y \exp\left(\frac{\theta}{1 + \theta/\gamma}\right) \quad (1)$$

$$\frac{\partial \theta}{\partial \tau} = \frac{\partial^2 \theta}{\partial x^2} + \frac{a}{x} \frac{\partial \theta}{\partial x} + \Phi^2 \gamma \beta y \exp\left(\frac{\theta}{1 + \theta/\gamma}\right) \quad (2)$$

where

$$a = \begin{cases} 0 & \text{for slab geometry} \\ 1 & \text{cylindrical} \\ 2 & \text{spherical} \end{cases}$$

subject to boundary conditions

$$x = 1: \quad y = 1 - \frac{1}{\text{Sh}} \frac{\partial y}{\partial x}$$

$$\theta = -\frac{1}{\text{Nu}} \frac{\partial \theta}{\partial x} \quad (3)$$

$$x = 0: \quad \frac{\partial y}{\partial x} = \frac{\partial \theta}{\partial x} = 0.$$

The present article is not designed to review the work devoted to theoretical treatment of multiple steady states and oscillatory activity predicted from these equations, for those readers who seek a profound review the texts by Aris (*1*), Schmitz (*3*), and others (*7,9*) are recommended.

Below we outline the quantitative results that have been gained by the theoreticians without recourse to rigorous analytical treatment of governing equations. The conditions under which a porous catalyst particle may exhibit multiple steady states have been conceived by Aris (*2*), Luss (*8*), and others (*9*):

Condition 1: The necessary and sufficient condition for unicity of the solution is $\gamma\beta < (\gamma\beta)^*$.

Condition 2: The necessary condition for multiplicity is given by $(\gamma\beta) > (\gamma\beta)^*$.

Condition 3: The necessary and sufficient condition of multiplicity is $\gamma\beta > (\gamma\beta)^*$ and $\Phi_{min} < \Phi < \Phi_{max}$. For instance, for the first-order reaction the critical value of $\gamma\beta$ is

$$(\gamma\beta)^* = \frac{4\gamma}{\gamma - 4}. \qquad (4)$$

The periodic activity of a porous catalyst pellet is governed by similar criteria which have been developed recently (10). The results are reported in Table I.

For $\gamma\beta \leq 1$, one steady state exists and the regime is globally stable for all values of the Lewis number Lw. For $1 < \gamma\beta < (\gamma\beta)^*$ and for sufficiently low values of Lewis number the system is again globally stable. Evidently for these conditions only one steady state occurs. For Lw > Lw*, undamped oscillations exist. For supercritical values of $\gamma\beta$, $\gamma\beta > (\gamma\beta)^*$, and $\Phi < \Phi_{min}$, a single steady state is stable or unstable according to the value of Lewis number. In the domain $\Phi_{min} < \Phi < \Phi_{max}$, three steady states occur, here the middle one is always unstable. Depending on the Lewis number the lower and upper steady states may be stable or unstable or limit cycle may be expected. Finally, for

TABLE I
Classification of the Stability Regions

$\gamma\beta$	ϕ	Number of steady states	Stability
$\gamma\beta < 1$	—	1	Stable
$1 < \gamma\beta < (\gamma\beta)^*$	—	1	Stable for Lw < Lw*
$\gamma\beta > (\gamma\beta)^*$	$\phi < \phi_{min}$	1	Stable for Lw < Lw*
	$\phi_{min} < \phi < \phi_{max}$	3	The lower state, stable for Lw < Lw* The middle state, always unstable The upper state, stable for Lw < Lw*

$\Phi > \Phi_{max}$ (one steady state), stability is governed by the Lewis number. Calculations have revealed that for the case where external concentration and temperature gradients are neglected, the critical value of the Lewis number is $Lw^* > 1$.

These criteria are, of course, a valuable guide in assessing multiplicity and periodic activity for a single catalyst particle. For the sake of comparison a list of values of γ, $\gamma\beta$, Lw and Φ is reported for some typical exothermic reactions (cf. Table II).

Since the critical values of $\gamma\beta^*$ and Lw^* are $\gamma\beta^* = 4$ and, $Lw^* > 1$ respectively, then referring to the results reported in Table II, it seems highly unrealistic to expect multiple steady states and periodic activity for a single catalyst particle resulting from intraparticle heat and mass transfer alone.

So far we have considered an infinite value of the gas-to-particle heat and mass transfer coefficients. One may encounter, however, an imperfect access of heat and mass by convection to the outer geometrical surface of a catalyst. Stated in other terms, the surface conditions differ from those in the bulk flow because external temperature and concentration gradients are established. In consequence, the multiple steady-state phenomena as well as oscillatory activity depend also on the Sherwood and Nusselt numbers. The magnitudes of the Nusselt and Sherwood numbers for some strongly exothermic reactions are reported in Table III (*11*). We may infer from this table that the range of Sh/Nu is roughly $Sh/Nu \in \langle 1.0, 10^4 \rangle$.

TABLE II
Parameters of Catalytic Exothermic Reactions

Reaction	γ	$\gamma\beta$	Lw	ϕ
Synthesis of NH_3	29.4	0.0018	0.00026	1.2
Synthesis of higher alcohols from CO and H_2	28.4	0.024	0.00085	—
Oxidation of CH_3OH to CH_2O	16.0	0.175	0.0045	1.1
Synthesis of vinylchloride from acetylene and HCl	6.5	1.65	0.1	0.27
Hydrogenation of ethylene	23–27	1–2.7	0.11	0.2–2.8
Oxidation of H_2	6.75–7.52	0.21–2.3	0.036	0.8–2
Oxidation of ethylene to ethylenoxide	13.4	1.76	0.065	0.08
Decomposition of N_2O	22.0	1.0–2.0	—	1–5
Hydrogenation of benzene	14–16	1.7–2.0	0.006	0.05–1.9
Oxidation of SO_2	14.8	0.175	0.0415	0.9
Decomposition of NH_3	24.3	0.26	0.0003	—

TABLE III
Dimensionless Ratio Sh/Nu for Some Heterogeneous Reactions

Reaction	Sh/Nu
Ethylene hydrogenation	580
Benzene hydrogenation	150
Methane oxidation	310
Ethylene oxidation	45
Naphthalene oxidation	2.7
$SO_2 \rightarrow SO_3$	105
Acrylonitrile synthesis	4260

To assess the influence of the Nusselt and Sherwood number on multiplicity explicit condition (5) may be consulted (*11*):

$$\gamma\beta \geq \left(\frac{\rho_1}{\beta_1}\right)^2 \frac{4\gamma}{\gamma - 4}. \tag{5}$$

Here ρ_1 and β_1 are the first roots of a certain transcendental equation [see Hlaváček and Kubíček (*11*)]. Evidently after inspection of Eq. (5) and Tables II and III the value of $(\gamma\beta)^*$ is essentially reduced so that Eq. (5) can be satisfied for a number of catalytic exothermic systems.

If we consider the effect of the real values of the parameter Sh/Nu on the region of instabilities, then the value Lw* decreases; however, the inequality Lw* > 1 is still valid (*12*). From the material presented it is apparent that the occurence of unstable steady states cannot be explained in terms of thermokinetic theory (*12*).

So far we have approximated the external heat and mass transfer coefficients by virtue of an average value; however, precise measurements have indicated that the local value may substantially change along the catalyst surface. Moreover, for a highly exothermic reaction a very strong gas phase concentration and temperature gradients may occur so that the pellet is bathed by a nonuniform external field (*13, 14*).

A. Experimental Observation of Multiple Steady-State Phenomena and Periodic Activity

So far, there have been published only a few papers devoted to experimental investigation of multiplicity and oscillatory activity of a single catalyst particle. Observations of multiple steady states and/or oscillations for a single catalyst particle are reported in Table IV. Evidently three types of strong exothermic reactions have been investigated:

TABLE IV
Experimental Observations of Multiply Steady States and Periodic Activity for a Single Catalyst Particle

Experimental system	Refs.	Oscillations	Multiplicity
i. Oxidation of hydrogen on suspended Pt/silica–alumina particles	Beusch et al. (15, 16)	X	X
ii. Oxidation of hydrogen on a single particle imbedded in a layer of inert particles		X	X
iii. Oxidation of CO on a suspended Pt/Al$_2$O$_3$ pellet		X	X
Ethylene hydrogenation on a suspended pellet of Adkin's catalyst	Furusawa and Kunii (24)		X
Oxidation of hydrogen on a suspended Pt/Al$_2$O$_3$ catalyst particle	Horák and Jiráček (21)		X
Oxidation of hydrogen on a Pt/alumina pellet	Jiráček et al. (22)		X
Oxidation of hydrogen on a Pt/alumina particle in a batch recycle-flow reactor	Jiráček et al. (23)		X
Oxidation of hydrogen on a Ni catalyst	Beljajeṽ et al. (18–20)	X	

oxidation of CO and H$_2$ and hydrogenation of ethylene. The laboratory studies have been conducted with a freely suspended catalyst pill or with a particle imbedded in a layer of inert material. To simulate the conditions in a flow tubular reactor for a sequence of pellets a single-pellet batch reactor with recirculation has been devised (see Ref. 23).

To elucidate the problem of multiplicity and instability the results of Beusch et al. (15, 16) will be discussed in detail. The exothermic catalytic reaction between H$_2$ and O$_2$ may exhibit both the multiple steady states and periodic activity. Hysteresis loop and the undamped oscillation of temperature in a single catalyst particle are presented in Figs. 1 and 2, respectively. For a high gas temperature ($t = 139°C$) an increasing H$_2$ concentration results in a slight overheating of catalyst pellet which finally for 5.5% H$_2$ gives rise to an abrupt temperature jump. On the other hand, for an ignited steady state a gradual decrease of the H$_2$ concentration gives rise to an extinction process. At a higher inlet gas

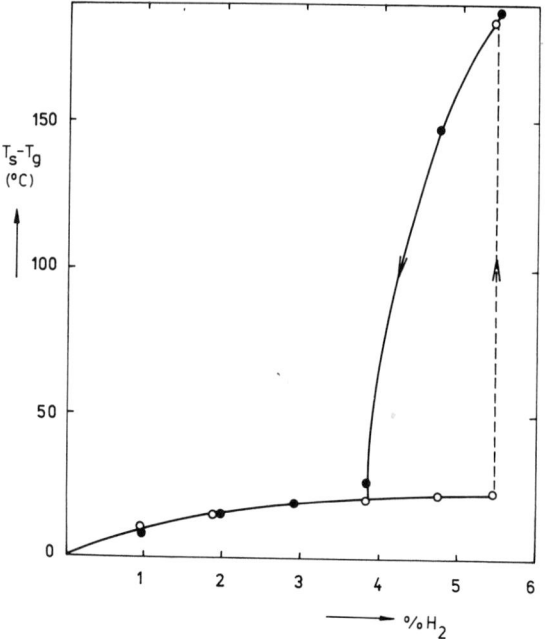

FIG. 1. Hysteresis loop for H_2 oxidation on a single catalyst particle in a packed bed of inactive pellets. Linear gas flow rate $w = 1.7$ cm/sec. $\bigcirc$:H_2 percentage increasing, $\bullet$:H_2 percentage decreasing (15). (Reprinted with permission from Advances in Chemistry Series. Copyright by the American Chemical Society.)

temperature the ignition process starts at lower values of H_2 concentration, for a further increase of H_2 concentration the hysteresis loop disappears.

For low values of the inlet gas temperatures (30°–80°C) oscillatory states within the particle have been observed. The periodic activity of the pellet has been observed both for the freely suspended and imbedded particles. While the temporal average temperature was 170°C, the temperature oscillations amounts to ±60°C. The frequence of oscillations is very low; the period is almost 1 hr. The temperature of the gas and solid oscillated in phase with the reaction rate.

While the multiple steady-state phenomena may be, at least qualitatively, explained in terms of a simple one-step kinetic mechanism and interactions of the intraphase and interparticle heat and mass transfer (thermokinetic model), there is no acceptable explanation for the periodic activity (12). Since the values of the Lewis number are at least by a factor of 10 lower than those necessary to produce undamped oscillations, there is no doubt that the instability cannot be viewed in terms of mutual

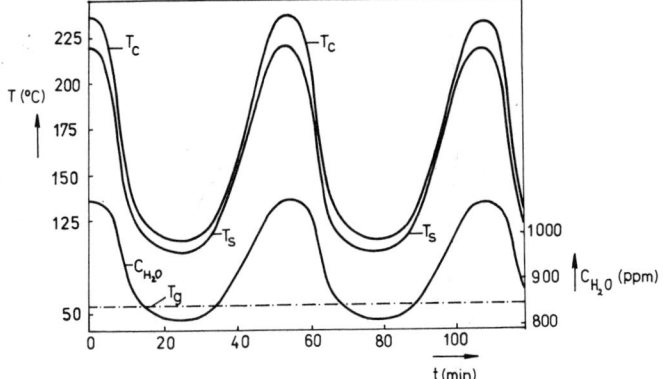

FIG. 2. Temperature oscillations of a single catalyst particle. Inlet temperature $T_0 = 70°C$, linear gas velocity $w = 36$ cm/sec, 3.14 vol% H_2. T_c center temperature, T_s surface temperature, C_{H_2O} effluent water concentration (15). (Reprinted with permission from Advances in Chemistry Series. Copyright by the American Chemical Society.)

interaction of heat and mass transfer within and outside a catalyst particle and a one-step kinetic mechanism. Evidently, the speculations to approximate the kinetic mechanism by an integer-power or Langmuir–Hinshelwood kinetic expressions are vastly oversimplified. The treatment may be rendered virtually dependent on the *kinetic term* if, for instance, branching schemes involving chain interactions are contemplated by the model (17).

Recently, Beljajev et al. (18–20) reported oscillations on a Ni catalyst for the hydrogen oxidation. They used a catalyt slab of the dimension 600 × 1 × 0.0025 mm. The reaction was followed for following parameters: temperature, 150°–300°C; hydrogen partial pressure, 40–760 torr. The frequence of oscillations observed was 1–20 min^{-1}.

The second type of exothermic reactions studied by Wicke in a single-pellet reactor is the catalytic oxidation of CO on Pd or Pt. Obviously the catalytic oxidation of CO on Pd or Pt is associated with strong chemisorption effects which essentially inhibit the reaction rate at higher CO concentrations; the dependence of the reaction rate on the CO partial pressure resembles a Langmuir–Hinshelwood kinetic expression with a high adsorption term (Fig. 3). This figure reveals that the dependence of the reaction rate versus partial pressure of CO measured for isothermal conditions is not a smoothed curve and that the ignition–extinction phenomena occur (Fig. 4). The hysteresis loop is almost independent of the gas flow rate. When the pellet is run at the upper steady state and the gas temperature is lowered the reaction rate may begin to oscillate. From the Wicke observations it can be inferred that the order of magnitude of the

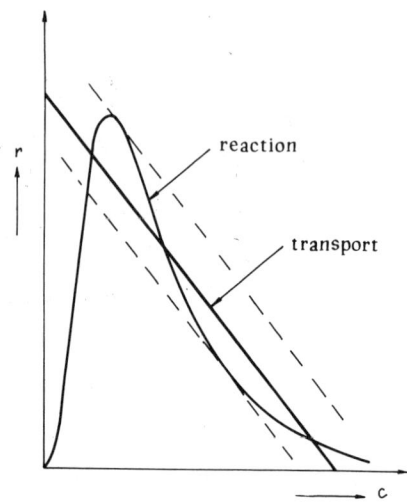

FIG. 3. A schematic picture of the dependence of reaction rate on partial pressure of CO. A case with a strong adsorption (15). (Reprinted with permission from Advances in Chemistry Series. Copyright by the American Chemical Society.)

period of oscillations is expressed in minutes. It is obvious that this observation cannot be reflected in terms of external heat and mass transport because the gas-to-solid gradients are absent. According to Wicke, several states of chemisorbed molecules participate in these phenomena.

Experiments with the system H_2–O_2 were performed also by Horák et al. (21–23) who observed pronounced ignition–extinction phenomena. They were able to construct a reliable mathematical model based on the heat and mass balances describing the gas-to-solid heat and mass transfer. Their general conclusion is that the multiplicity phenomena may be explained in terms of thermokinetic theory. However, on the other hand, because of the high thermal capacity of the pellet, the oscillations cannot be described by this mechanism (12). Obviously we should examine a more detailed kinetic mechanism to be able to analyze successfully this phenomenon (25).

In the homogeneous systems at low temperatures, oscillation phenomena were also observed for this reaction system. Similar observations were made for the CO and O_2 system. In order to explain the oscillation mechanism mathematically, a complicated kinetic scheme was devised which is capable of predicting the sustained oscillations [Yang (26, 27)]. Apparently a similar complicated kinetic mechanism governs the undamped oscillations in the heterogeneous system, however, a pertinent analysis has not been performed in the literature so far.

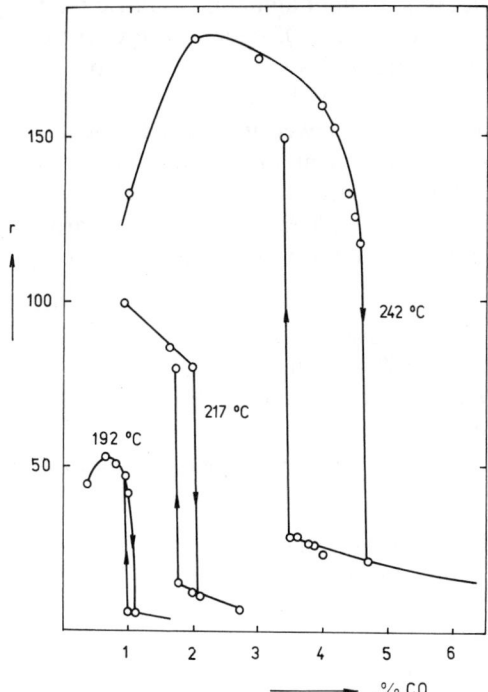

FIG. 4. Transition between the lower and upper steady state. CO oxidation on a single catalyst suspended freely in the gas stream at different temperatures (15). (Reprinted with permission from Advances in Chemistry Series. Copyright by the American Chemical Society.)

Lugovskoi et al. (28) in a qualitative study experimentally observed the ignition process on suspended catalyst particles ($H_2 + O_2$) and showed the relations between the local values of the transport coefficients and spots where the ignition process starts.

III. Catalytic Wire Problem

When a catalytic reaction takes place on a nonporous metal surface the coupling between the exothermic chemical reaction and the transport effects may also give rise to multiple steady states. Apparently, in the realm of chemical reaction engineering the first experimental observation of multiple steady states was done just for the catalytic wire problem [see Tamman (29), Davies (30), and Buben (31)]. Catalytic gauzes consisting of wire screens or layers of metal pills (e.g., the silver crystals) are used for a number of industrially important catalytic reactions as, e.g., synthesis of hydrogen cyanide, oxidation of ammonia to nitric oxide, and oxidation of certain alcohols to aldehydes. Of course, the occurrence of multiplicity phenomena is of importance during the start-up and control policy of these processes. Research in this area was strongly influenced by the classical Russian papers reviewed in Frank-Kamenetskii (32). The most general and rigorous treatment of the multiplicity of steady states of catalytic wires is that by Cardoso and Luss (33).

After rewriting the Frank-Kamenetskii (32), Cardoso and Luss (33), and Aris (2) results the conditions governing the multiplicity phenomena are:

Condition 1: The necessary and suficient condition for the occurrence of one steady state is $\gamma\beta < 4\text{Nu/Sh}$.

Condition 2: The necessary condition for multiple steady states is given by $\gamma\beta > 4\text{Nu/Sh}$.

Condition 3: The necessary and sufficient condition for multiplicity is $\gamma\beta > 4\text{Nu/Sh}$, $\Phi_{min} < \Phi < \Phi_{max}$.

Recently Ray (34) has shown that the periodic activity caused by thermal effects is impossible because of high thermal capacity of the metal catalysts.

A. Experimental Observations of Multiplicity and Periodic Activity

Tamman (29) observed in 1920 that for electrically heated catalytic wire multiple steady states occur for a certain value of electrical current. Similar experimental observations were done by Buben (31) and Davies (30) and recently by Rader (35), Barelko (36), and Cardoso and Luss (33). The Luss results involving oxidation of butane and carbon monoxide on a platinum wire will be discussed in detail.

Figure 5 shows the results obtained for three different butane concentrations. We may infer from this figure that for each butane concentration a region of gas temperatures exist where multiple steady states occur. For instance, for 1.5% of butane two stable steady states occur in the region $80° < T_S < 264°C$. The difference of surface temperatures for both steady states amounts to 600°C. The figure reveals that for higher butane concentration the domain of multiple steady states is enlarged. The ignition and extinction temperatures move to lower values. It was observed that the ignition jump occurs very rapidly and a sudden quick rise of temperature results. However, on the other hand, the extinction process is quite slow and the temperature decreases in a step-wise fashion. Figure 6 shows the effect of the mass flow rate on the

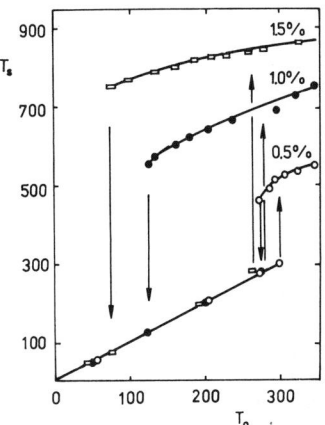

FIG. 5. Steady-state surface temperature T_s for various gas temperatures T_G and butane concentrations. Mass velocity $G = 25.6$ g/cm^{2}min (33). (Copyright by Pergamon Press. Reprinted with permission.)

dimension of the hysteresis loop. We can notice that for lower flow rates ignition and extinction temperatures decrease. The hysteresis loop for the carbon monoxide oxidation is displayed in Fig. 7. Here, in contradistinction to butane oxidation, for higher CO concentration the values of ignition temperature increase. Evidently, to reach the high temperature steady state different start-up policies must be used, i.e., to reach the upper steady state at gas temperatures below the ignition point it is necessary to start-up with a low carbon monoxide concentration.

Some other pathological phenomena connected with the existence of hysteresis loop have been reported in the literature. Frank-Kamenetskii (32) described the Buben experimental results for difference of surface temperatures for both steady states for the reaction between hydrogen and air on a Pt wire. These observations indicate a difference up to 1000°C if hydrogen with an excess of air was used, while the maximum temperature difference amounts to 250°C for air in excess of hydrogen. Frank-Kamenetskii explained this phenomenon by the thermal diffusion effects.

Another very strange observation associated with the hysteresis phenomena is that of Barelko (36) for ammonia oxidation on a Pt wire in a great excess of oxygen. He has observed a complicated shape of the hysteresis loop (cf. Fig. 9 for NH_3 concentration 2.16 vol %). If the extinction process is temporarily stopped by a sudden increase of the electrical current a new steady state may be adjusted. A slight increase of the wire temperature results in a new different ignition point. The course of this hysteresis loop is evident from Fig. 8. We may note that in a certain range of wire temperatures three stable steady states may exist, however,

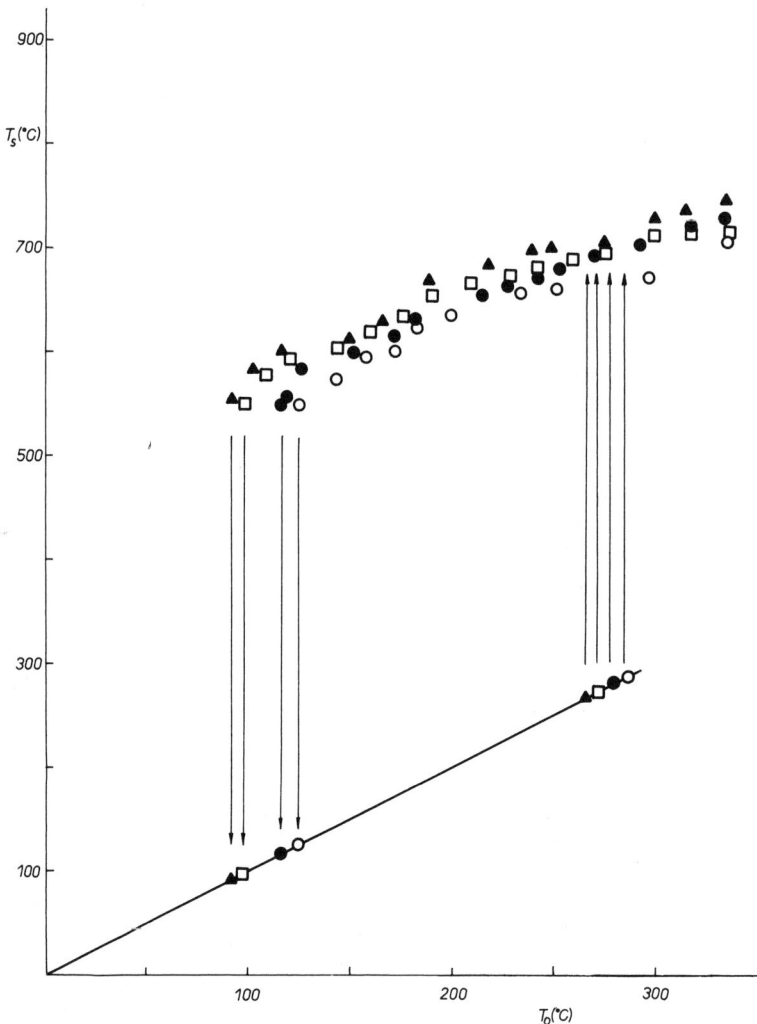

FIG. 6. Effect of the mass velocity G on the steady-state temperatures (1% butane). Mass velocity (g/cm²min): ○, 25.6; ●, 19.2; □, 12.8; ▲, 6.4 (*33*). (Copyright by Pergamon Press. Reprinted with permission.)

the middle stable steady state can be adjusted only as a result of a rather sophisticated manipulation with the system. Hence the cycle ignition–extinction–interruption of extinction–first steady state level–second steady state level–diffusion regime exhibits a double hysteresis loop.

Though the oscillations phenomena seem to be impossible for the

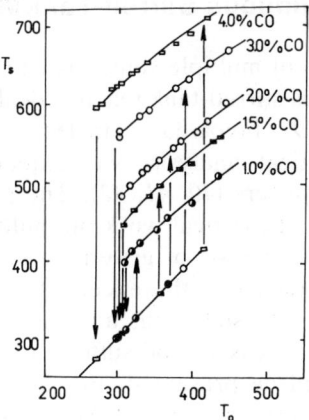

FIG. 7. Steady-state surface temperature T_s for various gas temperatures T_0 and carbon monoxide concentrations. Mass velocity $G = 25.6$ g/cm²min (33). (Copyright by Pergamon Press. Reprinted with permission.)

"reacting gas–catalytic wire" system, Hugo (37, 38) reported an experimental observation of the undamped oscillations for the CO oxidation on a Pt screen. Apparently these oscillations may be explained on the basis of a complicated reaction mechanism involving the concentrations of the activated molecules. However, a quantitative explanation of both aforementioned effects is missing so far.

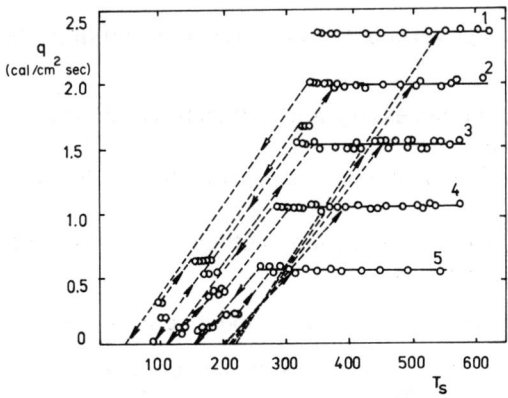

FIG. 8. Dependence of heat liberation q on the wire temperature T_s. Linear gas velocity $w = 15$ cm/sec. Ammonia concentration: (1) 2.7 vol%, (2) 2.16 vol%, (3) 1.67 vol%, (4) 1.1 vol%, and (5) 0.6 vol%. (Copyright by Publishing House Nauka. Reprinted with permission.)

IV. Continuous Stirred-Tank Reactor

The classical problem of multiple solutions and undamped oscillations occurring in a continuous stirred-tank reactor, dealt with in the papers by Aris and Amundson (39), involved a single homogeneous exothermic reaction. Their theoretical analysis was extended in a number of subsequent theoretical papers (40, 41, 42). The present paragraph does not intend to report the theoretical work on multiplicity and oscillatory activity developed from analysis of governing equations, for a detailed review the reader is referred to the excellent text by Schmitz (3). To understand the problem of oscillations and multiplicity in a continuous stirred-tank reactor the necessary and sufficient conditions for existence of these phenomena will be presented. For a detailed development of these conditions the papers by Aris and Amundson (39) and others (40) should be consulted.

The dimensionless mass and heat balances are

$$\frac{dy}{d\tau} = 1 - y - \text{Da} y \exp\left(\frac{\theta}{1 + \theta/\gamma}\right) \tag{6}$$

$$\chi \frac{d\theta}{d\tau} = -\theta + \text{Da} By \exp\left(\frac{\theta}{1 + \theta/\gamma}\right) - \beta(\theta - \theta_c) \tag{7}$$

subject to the initial conditions

$$\tau = 0: \quad \theta = \theta_0 \quad y = y_0. \tag{8}$$

The conditions governing the occurrence of multiple steady states are (40):

Condition 1: The necessary and sufficient condition for unicity of the solution is $B < B^*$.

Condition 2: The necessary condition for multiplicity is given by $B > B^*$.

Condition 3: The necessary and sufficient condition of multiplicity is $B > B^*$, $\text{Da}_{\min} < \text{Da} < \text{Da}_{\max}$.

The conditions which govern the periodic activity are

Condition 1: The necessary and sufficient condition for stability of a single steady state is $B < B^{**}$.

Condition 2: The necessary condition for instability of a single steady state is given by $B^{**} < B < B_{\max}$.

Condition 3: The necessary and sufficient condition of instability of a single steady state is $B^{**} < B < B_{max}$; $Da_{min} < Da < Da_{max}$.

The complicated situations of simultaneous occurrence of multiplicity a limit cycles will not be considered in this text.

So far there is a number of experimentally oriented papers dealing with steady-state multiplicity and stability, however, the majority of them are devoted to homogeneous systems. The important papers dealing with heterogeneous systems are discussed below.

A. Experimental Observation of Multiplicity and Periodic Activity

The two-phase (gas–solid) continuous stirred-tank reactors are represented by laboratory reactors as, for instance, one-pass differential reactors, reactors with forced recirculation, one-pellet reactors, etc. The industrial applications are the fluidized beds.[2] Table V presents a list of experimental studies along with a very brief description of each system studied.

Eckert et al. (43, 44) observed both oscillations and multiplicity in a laboratory recycle reactor for a CO oxidation on a Pt catalyst. Their experimental data are shown in Figs. 9 and 10. In Fig. 9, the behavior of the system at the inlet temperature $T_0 = 420°K$ is depicted. The lower part of the curve corresponds to the kinetic regime, the upper part to the regime influenced by diffusion. The dashed line shows the behavior of the reactor at the transition from one branch to the other. For higher inlet temperature, $T_0 = 430°K$, sustained oscillations of concentration have been observed (cf. Fig. 10). The dependence "exit conversion–inlet concentration" shows again the kinetic and diffusion regimes; within a certain region of feed concentrations the oscillations of temperature and concentration exist alone. At the further increase of the inlet temperature the oscillations disappeared and in the whole range of feed concentrations a unique steady state resulted.

Recently, very interesting results have been reported by Horák and Jiráček (45) for the catalytic exothermic reaction between hydrogen and

[2] Apparently, a number of experimental measurements of reaction rate data for strong exothermic reactions reported in the literature are carried out in the region of multiple steady states, however, at the lower (kinetic) branch. The measurements for a particular run are carried out as long as the experimental domain coincide with the kinetic region, after crossing the ignition point the measurements are stopped and a new run starts. Unfortunately, the bounds of the kinetic regime are not of interest in the kinetic studies and frequently are not reported since the differential reactor does not yield any more "kinetic data."

TABLE V
Experimental Observation of Multiple Steady States and Periodic Activity for a Continuous Stirred-Tank Reactor

Experimental system	Refs.	Oscillations	Multiplicity
Decomposition of N_2O on a copper oxide catalyst in adiabatic circulating reactor	Hugo (46)	X	
Decomposition of N_2O on a catalyst and isothermal oxidation of CO on platinum catalyst in circulating reactor	Hugo (37)	X	
Hydrogen oxidation on Pt catalyst in adiabatic recirculating reactor	Horák and Jiráček (45)	X	X
Isothermal oxidation of CO on Pt catalyst in a recirculating reactor	Hugo and Jakubith (38)	X	X
Oxidation of CO on CuO/Al_2O_3 catalyst in adiabatic circulating reactor	Eckert et al. (43, 44)	X	X

oxygen. They have used a single pellet reactor with a forced recirculation provided with an extra external reservoir (with a variable volume) included in the recirculation loop. As a result, for a given amount of catalyst in the reactor, they can vary the relation between the heat and mass capacity of the system (i.e., the parameter χ) if a reservoir of certain

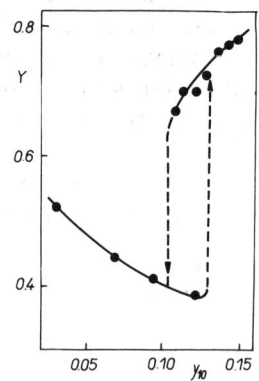

FIG. 9. Dependence of conversion on the feed CO concentration. The case of multiplicity ($T_0 = 420°K$) (43). (Copyright by Gordon and Breach. Reprinted with permission.)

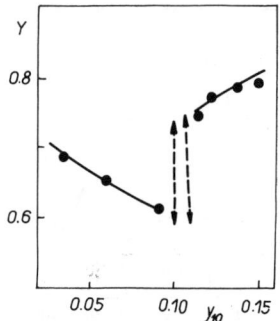

FIG. 10. Dependence of conversion on the feed CO concentration. The case of periodic activity (T_0 = 430°K) (43). (Copyright by Gordon and Breach. Reprinted with permission.)

volume is included in the recirculation loop. Evidently, since the stability of the steady states is affected by this ratio they are able to change the stability of the system. The experimental results for a case of one steady state are depicted in Figs. 11 and 12. For the reactor volume $V = 0.93$ liter, a monotonous approach to the steady state results, while for $V = 2.08$ liters, a slight overshooting has been observed. A further increase to $V = 4.48$ liters gives rise to slowly damped oscillations; finally, for $V = 23.03$ liters undamped oscillations (i.e., a limit cycle) have been observed (see Fig. 12). The phase plane portrait for multiple steady states is depicted in Fig. 13. Here the effect of the reactor volume on the stability of the system is shown. If the reactor is run at the upper steady state (e.g., for $V = 0.93$ liter), then a sudden increase of V to $V = 23.03$ liters destabilizes the upper steady state and a jump to the lower steady state results.

For an exothermic decomposition of N_2O Hugo (37, 47) observed limit cycles with a frequence ≈ 3 min and amplitudes ≈ 30°C.

So far, an experimental evidence for occurrence of multiple steady states in fluidized beds has not been reported.

V. Tubular Fixed-Bed Reactor

In a chemical packed-bed reactor in which a highly exothermic reaction is taking place conditions may be encountered under which, for a given set of input conditions (feed rate, temperature, concentration), the exit conversion is either high or low. To the experimental study of conditions connected with the existence of multiple steady states in packed catalytic reactors has not been paid as much attention in the past as to the study of

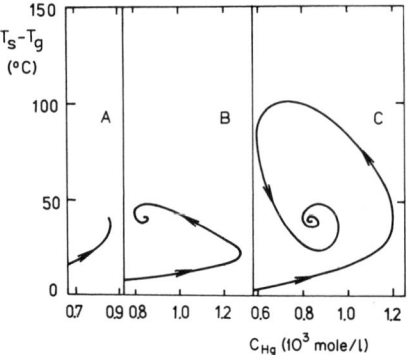

FIG. 11. The effect of reactor volume V on the stability of steady states: A, $V = 0.93$ liter; B, $V = 2.08$ liters; C, $V = 4.48$ liters. One steady state *(45)*. (Copyright by Pergamon Press. Reprinted with permission.)

other phenomena taking place in heterogeneous catalytic reactors. It is interesting to note that multiple states were previously described by researchers studying kinetic reactions catalyzed by various catalysts in laboratory apparatus. Usually they were looking for conditions under which the region of temperatures and concentrations was such that the reactor operated under so-called *kinetic conditions,* i.e., with low temperature and concentration gradients between the solid phase and gas and low degree of exit conversion as well. The fact that for a highly exothermic reaction the reactor may operate in two steady states has been known for a long time but a thorough investigation of this phenomenon was not usually included in experimental studies. Only in the last decade,

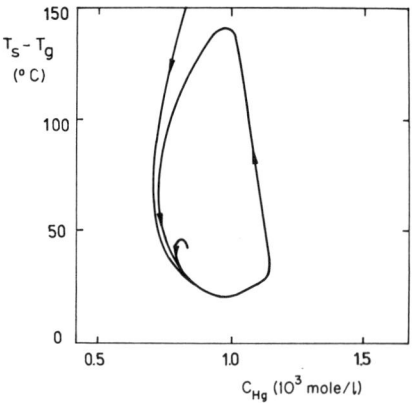

FIG. 12. The effect of the reactor volume V on the stability of steady states. $V = 23.03$ liters. Occurrence of a limit cycle. One steady state *(45)*. (Copyright by Pergamon Press. Reprinted with permission.)

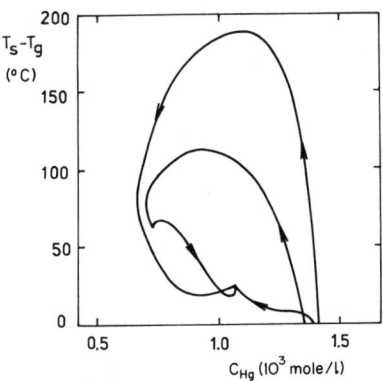

FIG. 13. The effect of the reactor volume V on the stability of steady states. Multiple steady states (45). (Copyright by Pergamon Press. Reprinted with permission.)

as the result of progress in experimental and calculation techniques, experimental study of such problems has become popular. A detailed study has shown that a number of industrial packed bed reactors (large-scale catalytic oxidation and hydrogenation, purification of exhaust gases) may operate in the region of multiplicity.

In this paragraph, based on experimental observations, we are going to review a number of observed facts which may help to elucidate the lows governing the occurrence of multiple steady states in tubular packed catalytic reactors.

The differential equations governing heat and mass transfer in tubular catalytic reactor are

$$\chi \frac{\partial \theta}{\partial t} = \frac{1}{Pe_H} \frac{\partial^2 \theta}{\partial \xi^2} - \frac{\partial \theta}{\partial \xi} - \beta(\theta - \theta_c) + B Da y \exp\left(\frac{\theta}{1 + \theta/\gamma}\right) \quad (9)$$

$$\frac{\partial y}{\partial t} = \frac{1}{Pe_M} \frac{\partial^2 y}{\partial \xi^2} - \frac{\partial y}{\partial \xi} - Da y \exp\left(\frac{\theta}{1 + \theta/\gamma}\right) \quad (10)$$

subject to the boundary conditions $\xi = 0$, $t > 0$:

$$Pe_H \theta = \frac{\partial \theta}{\partial \xi} \qquad Pe_M(y - 1) = \frac{\partial y}{\partial \xi}. \quad (11)$$

$\xi = 1, \tau > 0$:

$$\frac{\partial \theta}{\partial \xi} = \frac{\partial y}{\partial \xi} = 0. \quad (12)$$

To understand the experimental observations we outline in brief the quantitative theoretical results without any analytical investigation of the governing equations (47).

Condition 1: The necessary and sufficient condition for unicity of the steady states is $B < B^*$.

Condition 2: The necessary condition for multiplicity of steady states is given by $B > B^*$.

Condition 3: The necessary and sufficient condition for multiplicity of steady states is $B > B^*$, $Pe < Pe^*$, $Da \in \langle Da_{min}, Da_{max} \rangle$.

The conditions presented above have been formulated for the case that the values of the Peclet numbers for heat and mass dispersion in a bed are indentical. However, recent experimental research on axial heat dispersion in packed beds has indicated that the values of the Peclet number for heat transfer may be different from those for mass transfer (48). Fortunately, the aforementioned conditions are still valid, however, the critical values of B^* and Pe^* as well as the bounds Da_{min} and Da_{max} are dependent on the ratio $q = Pe_H/Pe_M$. From the theoretical results may be inferred that for highly exothermic reactions multiple steady states may

TABLE VI
Experimental Observations of Multiple Steady States in Tubular Catalytic Adiabatic Reactors

Experimental system	Refs.	Multiplicity	
		Two stable steady states	Three stable steady states
Oxidation of CO on Pt/Al$_2$O$_3$ catalyst	Padberg and Wicke (51)	X	
Oxidation of ethane on Pd/alumina catalyst	Wicke et al. (52)	X	
Oxidation of CO on Pt/alumina catalyst			
Oxidation of CO on Pd/Al$_2$O$_3$, Pt/Al$_2$O$_3$, and CuO/Al$_2$O$_3$ catalyst	Hlaváček and Votruba (54, 55)		
Oxidation of CO on Pt/Al$_2$O$_3$ catalyst Heteregeneous dilution of the bed by inert particles	Hlaváček and Votruba (unpublished results)	X	X
Oxidation of CO on Pt/Al$_2$O$_3$ catalyst Heterogeneous mixture of the catalysts	Hlaváček and Votruba (unpublished results)	X	X

occur in a broad range of the Damköhler numbers also for high values of the Peclet number. As a result the ignition–extinction phenomena may occur also in long tubular reactors (49) and the sudden jump in temperature which is usually referred to as the parametric sensitivity phenomena can also be explained by ignition–extinction phenomena.

Numerical analysis has shown that the periodic activity in integral tubular reactors is restricted to quite low Peclet numbers (50). As a result such reactors may be rather considered as differential reactors. Unfortunately, so far, the detailed parametric analysis of the oscillations in tubular reactors is missing.

A. Experimental Observations

Experimental data on multiple steady-state profiles in tubular packed bed reactors have been reported in the literature by Wicke et al. (51–53) and Hlaváček and Votruba (54, 55) (Table VI). The measurements have been performed in adiabatic tubular reactors. In the following text the effects of initial temperature, inlet concentration, velocity, length of the bed, and reaction rate expression on the multiple steady state profiles will be studied.

1. Effect of Inlet Temperature

The inlet temperature strongly affects the exit conversion because of exponential dependence of the reaction rate constant on temperature. For a strong exothermic reaction occurring in a tubular adiabatic system hysteresis phenomena can be observed. Figure 14 graphically compares the hysteresis loops measured for CuO, Pd, and Pt catalysts, the inlet CO concentration was 3%. The circles are the measured steady states, the arrows show the orientation of the hysteresis loop, and the dashed lines are the ignition and extinction transitions. The hysteresis loop for the CuO catalyst is short (15°–30°C) while the platinum catalysts are more apt to result in multiple steady states. Figure 15 shows two steady-state profiles measured experimentally for a CuO catalyst in the region of the multiple steady states.

2. Effect of Inlet Concentration of Carbon Monoxide

In several experiments we have followed the exit conversion as a function of the inlet CO concentration. For a Pd catalyst, decreasing inlet CO concentration increases the reaction rate resulting in a jump to the upper steady state. Obviously the lower steady state disappears. On the

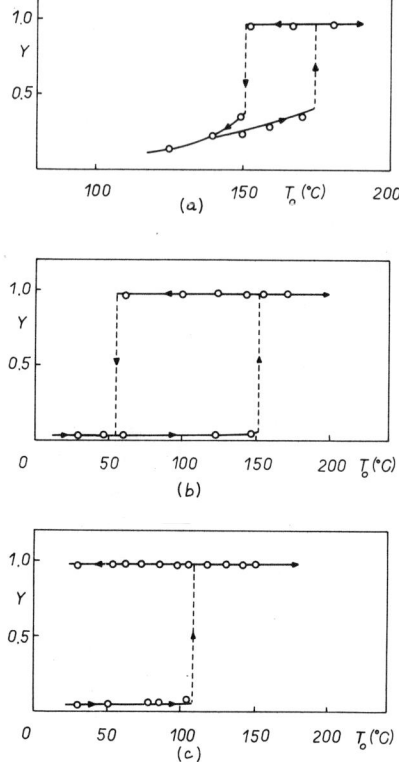

FIG. 14. Dependence of exit conversion Y on inlet temperature: (a) CuO/Al_2O_3 (3% CO, $L = 8.9$ cm, Re = 40); (b) Pd/Al_2O_3 (3% CO, $L = 6.8$ cm, Re = 75); (c) Pt/Al_2O_3 (3% CO, $L = 10.3$ cm, Re = 58) (55). (Reprinted with permission from Advances in Chemistry Series. Copyright by the American Chemical Society.)

other hand, for a CuO catalyst, for a lower inlet CO concentration the upper steady state is extinguished and a lower steady state results. The results are shown in Fig. 16.

3. *Effect of Inlet Gas Velocity*

The gas velocity affects heat and mass transfer between the particles and flowing gas as well as the axial dispersion and heat conduction phenomena. For a reactor operating near at the extinction boundary, an increase of inlet velocity results in a sudden decrease of exit conversion. Sometimes this effect is called the *blow-out phenomenon* (Fig. 17). On the other hand, for a very active catalyst a decrease of inlet velocity leads to an ignited upper steady state.

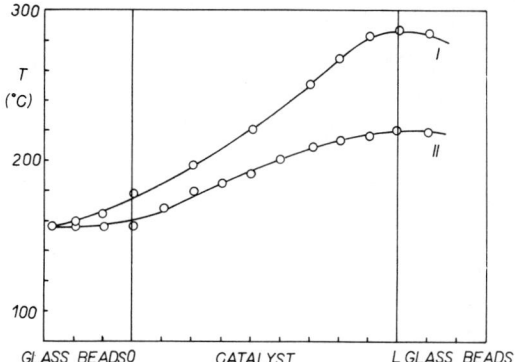

FIG. 15. Typical temperature profiles in the bed. Region of multiplicity (CuO/Al$_2$O$_3$, 4% CO, Re = 26, T_0 = 166°C) (55). (Reprinted with permission from Advances in Chemistry Series. Copyright by the American Chemical Society.)

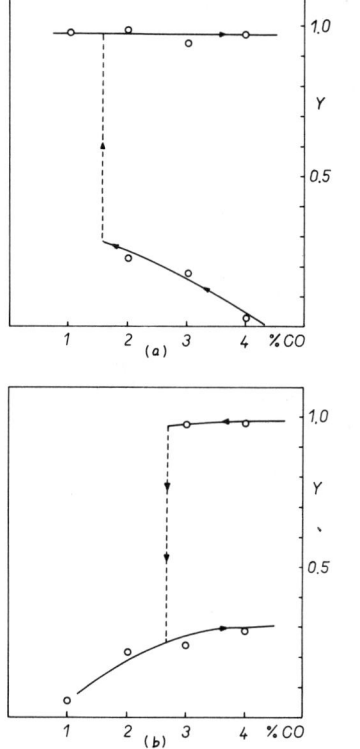

FIG. 16 Dependence of exit conversion on CO inlet concentration: (a) Pd/Al$_2$O$_3$ (Re = 100, L = 32 cm, T_0 = 124°C); (b) CuO/Al$_2$O$_3$ (Re = 34, L = 8.9, T_0 = 148°C) (55). (Reprinted with permission from Advances in Chemistry Series. Copyright by the American Chemical Society.)

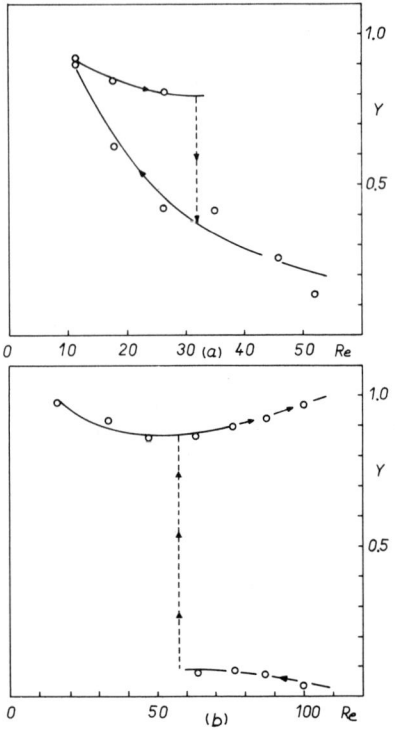

FIG. 17 Dependence of exit conversion, Y, on Reynolds number; (a) CuO/Al_2O_3 (4% CO, $L = 6.8$ cm, $T_0 = 145°C$) (55). (Reprinted with permission from Advances in Chemistry Series. Copyright by the American Chemical Society.)

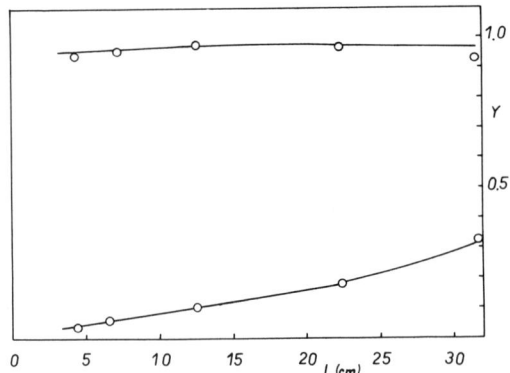

FIG. 18. Dependence of exit conversion Y on reactor length. Pd/Al_2O_3 catalyst (3% CO, $Re = 63$, $T_0 = 124°C$) (55). (Reprinted with permission from Advances in Chemistry Series. Copyright by the American Chemical Society.)

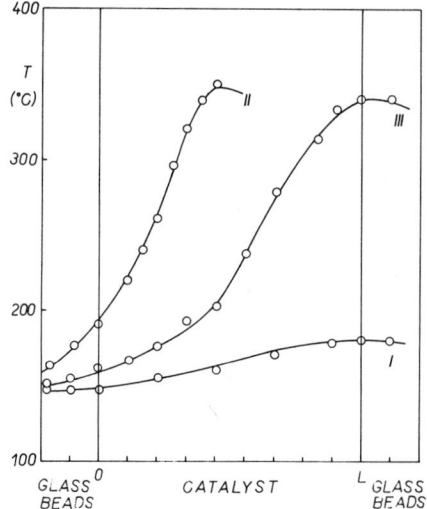

FIG. 19. Three steady-state temperature profiles in packed bed. CuO/Al_2O_3 catalyst (4% CO, L = 8.9 cm, Re = 52, T_0 = 148°C): I, Y = 0.25; II, Y = 0.99; III, Y = 0.99 (55). (Reprinted with permission from Advances in Chemistry Series. Copyright by the American Chemical Society.)

4. *Effect of Reactor Length*

Recent theoretical studies have shown that for higher Peclet numbers (i.e., for longer beds) multiple solutions disappear. This conclusion is supported by our experimental observations. For a bed packed with a CuO catalyst, multiple solutions disappeared with a bed length higher than 12 cm, while for a Pt and Pd catalysts multiple steady state were observed for a bed of 30 cm (i.e., for $Pe_M \sim 180$) (Fig. 18). Extrapolation of this observation indicates that multiple steady states occur also for high Peclet numbers. This agrees with our recent theoretical findings (49).

5. *Occurrence of Three Steady States in Packed Bed*

A detailed experimental exploration of temperature profiles in the reactor packed with the CuO catalyst showed near at the extinction boundary three steady-state axial temperature profiles which were easily reproducible (Fig. 19). There is no simple explanation of these effects so far 55).

6. *Effect of Catalyst Bed Dilution*

In the afterburner reactors the Pt catalyst may be diluted by inert particles because of better distribution of gas in the catalyst layer and de-

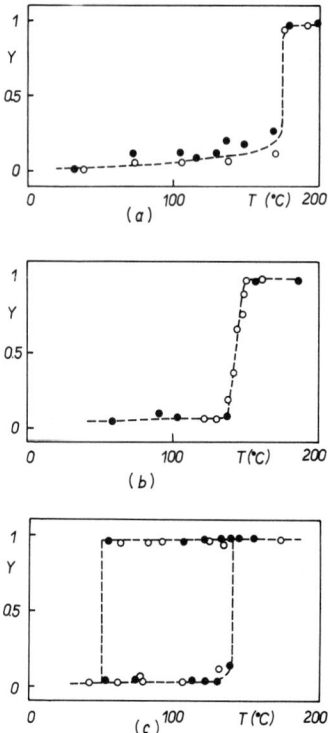

FIG. 20. Dependence of exit conversion on inlet temperature. Platinum catalyst diluted with inert particles (homogeneous mixing); linear velocity $w = 5.3$ cm, bed length $L = 6$ cm.

crease of Pt catalyst consumption per reactor as well. In Fig. 20, the dependence of exit conversion versus inlet temperature is drawn. The results reveal that for different dilution ratios and inlet concentrations of CO these dependences are almost identical both for the unique and multiple steady-state operations ("homogeneous" dilution). Some experiments have been carried out with combination of beds consisting of active particles and inert material ("heterogeneous" dilution). In the region of multiple steady states three stable steady states have been observed (Fig. 21).

7. Mixing of Catalysts

In the afterburner reactors the Pt catalyst may be diluted also by some other catalysts with lower activity, e.g., with a CuO catalyst. Again both

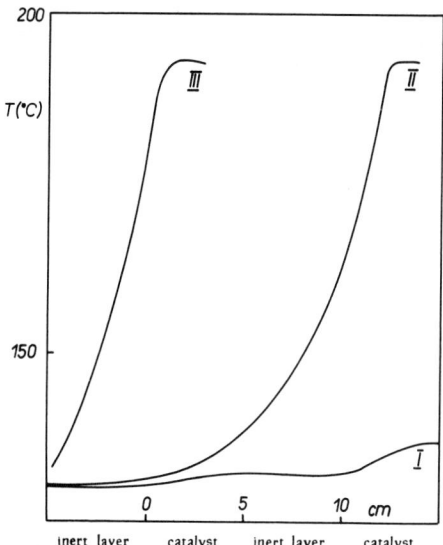

FIG. 21. Three steady states in a catalyst bed. Pt catalyst diluted with inert particles (heterogeneous mixing).

homogeneous and heterogeneous dilution may be considered. In Fig. 22, the dependence of exit conversion on inlet temperature is drawn. We can see that if the Pt catalyst is placed at the reactor inlet the ignition temperature is low; however, the reactor may be extinguished at approximately 50°C. On the other hand, if the Pt catalyst is placed at the reactor outlet, the ignition temperature is higher and the extinction temperature is shifted to lower values. The hysteresis curve for a homogeneously mixed CuO–Pt catalysts is shown in Fig. 22(c). For a catalyst layer combination shown in Fig. 23, three steady states have been observed.

8. Periodic Activity

In the last decade we have performed some thousands of experiments in packed adiabatic tubular reactors (CO oxidation), however, we have never observed oscillations. If on certain catalysts (e.g., Pt/Al_2O_3) the oscillation are caused by the kinetic mechanism then, apparently, the interactions of heat and mass transfer with chemical reaction suppress the occurence of periodic activity in tubular reactors.

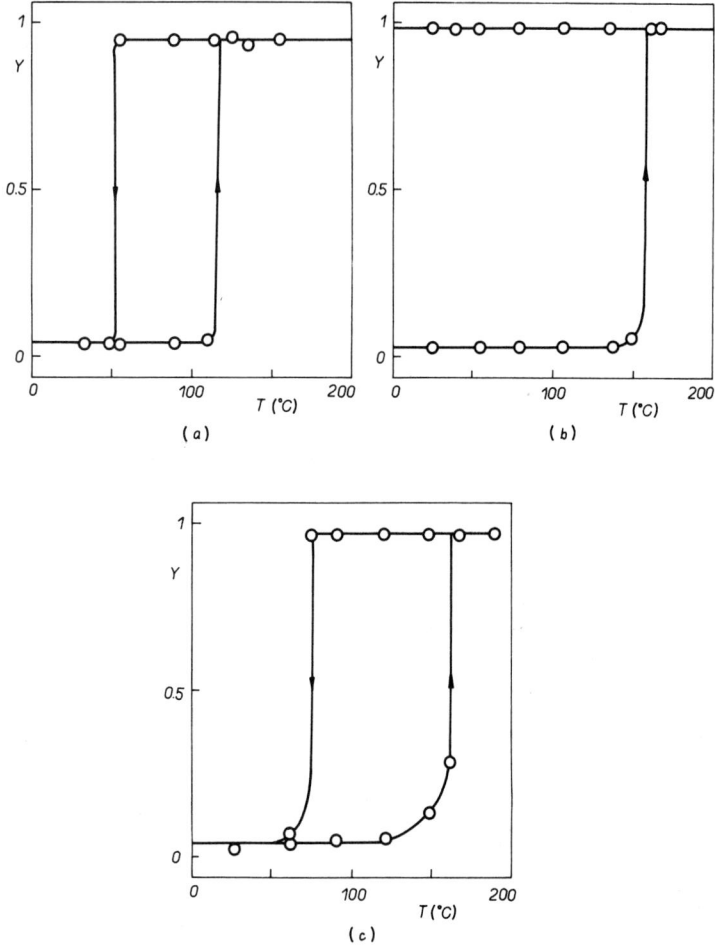

FIG. 22. Dependence of exit conversion on inlet temperature. Pt catalyst (12.5%) diluted with CuO particles (87.5%), 3% CO. (a) Pt catalyst located at reactor inlet (heterogeneous mixing), (b) Pt catalyst located at reactor outlet (heterogeneous mixing), and (c) homogeneous mixing of Pt and CuO catalyst.

VI. Monolithic Catalyst

Since the later 1960s there is growing awareness of the potential utility of monolithic and composite supports in the air pollution control. Prominent among the air pollution problems that cause immediate public annoyance are those of auto exhaust emissions, exhaust malodorous gases from some industrial processes, and finally, inorganic or organic constituents in waste gases of ecological danger. To combat these difficulties ceramic and/or metal honeycomb structures and composite supports have been recognized as a powerful technique that meets the aforementioned requirements. Since the reactions under question are strongly exothermic with high activation energy, multiple steady-state profiles may be expected. To guarantee a complete conversion of the pollutants, the reaction must be carried out at the upper steady state. Therefore, the knowledge of the hysteresis phenomena is for honeycomb structures also important from the industrial point of view. Unfortunately, the body of information on multiple steady states in honeycomb structures is very limited so far.

The significance of the parameters that govern heat and mass transfer in a monolith may be illustrated most effectively by presentation and manipulation of the differential equations which are supposed to describe the system (56). The governing mechanism for heat and mass transfer within a monolithic structure are:

Convective heat and mass transfer in the holes of the structure.
Longitudinal heat conduction within the solid structure.
Radiation effect.
Interphase heat and mass transfer between the flowing gas and solid structure.

Formulation of the transport equations in the terms of the aforementioned mechanism leads to a set of integro-differential and differential equations. The numerical solution of these equations is associated with serious computational difficulties (57).

While we have for heat and mass transfer in a porous catalyst explicit relations for parameters giving rise to multiple steady states, there is nothing similar developed for the monolithic catalysts so far. Hence we are forced to investigate, for particular conditions, the region of multiple steady states numerically.

A. Experimental Observation of Multiple Steady-State Phenomena

So far the multiple steady-state phenomena in the flow of reactants through channels with catalytic walls have been described only by Hlaváček et al. (55, 58). The behavior of the honeycomb catalyst is similar to that of the packed bed; however, the domain of multiple steady state

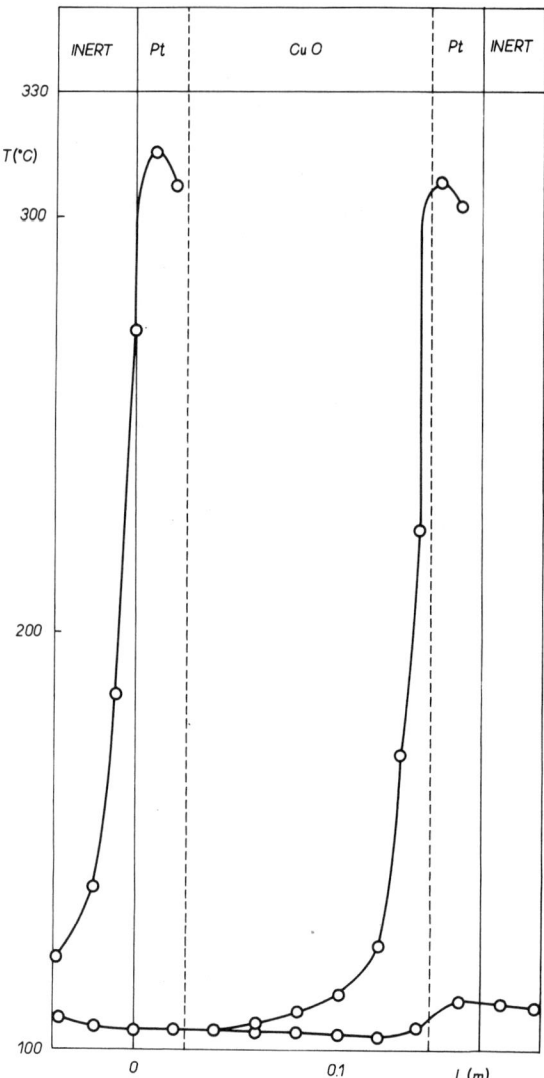

FIG. 23. Three steady states in a catalyst bed. Pt catalyst diluted with CuO particles (heterogeneous mixing).

is wider. Apparently the effect of axial heat transfer is more pronounced because of the higher thermal conductivity and the radiation effect as well.

To elucidate the problem of multiplicity our experimental observations will be discussed in detail. Three different catalyst have been tested: Pt, $CuO-MnO_2$, $Pt-CuO-MnO_2$. Figure 24 shows the dependence of the exit conversion on the inlet temperature for Re = 152, MnO_2-CuO catalyst and for the inlet concentration of CO 2,4 and 6 vol %. For low concentration of CO multiple profiles do not exist. With higher inlet CO concentration the domain of multiple steady states enlarges, i.e., the ignition temperature moves to higher values while the extinction temperature shifts to lower values. The dependences exit conversion–inlet temperature measured for a Pt honeycomb catalyst are presented in Fig. 25. Finally, the behavior of the mixed $Pt-CuO-MnO_2$ catalyst is displayed in Fig. 26. We may see that the Pt component is capable of causing multiple steady states; however, the ignition temperature moves to higher values. The effect of the Reynolds number on the region of multiple steady states is shown in Fig. 27 (1% of CO in the inlet gas, Pt catalyst). We may note that the domain of multiple steady states is rather wide, the multiple profiles can exist up to the value of Reynolds number Re ≈ 900. Of course, the domain of multiple profiles for a $CuO-MnO_2$ catalyst is essentially smaller. The shape of this domain is in a qualitative agreement with our observations for a packed bed multiplicity.

The monolithic structures exhibit higher longitudinal thermal conduc-

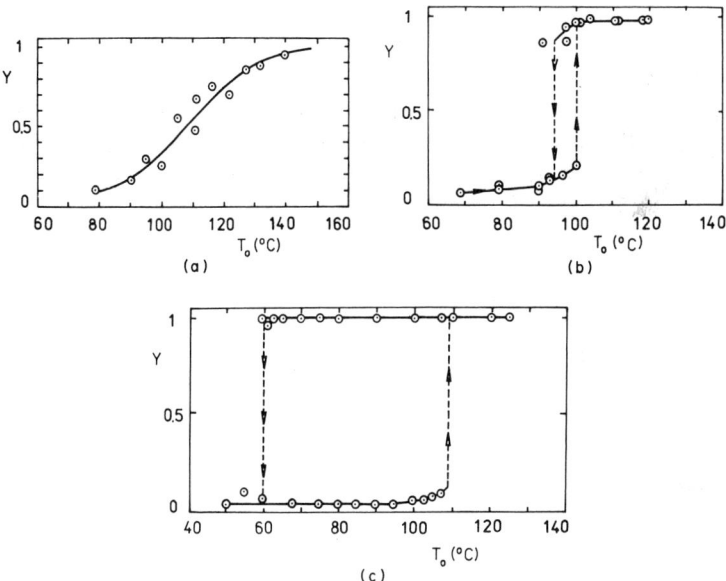

FIG. 24. Dependence of exit conversion. Y, on inlet temperature, T_0. Honeycomb structures, CuO/MnO_2 catalyst, Re = 152, (a) 2% CO, (b) 4% CO, and (c) 6% CO (58). (Copyright by Verlag Chemie GmbH. Reprinted with permission.)

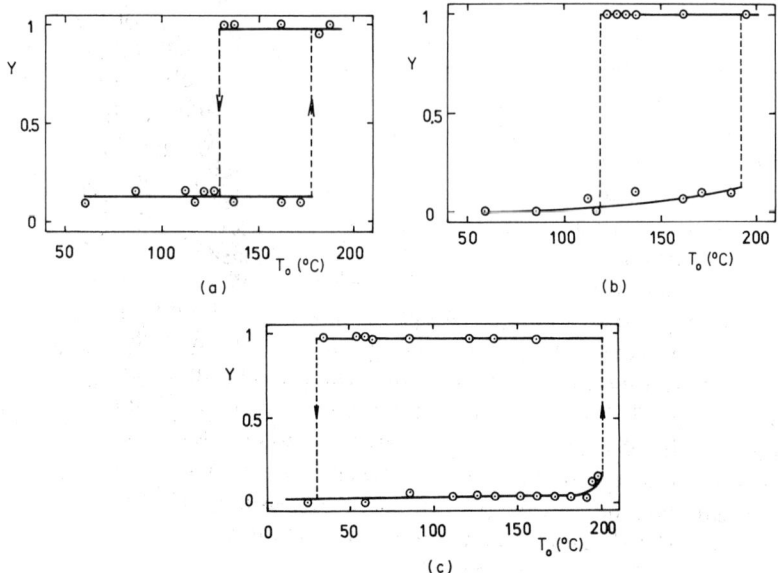

FIG. 25. Dependence of exit conversion, Y, on inlet temperature, T_0. Honeycomb structures, CuO/MnO_2 catalyst, Re = 152, (a) 2% CO, (b) 4% CO, and (c) 6% CO (58). (Copyright by Verlag Chemie GmbH. Reprinted with permission.)

tivity than packed catalytic reactors and accordingly better ignition conditions result. This is important especially in the region of low Reynolds number and, moreover, the ignited steady state can be reached at lower inlet temperature. The effect of longitudinal heat transfer is important for honeycomb structures with $Pe_s < 200$. For higher values of

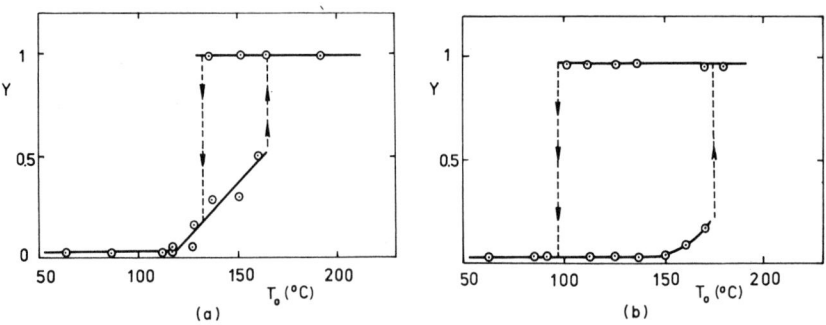

FIG. 26. Dependence of exit conversion, Y, on inlet temperature, T_0. Honeycomb structures, $Pt-MnO_2-CuO$ catalyst, Re = 356; (a) 1% CO and (b) 4% CO. (Copyright by Verlag Chemie GmbH. Reprinted with permission.)

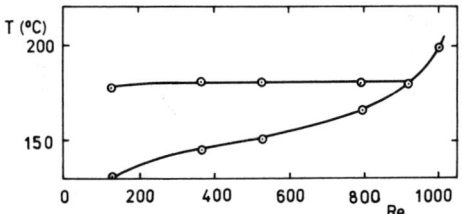

FIG. 27. Region of multiple steady states. Honeycomb structures, Pt catalyst, 1% CO (58). (Copyright by Verlag Chemie GmbH. Reprinted with permission.)

Pe_s the thermal conductivity of the solid phase need not be considered. So far the effect of radiation (which is very important) on the occurence of multiple steady states is not completely known.

VII. Other Catalytic Systems

Multiple steady states may be expected also in other catalytic systems. Two important cases should be mentioned: (i) catalytic systems with some kind of heat recirculation and (ii) trickle-bed reactors.

A. Catalytic Systems with Heat Recirculation

Tubular reactors where an exothermic reaction occurs with heat recirculation may give rise to multiple steady states, e.g., reactors with external heat exchangers, reactors with internal heat exchangers, and countercurrently cooled tubular reactors. A beautiful, industrially very important, example of multiplicity of this type is a TVA reactor for ammonia synthesis (reactor with external heat exchanger) [Baddour et al. (59)]. Though there have been observed multiple steady states for a reactor with internal heat exchangers [Kagan et al. (60)] and for a reactor with countercurrent cooling in homogeneous systems [Luss and Medellin (61)] the corresponding effects in catalytic systems have not been reported in the literature so far.

B. Trickle-Bed Reactors

Hanika et al. (62) and Germain et al. (63) observed multiple steady states in the trickle-bed reactors for the hydrogenation of cyklohexane and hydrogenation of α-methylstyrene, respectively. One possible explanation of this effect is an abrupt increase of the reaction rate arising from temperature gradients within the bed and in the gas film surrounding the

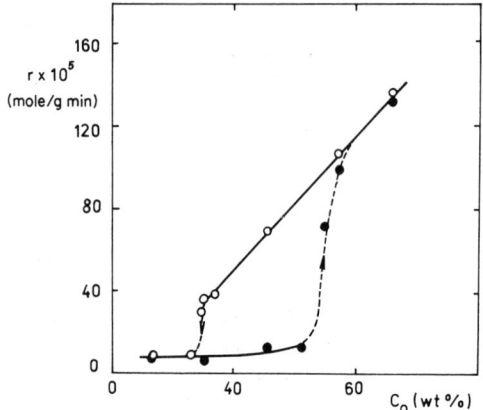

FIG. 28. The effect of cyklohexene concentration (wt %) on the reaction rate (mol/min g), trickle-bed reactors (62). (Copyright by Gordon and Breach. Reprinted with permission.)

catalyst pellet during the transition from the liquid to the gas phase regime. The transition is accompanied by the change of apparent kinetics as well as by a change of regime and operation of the pellet. The measured hysteresis loop for hydrogenation of cyklohexene is presented in Fig. 28.

VIII. Concluding Remarks

While the trend of the research activity in the area of multiplicity and periodic activity in the 1960s has been focused on theoretical investigation, the recent development has indicated an increase of experimental information. However, the number of experimental papers in comparison with the theoretical studies is still low and the need for additional laboratory studies is obvious. We have tried in this report to focus only on experimental papers and on behavior of "real" systems. We hope that a qualitative description of multiplicity and oscillations phenomena presented here will catalyze the research oriented to more detailed investigation of fundamental laws governing transport phenomena in chemically reacting systems.

Notation

a	coefficient describing geometry of a catalyst
B	dimensionless adiabatic temperature rise
t	dimensionless time
T_0	inlet temperature
x	dimensionless coordinate (catalyst particle)

y	dimensionless concentration
Y	conversion
β	dimensionless parameter of heat evolution
γ	dimensionless activation energy
θ	dimensionless temperature
θ_c	dimensionless temperature of cooling medium
χ	dimensionless parameter expressing ratio of thermal capacity of heterogeneous and homogeneous system
ξ	dimensionless coordinate (tubular reactor)
ϕ	Thiele modulus
Da	Damköhler number
Lw	Lewis number
Nu	Nusselt number
Pe_H, and Pe_M	Peclet number for heat and mass transfer, respectively
Pe_s	Peclet number related to the solid phase
Sh	Sherwood number
*	critical value (multiplicity)
**	critical value (periodic activity)
min	lower bound
max	upper bound

REFERENCES

1. Aris, R., "The Mathematical Theory of Diffusion and Reaction in Permeable Catalysts," Vols. 1 and 2. Oxford Univ. Press (Claredon), London and New York, 1975.
2. Aris, R., *Chem. Eng. Sci.* **24,** 149 (1969).
3. Schmitz, R. A., *Int. Symp. Chem. React. Eng.*, 3rd, Evanston, Ill., 1974.
4. Ray, W. H., *Chem. React. Eng.*, *Proc. 5th Eur./2nd Int. Symp. Chem. React. Eng.* p. A-8-1 (1972).
5. Weisz, P. B., and Hicks, G. S., *Chem. Eng. Sci.* **17,** 265 (1962).
6. Hlaváček, V., and Marek, M., *Proc. Eur. Symp. Chem. React. Eng.*, 4th, Brussels, 1968 p. 107 (1971).
7. Hlaváček, V., *Sci. Pap. Inst. Chem. Technol., Prague* **K9,** 149 (1974).
8. Luss, D., *Chem. Eng. Sci.* **26,** 1713 (1971).
9. Hlaváček, V., *Can. J. Chem. Eng.* **48,** 656 (1972).
10. Hlaváček, V., Kubíček, M., and Marek, M., *J. Catal.* **15,** 17 (1969).
11. Hlaváček, V., and Kubíček, M., *Chem. Eng. Sci.* **25,** 1537 (1970).
12. Hlaváček, V., and Kubíček, M., *J. Catal.* **22,** 364 (1971).
13. Hlaváček, V., and Kubíček, M., *Chem. Eng. Sci.* **25,** 1527 (1970).
14. Copelowicz, I., and Aris, R., *Chem. Eng. Sci.* **25,** 885 (1970).
15. Beusch, H., Fieguth, P., and Wicke, E., *Adv. Chem. Ser.* No. 109, p. 615 (1972).
16. Beusch, H., Fieguth, P., and Wicke, E., *Chem.-Ing.-Tech.* **44,** 445 (1972).
17. Gray, P. F., *Trans. Faraday Soc.* **66,** 1118 (1970).
18. Beljajev, V. D. et al., *Kinet. Katal.* **14,** 810 (1973).
19. Beljajev, V. D. et al., *Dokl. AH CCCP* **214,** 1098 (1974).
20. Beljajev, V. D., Slinko, M. G., and Timoschenko, V. I., *Kinet. Katal.* **16,** 555 (1975).
21. Horák, J., and Jiráček, F., *Chem. Tech. (Leipzig)* **22,** 393 (1970).
22. Jiráček, F., Havlíček, M., and Horák, J., *Collect. Czech. Chem. Commun.* **36,** 64 (1971).

23. Jiráček, F., Horák, J., and Hájková, M., *Collect. Czech. Chem. Commun.* **38,** 185 (1973).
24. Furusawa, T., and Kunii, D., *J. Chem. Eng. Jpn.* **4,** 274 (1971).
25. Linett, J. W., Reuben, B. G., and Wheatley, T. F., *Combust. Flame* **12,** 325 (1968).
26. Yang, C. H., *Combust. Flame* **23,** 97 (1974).
27. Yang, C. H., and Berald, A. L., *J. Chem. Soc., Faraday Trans. 1* **70,** 1661 (1974).
28. Lugovskoi, B. I., Matros, J. S., Kirillov, V. A., and Slinko, M. G., *Teor. Osn. Khim. Tekhnol.* **6,** 616 (1974).
29. Tamman, G., *Z. Anorg. Allg. Chem.* **111,** 90 (1920).
30. Davies, W., *Phil. Mag.* **17,** 233 (1934).
31. Buben, N., *Zh. Fiz. Khim.* **19,** 250 (1945).
32. Frank-Kamenetskii, D. A., "Diffusion and Heat Exchange in Chemical Kinetics." Plenum, New York, 1969.
33. Cardoso, M. A. A., and Luss, D., *Chem. Eng. Sci.* **24,** 1699 (1969).
34. Ray, W. H., Uppal, A., and Poore, A. B., *Chem. Eng. Sci.* **29,** 1330 (1974).
35. Rader, C. G., and Weller, S. W., *AICHE J.* **20,** 515 (1974).
36. Barelko, V. V., *Kinet. Katal.* **14,** 196 (1973).
37. Hugo, P., *Ber. Bunsengls. Phys. Chem.* **74,** 121 (1970).
38. Hugo, P., and Jakubith, M., *Chem.-Ing.-Tech.* **44,** 383 (1972).
39. Aris, R., and Amundson, N. R., *Chem. Eng. Sci.* **7,** 121, 132, 148 (1958).
40. Hlaváček, V., Kubíček, M., and Jelínek, J., *Chem. Eng. Sci.* **25,** 1441 (1970).
41. Poore, A. B., *Arch. Ration. Mech. Anal.* **52,** 358 (1973).
42. Uppal, A., Ray, W. H., and Poore, A. B., *Chem. Eng. Sci.* **29,** 967 (1974).
43. Eckert, E., Hlaváček, V., and Marek, M., *Chem. Eng. Commun.* **1,** 89 (1973).
44. Eckert, E., et al., *Chem.-Ing.-Tech.* **45,** 83 (1973).
45. Horák, J., and Jiráček, F., *Chem. React. Eng., Proc. 5th Eur./2nd Int. Symp. Chem. React. Eng.* p. B-3-47 (1972).
46. Hugo, P., *Proc. Eur. Symp. Chem. React. Eng., 4th, Brussels, 1968* p. 459 (1971).
47. Hlaváček, V., and Hofmann, H., *Chem. Eng. Sci.* **25,** 173, 1517 (1970).
48. Votruba, J., Hlaváček, V., and Marek, M., *Chem. Eng. Sci.* **27,** 1845 (1972).
49. Hlaváček, V., et al., *Chem. Eng. Sci.* **28,** 1897 (1973).
50. Hlaváček, V., and Hofmann, H., *Chem. Eng. Sci.* **25,** 1517 (1970).
51. Padberg, G., and Wicke, E., *Chem. Eng. Sci.* **22,** 1035 (1967).
52. Wicke, E., Padberg, G., and Arens, H., *Proc. Eur. Symp. Chem. React. Eng., 4th, Brussels, 1968* p. 425 (1971).
53. Fieguth, P., and Wicke, E., *Chem.-Ing.-Tech.* **43,** 604 (1971).
54. Votruba, J., and Hlaváček, V., *Int. Chem. Eng.* **14,** 461 (1974).
55. Hlaváček, V., and Votruba, J., *Adv. Chem. Ser.* No. 133, p. 545 (1974).
56. Votruba, J., et al., *Chem. Eng. Sci.* **30,** 117 (1975).
57. Sinkule, J., and Hlaváček, V., *Chem. Eng. Sci.* **33,** 839 (1978).
58. Mostecký, J., Hlaváček, V., and Votruba, J., *Erdoel Kohle* **27,** 261 (1974).
59. Baddour, R. F., et al., *Chem. Eng. Sci.* **20,** 281 (1965).
60. Kagan, J. B., et al., "Modelirovanie i Optimizacija Kaliticeskich Processov," p. 155. Nauka, Moscow, 1965.
61. Luss, D., and Medellin, P., *Chem. React. Eng., Proc. 5th Eur./2nd Int. Symp. Chem. React. Eng.* p. 134 (1972).
62. Hanika, J., et al., *Chem. Eng. Commun.* **2,** 19 (1975).
63. Germain, A. H., Lefebvre, A. G., and L'Homme, G. A., *Adv. Chem. Ser.* No. 133, p. 164 (1974).

Surface Acidity of Solid Catalysts

H. A. BENESI* AND B. H. C. WINQUIST

Shell Development Company
Westhollow Research Center
Houston, Texas

I.	Introduction	98
II.	Critique of Methods for Determination of Surface Acidity	99
	A. Aqueous Method	99
	B. Nonaqueous Indicator Methods	100
	1. Acid Strength	100
	2. Titration of Surface Acidity	104
	C. Adsorption of Gaseous Bases	107
	D. Infrared Spectroscopic Method	110
	E. Model Reactions for Determination of Surface Acidity	112
	1. Number of Acid Sites from Reaction Rates	112
	2. Number and Strength of Acid Sites from Catalytic Titrations	114
	F. Recommendations	118
III.	Review of Acidity and Catalytic Activity of Specific Solids	120
	A. Silica Gel	120
	B. Alumina	123
	1. Fluorided Alumina	125
	2. Chlorided Alumina	126
	3. Other Promoted Aluminas	130
	C. Silica–Alumina	131
	D. Other Binary Oxides	136
	E. Zeolites	138
	1. Faujasitic Zeolites	139
	a. Hydrogen Forms	139
	b. Ultrastable and Aluminum Deficient Forms	154
	c. Group Ia Forms	159
	d. Alkaline and Rare Earth Forms	160
	e. Transition Metal Forms	165
	2. Mordenite	166
	F. Clays	168
	1. Structure	168
	2. Surface Acidity	169
	3. Synthetic Mica-Montmorillonite	174
	References	176

* Present address: Catalytica Associates, Inc., Santa Clara, California 95051.

I. Introduction

Although several excellent reviews dealing with the subject of surface acidity of solid catalysts have recently been published (1–3), the activity in this area of research is great enough to warrant updating—hence, the present review. In addition to updating, we have attempted to increase the usefulness of this review by supplying a critique of methods used to measure surface acidity in which special emphasis is placed on the relevance of the results for correlations with acid-catalyzed activities. For the sake of brevity, we have limited ourselves to a discussion of solid oxides that are now used or that could be of potential use as acidic catalysts for hydrocarbon reactions in which the key steps involve transformations of carbon skeletons. Such hydrocarbon reactions have been well summarized in previous reviews; they include cracking (4) and isomerization (5) of hydrocarbons, alkylation of paraffins (6) and aromatics (7) with olefins, and polymerization of olefins (8). It is generally accepted that these reactions occur by means of a carbonium ion mechanism that was proposed by Whitmore (9) more than 30 years ago.

We will not generally concern ourselves with additional classes of reactions that can also be catalyzed via carbonium ion intermediates, such as hydrogen–denterium exchange in hydrocarbons or double-bond migration in olefins. These are facile reactions that can also occur over metals or strong bases via free-radical or carbanion intermediates and are therefore more likely to give anomalous behavior. Nor will we deal with reactions such as the catalyzed hydration of olefins (or the dehydration of alcohols) that involve the removal or formation of water, for water profoundly affects surface acidity in most cases. Though the role of surface acidity in the above-mentioned reactions can obviously be relevant, it would require a much lengthier review to include this area of catalytic research.

In the following section, we will critically review representative methods for measuring surface acidity of solid catalysts. Recommendations will then be made of the most appropriate methods from the standpoint of the needs of the investigator. The final section is devoted to updating research activities dealing with individual solid catalysts. Particular attention will be devoted to studies of acidities of unusually active catalysts such as crystalline zeolites, synthetic clays, and chlorinated aluminas.

II. Critique of Methods for Determination of Surface Acidity

Before we review the methods used to determine surface acidity, we wish to define the type of acidity that should be measured. *An acid is an electron-pair acceptor.* In our opinion, the term *acid* should be limited to this definition rather than broadening the term to include oxidizing agents as well. We agree with Flockhart and Pink (*10*) who suggest a clear distinction be made between Lewis acid–Lewis base reactions (which involve coordinate bond formation) and oxidation–reduction reactions (which involve complete transfer of one or more electrons).

There are thus two classes of acids on surfaces of metal oxides: *Lewis* acids and *Brønsted* acids (which are also termed *proton* acids). The weight of evidence (*1–8*) shows that strong Brønsted acids are the primary seat of catalytic activity for skeletal transformations of hydrocarbons. In the solids under review, they consist of protons associated with surface anions.

On the other hand, Lewis acids are catalytically inactive for skeletal transformations unless proton donors are available at the same locality.[1] Lewis acids consist of incompletely coordinated surface ions; aluminum ion is the most frequently cited example. Since the relevance of acidity measurements for the prediction of catalytic activity is what we are trying to emphasize, we have concentrated on the determination of Brønsted acidity in this critique. The problem of finding the most relevant method for acidity measurement has therefore been treated as an evolutionary process in which successive methods have been used more and more successfully for the characterization of a relatively small number of strong Brønsted acids that are frequently accompanied by a multitude of other surface acids.

A. Aqueous Methods

In essence, aqueous titration of surface acidity is an ion-exchange process in which hydrated surface protons are replaced by other hydrated cations (e.g., Na^+, NH_4^+) during the course of the titration. The procedure is straightforward. It usually consists of the direct titration of an aqueous suspension of the sample of powdered solid with a dilute base (e.g., sodium hydroxide) to a neutral endpoint. Another commonly used procedure consists of noting the pH of an appropriate salt solution (e.g., ammonium acetate), adding the sample, and measuring the amount of di-

---

[1] We know of only one possible exception to this rule: the case of pure alumina (p. 125).

lute base required to raise the pH of the suspension to its initial value. Examples of such determinations have been listed in previous reviews (*1–3, 11*).

The aqueous titration method is the least suitable method discussed in this review for measuring the seat of activity of acidic solids. The reason for its unsuitability is that the state of the surface of a solid catalyst in a water suspension is radically different from its state during use as an acidic catalyst. In the first place, water has a leveling effect; that is, it will form H_3O^+ by reaction with all Brønsted acids having an acid strength higher than that of H_3O^+. It therefore becomes impossible to discriminate among acid strengths beyond a given threshold value. Second, water creates new Brønsted sites when it reacts with incompletely coordinated metal atoms in near anhydrous oxides. In addition to these two general effects, water can drastically alter structural properties of solid catalysts. Colloidal aerogels and "decationated" zeolites can undergo structural collapse; on the other hand, montmorillonite clays undergo swelling (and thus expose previously inaccessible crystal surfaces) when water is added to such solids.

B. Nonaqueous Indicator Methods

Nonaqueous methods for the determination of surface acidity represent a considerable improvement over aqueous methods because the solvents used (e.g., benzene, iso-octane) do not react with catalyst surfaces as previously described in the case of water. Of the available types of nonaqueous methods (*1–3*), the simplest is that employing adsorbed indicators. It can be used to determine acid strengths and amounts of surface acids as described in the following section.

1. *Acid Strength*

Following Walling (*12*), the acid strength of a solid surface can be defined as its *proton-donating ability*—quantitatively expressed by Hammett and Deyrup's H_0 function (*13, 14*). H_0 is defined by

$$H_0 = -\log \frac{a_{H^+} f_B}{f_{HB^+}} \tag{1}$$

where a_{H^+} is the proton activity of the solid surface, and f_B and f_{HB^+} are activity coefficients of the basic and acid forms, respectively, of the adsorbed indicators (henceforth called *Hammett indicators*). Such indicators, B, react with a Brønsted acid, HA, according to the equation

$$B + HA = HB^+ + A^- \tag{2}$$

We emphasize that the acid strength definition embodied by Eq. (1) applies only to Brønsted acids and therefore differs from that orginally proposed by Walling (12), who suggested that H_0 functions could be used to denote acid strengths of Lewis as well as Brønsted acids. Hammett and Deyrup's treatment is not applicable to Lewis acids because relative strengths of such acids are strongly influenced by steric requirements as well as by intrinsic coordinating abilities (15, 16). Since comparison of acid strengths for a given pair of Lewis acids depend on the particular base used, strengths of this class of acids cannot be expressed in terms of a general H_0 function.

Colors of adsorbed Hammett indicators can be used to bracket the H_0 of a solid surface in the same way that the colors of more conventional acid-base indicators are used to bracket the pH of an aqueous solution. Thus, when an "acid" color is observed for the adsorbed indicator, the H_0 of the solid surface is equal to or lower than the pK_a of the indicator. As an example, a solid surface that has an H_0 that lies between $+1.5$ and -3.0 gives an acid color with benzeneazodiphenylamine (with a pK_a of $+1.5$) and a basic color with dicinnamalacetone (pK_a of -3.0). Color tests are made by adding 3–5 ml of dry solvent (e.g., benzene) to roughly 0.1 g of dried, powdered solid in a test tube, adding a few drops of a 0.1% solution of indicator in benzene, mixing the resulting suspension, and noting the color formed on the powdered solid.

Weil-Malherbe and Weiss (17) were the first to note that weakly basic indicators adsorbed on clays gave the same colors as those formed when such indicators were added to concentrated sulfuric acid. Walling (12) suggested that such color tests could be used to measure acid strengths of solid surfaces in terms of the Hammett acidity function; he used this type of test to characterize a variety of solid salts and oxides. This approach was extended by Benesi (18). Table I lists the Hammett indicators he used together with their color changes and the sulfuric acid composition corresponding to the midpoint of each indicator transition; acid strengths of a variety of acidic solids measured by means of such indicators are listed in Table II.

The indicator method is by far the easiest and quickest way of screening surface acidities of solid catalysts, but it has at least two drawbacks. First of all, the number of suitable indicators is limited because of the visual requirement that the color of the acid form mask that of the basic form. Second, the acid color of many of the Hammett indicators can be produced by processes other than simple proton addition. The first of these drawbacks can be overcome by using absorption spectroscopy to measure

TABLE I
Hammett Indicators Used for Visual Measurements of Acid Strength[a]

Indicator	Basic color	Acid color	pK_a	H_2SO_4 (% wt.)
Natural red	Yellow	Red	+6.8	8×10^{-8}
Phenylazonaphthylamine	Yellow	Red	+4.0	5×10^{-5}
Butter yellow	Yellow	Red	+3.3	3×10^{-4}
Benzeneazodiphenylamine	Yellow	Purple	+1.5	0.02
Dicinnamalacetone	Yellow	Red	−3.0	48
Benzalacetophenone	Colorless	Yellow	−5.6	71
Anthraquinone	Colorless	Yellow	−8.2	90

[a] From Benesi (18). Reprinted with permission of the American Chemical Society.

colors of adsorbed indicators (19–21). Since this tool enables the quantitative measurement of both the acid and basic forms of adsorbed indicators, the list of useful indicators can be greatly expanded and the H_0 of an acidic surface can therefore be more closely bracketed. Table III contains a list of Hammett indicators used by Leftin and Hobson (19) in their spectrophotometric studies of indicators adsorbed on acidic surfaces.

The second drawback—that "acid" colors can be produced by processes other than simple proton addition—is more serious. The case of alumina is an example. Several of the indicators listed in Table I give faint acid colors with samples of pure, activated aluminas that have previously been shown to be inactive for an acid-catalyzed reaction such as cumene cracking (22). However, closer visual examination indicates that the acid

TABLE II
Acid Strengths of Various Solids[a]

Acidic solids	H_o
Original kaolinite	−3.0 to −5.6
Hydrogen kaolinite	−5.6 −8.2
Original montmorillonite	+1.5 to −3.0
Hydrogen montmorillonite	−5.6 to −8.2
Silica–alumina	< −8.2
Silica–magnesia	+1.5 to −3.0
Alumina–boria	<8.2
H_3BO_3/silica gel	+3.3 to +1.5
H_3PO_4/silica gel	+1.5 to −3.0
H_2SO_4/silica gel	−5.6 to −8.2
$HClO_4$/silica gel	−5.6 to −8.2

[a] From Benesi (18). Reprinted with permission of the American Chemical Society.

TABLE III
Hammett Indicators for Spectrophotometric Studies of Acid Strength[a]

Indicator	pK_a
Phenylazonaphthylamine	+4.0
Aminoazobenzene	+2.8
Benzeneazodiphenylamine	+1.5
p-Nitroaniline	+1.1
o-Nitroaniline	−0.2
p-Nitrodiphenylamine	−2.4
p-Nitroazobenzene	−3.3
2,4-Dinitroaniline	−4.4
Benzalacetophenone	−5.6
p-Benzoyldiphenyl	−6.2
Anthraquinone	−8.1
2,4,6-Trinitroaniline	−9.3

[a] From Leftin and Hobson (19).

colors obtained with alumina are not quite genuine; yellow colors obtained with the less basic indicators look washed out and the red or purple colors of the more basic indicators are not quite the same as the colors obtained in homogeneous acid solutions.

Spectrophotometric studies provide quantitative information regarding the unusual acid colors mentioned above. An example of this type of study is the work by Drushel and Sommers for anthraquinone on alumina (21) as shown in Fig. 1. Reference spectra for the basic and acid forms of

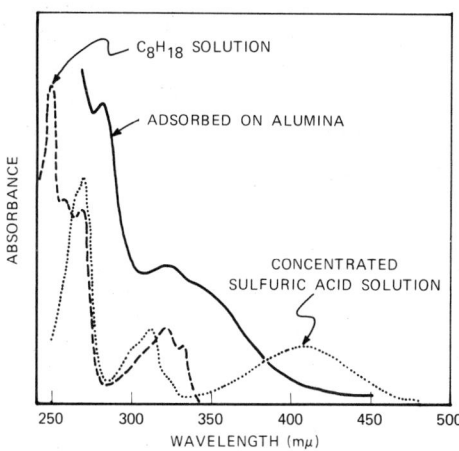

FIG. 1. Absorption spectra of anthraquinone (21). (Reprinted with permission of the American Chemical Society.)

antraquinone in homogeneous solutions are also given. These results indicate that the yellow color obtained for anthraquinone on alumina arises from the high-wavelength tail of a shifted, smeared-out absorption band arising from the basic form of this indicator. Drushel and Sommers suggest that this shift in the anthraquinone spectrum arises from physical adsorption. It seems more likely, however, that the shift arises from coordination of the oxygen groups in anthraquinone with Lewis acid sites that are known to be present in alumina surfaces (23). The physical adsorption process, in itself, does not seem to be sufficiently energetic to produce acid colors; for example, anthraquinone adsorbed on pure silica gel is colorless.

Determination of acid strengths by means of fluorescent indicators (21) and still another class of indicators, the arylcarbinols (24), has also been reported. Comparison of these indicators for the titration of surface acidity will be discussed in the following section.

2. Titration of Surface Acidity

After the acid strength of a catalyst surface has been bracketed by means of colors of adsorbed indicators, the next logical step in the determination of surface acidity is the measurement of the number of acidic groups. This is generally done by titrating a suspension of the catalyst with a solution of a suitable amine in an inert solvent; the previously described indicators are used to determine endpoints.

Johnson (25) was the first investigator to titrate surface acidities of solid catalysts in nonaqueous media. His method consisted of slowly adding a dilute benzene solution of n-butylamine to a benzene suspension of the powdered sample until the acid color (red) of the adsorbed indicator he selected (butter yellow) was completely discharged. Benesi (26) suggested the use of a complete set of Hammett indicators to yield titers of surface acids as a function of acid strength. He also developed a simplified titration technique which consisted of testing portions of catalyst samples with indicators *after* they had been equilibrated with varying amounts of n-butylamine in benzene. Equilibration was then repeated with intermediate amounts of n-butylamine until endpoints had been determined within sufficiently narrow limits. The preceding procedure required less operator time than the normal titration procedure and also enabled convenient addition of basic reagent to the catalyst without introducing traces of moisture.

Hirschler (24) suggested the use of arylcarbinol indicators (see Table IV) for the visual measurement of surface acidity. He pointed out that a variety of physical measurements (27–29) show that arylcarbinols (ROH)

TABLE IV
Arylcarbinol Indicators Used for Acid Strength Measurements[a]

Indicator	pK_{R+}	Acid strength (wt % H_2SO_4)
4,4′,4″-Trimethoxytriphenylmethanol	+0.82	1.2
4,4′,4″-Trimethyltriphenylmethanol	−4.02	36
Triphenylmethanol	−6.63	50
3,3′,3″-Trichlorotriphenylmethanol	−11.03	68
Diphenylmethanol	−13.3	77
4,4′,4″-Trinitrotriphenylmethanol	−16.27	88
2,4,6-Trimethylbenzyl alcohol	−17.38	92.5
1,1-Diphenylethylene	—	72

[a] From Hirschler (24).

react with strong Brønsted acids (HA) according to the equation

$$ROH + HA = R^+ + H_2O + A^- \qquad (3)$$

where R^+ is a colored carbonium ion. A new acidity function, denoted as H_R by Deno et al. (30), was used to define acid strength in terms of this class of indicators. Hirschler observed that the arylcarbinol indicators did not give acid colors with alumina and therefore showed a better correlation between acidity and catalytic activity than was the case with Hammett indicators.

The ultimate technique for the nonaqueous titration of surface acidity appears to be that described by Drushel and Sommers (21), who employed spectrophotometric measurements of adsorbed indicators to determine endpoints in the titration of silica gel, alumina, and silica–alumina cracking catalysts. Their procedure consisted of calcining the powdered sample *in situ* at 528°C in a quartz absorption cell, cooling the sample to room temperature, adding an iso-octane solution of the indicator to the sample through a septum, then monitoring the spectrum of the resulting slurry as small increments of diethylamine in iso-octane were added through the septum. The titer with a given indicator was obtained from the minimum amount of diethylamine required to remove the absorption band arising from the protonated form of the adsorbed indicator.

Results of three independent titration studies of silica–alumina cracking catalyst are compared in Fig. 2. Benesi (26) and Hirschler (24) used *n*-butylamine as the basic reagent and detected endpoints visually by means of Hammett and arylcarbinol indicators, respectively. Drushel and Sommers (21) used diethylamine as the basic reagent and measured end-

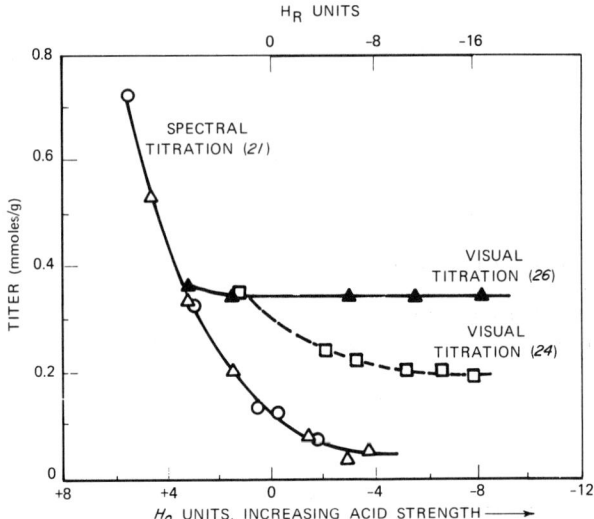

FIG. 2. Surface acidity of silica–alumina; visual and spectrophotometric titrations (21, 24, 26). ▲, △, Hammett indicators; ○, Fluorescent indicators; □, arylcarbinol indicators.

points spectrophometrically by means of fluorescent and Hammett indicators. The H_0 scale at the bottom of the figure refers to the fluorescent and Hammett indicators. The H_R scale at the top of the figure refers to the arylcarbinol indicators.

Figure 2 shows that the arylcarbinols give considerably lower visual titers at high acid strengths than those obtained with Hammett indicators. Part of this difference may arise from the previously discussed drawback of the more weakly basic Hammett indicators: an acid color can arise from the shifted absorption band of the basic form of the indicator. Differences in the behavior of the two classes of indicators may also arise from the fact that arylcarbinols produce water when they react with proton acids. (We have observed that addition of water to pure, activated alumina, followed by drying of the hydrated product, results in an alumina surface that no longer gives acid colors with Hammett indicators.) Both sets of titration results obtained by means of visual endpoints are much higher than those obtained spectrophotometrically. Drushel and Sommers conclude, quite correctly, that use of the visual method often results in over titration of catalyst acidity, especially in the case of the more weakly basic indicators.

Even in the case of the more exact determination of endpoints by spectrophotometry, doubts still remain regarding the accuracy of the indicator titration method. For example, there is always the possibility that certain

Lewis acid sites can coordinate with a given indicator molecule to produce an adsorption band identical in position with that produced through proton addition. Even if the indicators used are responsive only to Brønsted acids, most basic reagents used to titrate surface acidity (e.g., *n*-butylamine, pyridine) are strongly adsorbed on surface sites other than Brønsted acid sites. In this connection, a recent study indicates that adsorption equilibrium is not fully established during titration of silica–alumina with *n*-butylamine because of the irreversible attachment of amine molecules by adsorption sites at which they first arrive *(31)*.

In summary, the preceding considerations make it evident that amine titrations generally give acidity values higher than the number of strong Brønsted acid sites that constitute the primary seat of catalytic activity.

C. Adsorption of Gaseous Bases

Adsorption measurements of gaseous bases are often used for the determination of surface acidity. The chief advantage of this method over those discussed previously is that adsorption can be measured at or near temperatures at which catalyzed reactions occur. A variety of bases and techniques have been used as described in previous reviews *(1–3)*. Two representative approaches will be discussed in this critique: one consists of using chemisorption measurements of a suitable base at a given set of conditions to "count" acid sties on a catalyst surface; the other is a more comprehensive approach in which adsorption is measured at several temperatures to enable the determination of the thermodynamics of adsorption as a function of surface coverage.

The classical study of cracking catalysts by Mills *et al.* *(32)* illustrates the first approach. These investigators measured the adsorption of an amine such as quinoline or pyridine by weighing the catalyst sample in a perforated glass bucket as it was saturated with amine in a mixture of nitrogen and amine vapor. *Chemisorbed* amine was defined as the amount of amine retained by the presaturated catalyst after it was swept with pure nitrogen for a convenient length of time at the same temperature. This measurement really consists of the determination of a single point in the desorption isotherm—a point that corresponds to the amount of amine retained at the low residual pressure that is attained after excess amine is removed. As illustrated in Fig. 3, quinoline chemisorption correlates well with cracking activity for catalysts having a wide range of composition and activity.

In principle, this chemisorption method should enable the investigator to count surface sites that are catalytically active. In practice, this does not appear to be the case for most amines. Even in the case of a highly

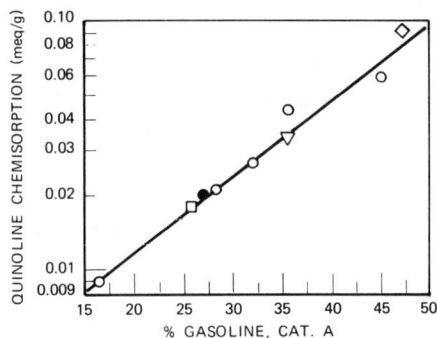

FIG. 3. Quinoline chemisorption at 315° as a function of activity for cracking light East Texas gas–oil (32). ○, SiO$_2$-Al$_2$O$_3$ Houdry type S); ●, SiO$_2$-1% Al$_2$O$_3$; □, clay catalyst (filtrol); ◇, SiO$_2$-MgO; ▽, SiO$_2$-ZrO$_2$. (Reprinted with permission of the American Chemical Society.)

effective poison such as quinoline, a considerable fraction of the chemisorbed amine is attached to inactive parts of a catalyst surface. This problem will be more fully discussed in Section II.E2.

The second approach can be illustrated with the comprehensive study of ammonia adsorption by Clark et al. (33). These investigators measured ammonia adsorption isotherms at 50° intervals in the range of 100° to 400°C for 11 silica–alumina catalysts ranging in composition from pure alumina to pure silica. Figure 4, taken from their study, contains plots of isosteric heats of adsorption of ammonia versus surface coverage for representative members of the silica–alumina series. (We focus attention on isosteric heats because this thermodynamic function is more easily interpreted than differential entropies or free energies of adsorption.) In the case of each catalyst in this series, isosteric heat drops rapidly as coverage increases. At low coverage, isosteric heats are highest in the case of alumina and fall to lower values as the silica content increases—an order very different from that for acid-catalyzed activity in this catalyst series. A subsequent study by two of the same authors (34) shows that pure alumina is inactive for a simple acid-catalyzed reaction such as isomerization of o-xylene (Fig. 5); maximum activity is obtained at a composition of 15% Al$_2$O$_3$, 85% SiO$_2$.

The lack of correlation between isosteric heats of adsorption and catalytic activity illustrates the point we have been stressing: the strength of attachment of a base to a catalyst surface is not a valid index of catalytic activity when a substantial portion of the base is strongly bonded to inactive portions of a catalyst surface. Putting it more explicitly, isosteric heats of adsorption are useful indexes for the prediction of acid-catalyzed

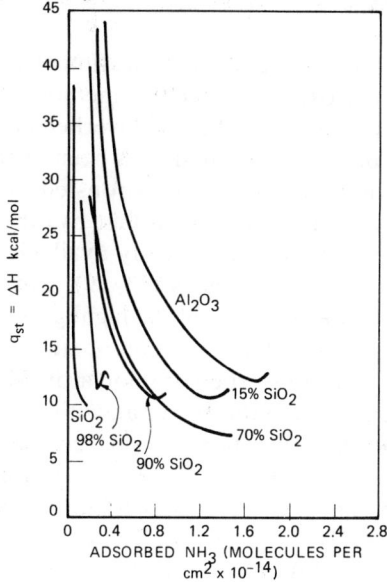

FIG. 4. Variation of isosteric heats of adsorption with coverage for ammonia on silica–alumina catalysts of various compositions (33).

activity only in those cases in which the basic reagent selectively reacts with strong Brønsted acids on a catalyst surface. Ammonia is a poor choice in this regard; infrared studies (35, 36) show that even in the case of active silica–alumina catalysts, only 20% of the chemisorbed ammonia has undergone proton addition to form NH_4^+.

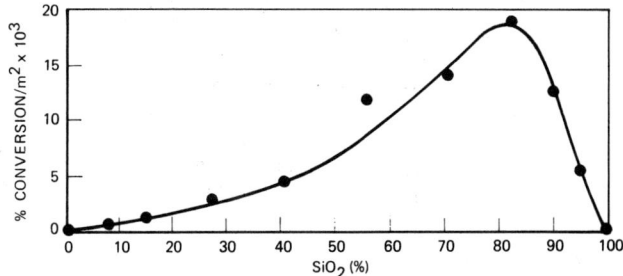

FIG. 5. Isomerization of o-xylene; activity versus silica–alumina composition (34).

D. INFRARED SPECTROSCOPIC METHOD

The most powerful tool for the characterization of acidic groups in solid surfaces is infrared spectroscopy. Since this method enables the measurement of the extent of protonation of a chemisorbed base, the chief handicap of the previously described methods has been overcome: the amount of basic reagent required to neutralize Brønsted acids can be distinguished from that attached to a catalyst surface by other types of chemical bonds. Several general reviews are already available that deal with the infrared spectroscopy of solid surfaces (37–39). Our discussions of this tool will therefore be limited to some of the highlights in the application of infrared spectroscopy for the determination of surface acidity.

The infrared study of chemisorbed ammonia by Mapes and Eischens (35) was the first to demonstrate the power and utility of the infrared spectroscopic method for determination of surface acidity. These investigators demonstrated that IR spectra of ammonia chemisorbed on cracking catalyst contained H—N—H-bending bands that arose from NH_4^+ and coordinated NH_3 (Fig. 6), a finding that constituted direct evidence for the existence of Brønsted and Lewis acids on the surface of silica–alumina catalyst. Parry (23) subsequently suggested the use of pyri-

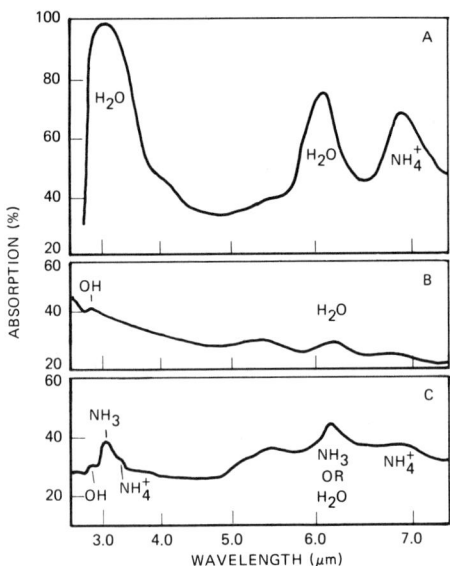

FIG. 6. Infrared spectra of silica–alumina catalyst: A, unheated; B, dried at 500°C under vacuum; C, dried catalyst with chemisorbed ammonia (35). (Reprinted with permission of the American Chemical Society.)

dine rather than ammonia for infrared studies of chemisorbed bases. Pyridine is preferred because it is a weaker base and therefore more selective for strong acid sites. Also, the pertinent pyridine absorption bands are comfortably remote from possible interference with bands arising from O—H-stretching or H—O—H-bending vibrations. A band at about 1545

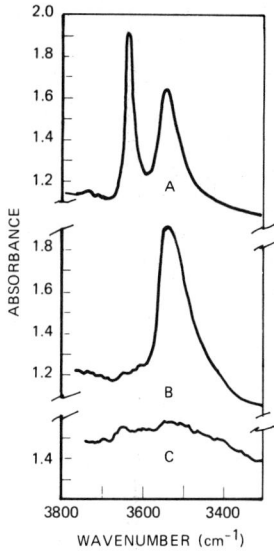

FIG. 7. Spectra of the hydroxyl stretching region of HY, illustrating the interaction of OH groups with amines: A, HY at 150°; B, HY treated with excess pyridine at 35°, then degassed at 150°C; C, HY treated with excess piperidine at 35° and degassed (41). (Reprinted with permission of the American Chemical Society.)

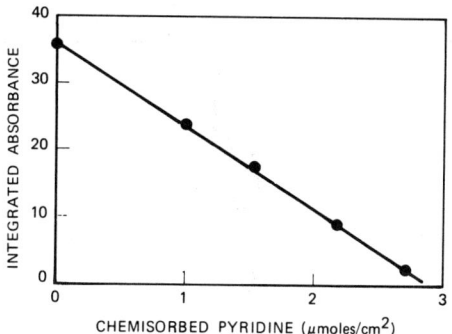

FIG. 8. Effect of pyridine addition on integrated absorbance of 3650 cm^{-1} OH band of HY (41). (Reprinted with permission of the American Chemical Society.)

cm^{-1} arises from pyridinium ions; bands in the 1440- to 1465-cm^{-1} region arise from coordinately bound pyridine.

A variety of quantitative infrared studies of amines chemisorbed on acidic catalysts have since been reported (40–44). The infrared study of Y zeolite by Hughes and White (41) exemplifies the type of information that can be obtained by these methods. Their equipment consisted of a vacuum-tight infrared cell that was permanently mounted in the sample compartment of a recording infrared spectrophotometer. Sample temperature could be adjusted within the range of −110° to 760°C. Infrared spectra of the zeolite sample were recorded *in situ* to follow spectral changes produced upon addition (or removal) of water or amines. The sample was in the form of a wafer (prepared by compressing NH$_4$Y powder) that had been heated to 300°C to form the hydrogen form of Y zeolite (HY). Hughes and White measured intensity changes in two OH-stretching bands (at 3650 and 3550 cm^{-1}) as measured amounts of amine were added to HY. Figure 7 shows that addition of excess piperidine resulted in the removal of both OH bands; excess pyridine (which is a weaker base than piperidine) removed the 3650-cm^{-1} band but not the 3550-cm^{-1} band. Thus the OH groups that produced the 3650-cm^{-1} band appear to be stronger or more accessible Brønsted acids than those giving rise to the 3550-cm^{-1} band. The progressive reaction of acidic OH groups with pyridine is illustrated in Fig. 8, in which the intensity of the 3650-cm^{-1} band is plotted as a function of added pyridine. The simultaneous appearance of an absorption band at 1540 cm^{-1} (and the absence of a band at 1450 cm^{-1}) demonstrated that pyridine is quantitatively converted to pyridinium ion and that this sample of Y zeolite does not contain appreciable concentrations of Lewis acid.

E. Model Reactions for Determination of Surface Acidity

1. *Number of Acid Sites from Reaction Rates*

To a catalytic chemist, the most relevant approach for the determination of surface acidity is from rate measurements of a relatively simple acid-catalyzed reaction (i.e., a "model" reaction). The theoretical justification for this approach is based on the Arrhenius relation

$$r = Ae^{-E/RT} \tag{4}$$

where r is the reaction rate, A is the frequency factor, and E is the energy of activation for a suitable acid-catalyzed reaction. Under the appropriate conditions, A will be directly proportional to the concentration of acid sites and E will include an energy term that represents acid strength. If the kinetics are sufficiently simple, the absolute number of active sites per unit area can be calculated from the frequency factor by means of absolute rate theory (45). For example, the rate r of an acid-catalyzed unimolecular reaction that exhibits zero-order kinetics will be given by

$$r = c_s(kT/h)e^{-E/RT} \tag{5}$$

where c_s is the concentration of active sites (number per unit area), k is the Boltzmann constant, and h is the Planck constant. Combining Eqs. (4) and (5) one obtains

$$c_s = hA/kT \tag{6}$$

by which the absolute concentration of acid sites can be calculated. In an extensive study of the kinetics of cumene cracking over silica–alumina catalyst, Prater and Lago (46) demonstrated that the necessary conditions for the applicability of Eq. (5) were satisfied at appropriate temperatures and partial pressures of cumene. From the frequency factor A, they calculated a value of 0.8×10^{17} sites/m² for the acid site concentration on a silica–alumina catalyst that contained 10% (w) alumina. The authors point out that this value is in remarkable agreement with the value of 1.3×10^{17} sites/m² obtained by Mills et al. (32) from quinoline chemisorption measurements. The application of absolute rate theory for the calculation of active site concentrations in a variety of solid catalysts has been extended by Maatman (47).

Rates of model reactions are more commonly used to determine relative rather than absolute surface acidities and a variety of acid-catalyzed reactions have been used for this purpose (1–3). Xylene isomerization is a particularly well-substantiated model reaction, thanks to work by Ward and Hansford (43). They demonstrated that the conversion of o-xylene to p- and m-xylenes over a series of synthetic silica–alumina catalysts increases as the alumina content is increased from 1 to 7%. The number of strong Brønsted acids in each member of the catalyst series was measured by means of infrared spectroscopy. Since conversion of o-xylene was found to be a straight-line function of the number of Brønsted acids (see Fig. 9), rate of xylene isomerization appears to be a valid index of the amount of surface acidity for this catalyst series. This correlation also indicates that the acid strengths of these silica–alumina preparations are roughly equivalent.

In the general case, however, both the number and strength of acid sites will vary from one catalyst to another. The determination of both of these parameters by means of model reactions is described in the following section.

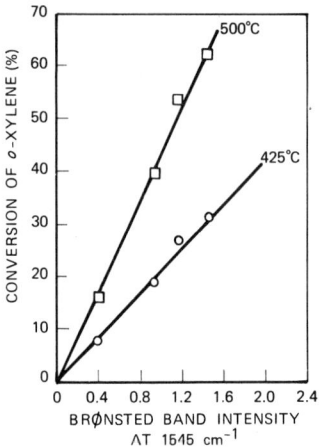

FIG. 9. Conversion of o-xylene as a function of Brønsted acidity for synthetic silica–alumina catalysts (43).

2. Number and Strength of Acid Sites from Catalytic Titrations

An apparently straightforward method for the determination of number and strength of acid sites consists of the determination of the amount of base required to poison catalytic activity for a model reaction. By means of plots of activity versus amount of added base, the number of acid sites is obtained from the threshold amount of base required to remove catalytic activity; acid strength is gauged from the slope of the titration curve. This method can therefore be called a *catalytic titration*.

The work by Mills et al. (32) includes an early example of catalytic titration behavior. Figure 10 taken from their study shows that cumene cracking at 425°C drops sharply as nitrogen bases are chemisorbed in increasing amounts on silica–alumina catalyst. Base effectiveness decreases in the order: quinaldine > quinoline > pyrrole > piperidine > decylamine > aniline.

A different type of catalytic titration was carried out by Stright and Danforth (48), who added varying amounts of lithium or potassium hydroxide to aqueous suspensions of cracking catalysts. The products were dried, calcined, and tested for cetane cracking at 500°C by means of a flow reactor. Plots of cetane conversion versus amount of added lithium hydroxide were used to determine titers for a variety of cracking catalysts (Fig. 11). This type of catalytic titration is not recommended for quantitative studies because it given high acidity values. In aqueous media, alkali

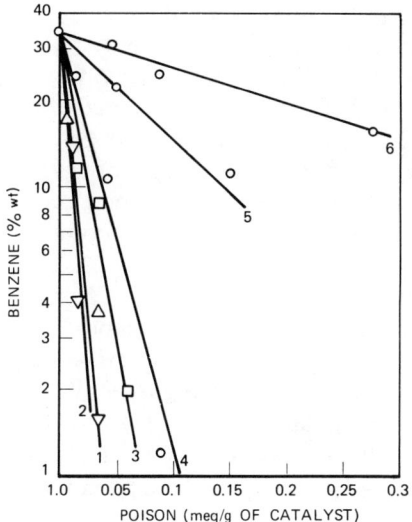

FIG. 10. Poisoning effect of amines on cumene cracking over silica–alumina catalyst: 1, quinoline; 2, quinaldine; 3, pyrrole; 4, piperidine; 5, decylamine; 6, aniline (32). (Reprinted with permission of the American Chemical Society.)

metal ions replace hydrogen ions that may originally have had a wide spectrum of acid strengths.

The advent of gas–liquid chromatography and the use of pulse reactors (49) led to the independent development in two laboratories (50, 51) of a simple, straightforward technique for carrying out catalytic titrations. It

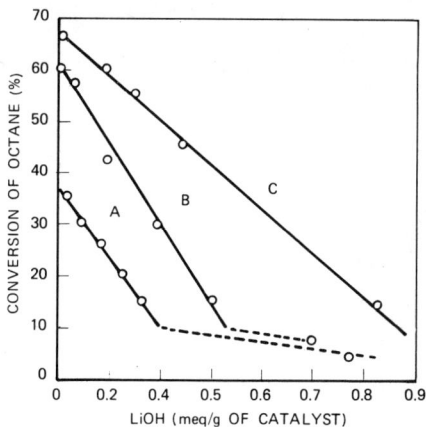

FIG 11. Conversion of cetane versus lithium hydroxide addition for several cracking catalysts: A, 19% Al_2O_3/81% SiO_2: B, 31% MgO/69% SiO_2; C, 12% B_2O_3/88% Al_2O_3 (48). (Reprinted with permission of the American Chemical Society.)

consists of the alternate addition of small pulses of a suitable reactant (e.g., cumene) and a strongly adsorbed base (e.g., quinoline) into an inert carrier gas that flows over the catalyst sample. Conversion of the reactant is measured by means of gas–liquid chromatography. Figure 12 taken from the study of Turkevich *et al.* (*50*) illustrates this type of catalytic titration. Upon adding quinoline to a sample of crystalline zeolite, conversion of cumene drops continuously until the catalyst is completely poisoned. The minimum amount of quinoline required to poison the catalyst is the catalytic titer.

Special care has to be taken, however, that the quinoline titer truly represents the *minimum* amount of catalyst poison. In most cases this type of base is adsorbed by inactive as well as active sites. Demonstration of indiscriminate adsorption is furnished by the titration results of Romanovskii *et al.* (*52*). These authors (Fig. 13) showed that introduction of a given dose of quinoline at 430°C in a stream of carrier gas caused the activity of Y-zeolite catalyst (as measured by cumene conversion) to drop with time, reach a minimum value, then slowly rise as quinoline was desorbed. The decrease in catalytic activity with time is direct evidence for the redistribution of initially adsorbed quinoline from inactive to active centers. We have observed similar behavior in carrying out catalytic titrations of amorphous and crystalline aluminosilicates with pyridine, quinoline, and lutidine isomers. In most cases, we found that the poisoning effectiveness of a given amine can be increased either by lengthening the time interval between pulse additions or by raising the sample temperature for a few minutes after each pulse addition.

An interesting type of catalytic titration procedure was reported by Goldstein and Morgan (*53*). They employed a flow technique in which

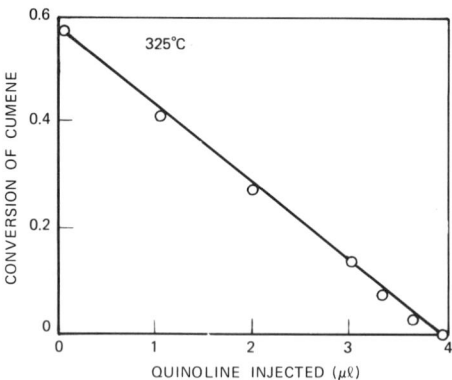

Fig. 12. Titration with poison of a catalyst surface (*50*). (Reprinted with permission of Elsevier/North-Holland Biomedical Press.)

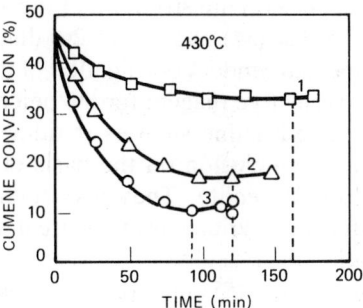

FIG. 13. Variation of catalytic activity after quinoline introduction at 430°: 1, 0.03; 2, 0.05; 3, 0.07 mmoles of quinoline/g of catalyst (52). (Reprinted with permission of Plenum Press.)

both the reactant (cumene) and a low concentration of the basic reagent (quinoline) pass continuously over the catalyst sample at a suitable reaction temperature while extent of cumene cracking is monitored as a function of time. Figure 14 shows an example of their results: the titration of CeY zeolite at 353°C. Cumene conversion is initially complete (100%) until enough active sites have been poisoned to affect the high activity of this type of catalyst. Cumene conversion then drops to zero with time. In the case cited, it took 55 min to poison the catalyst; the titer was 0.62 mmoles/g. The authors found that such titers decreased as reaction temperature increased—presumably because of the more effective distribution of quinoline at the higher temperatures.

Even after the poisoning effectiveness of an amine is optimized by redistribution treatments, a considerable fraction of the amine can remain coordinatively bound to Lewis acid sites that are inactive for skeletal rearrangements of hydrocarbons. However, advantage can be taken of the fact that the relative extent of reaction of an amine with Lewis and

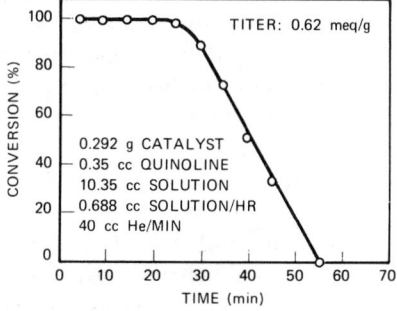

FIG. 14. Titration of CeY with quinoline at 353° (53).

Brønsted acid sites depends on the structure of the amine. Chemisorption data for amorphous oxides (54) show that 2,6-dimethylpyridine (which contains methyl groups that hinder coordination of the nitrogen atom with Lewis acids) is a more selective reagent for the determination of Brønsted acidity than an unhindered amine such as pyridine. Jacobs and Heylen (44) arrived at a similar conclusion on the basis of an infrared study of amines chemisorbed on Y zeolite. They also found that the poisoning effectiveness of 2,6-dimethylpyridine is much greater than that of pyridine for the catalytic titration of Y zeolite.

The work of Misono et al. (55) illustrates how acid strength distributions for silica–alumina catalyst can be deduced from catalytic titration measurements by use of an appropriate series of reactants. Surface concentration of amine, pyridine in this case, was adjusted by proper choice of amine partial pressure and desorption temperature while carrier gas flowed over the catalyst sample. At each level of chemisorbed pyridine, pulses of the reactants were passed over silica–alumina at 200°C and the products analyzed. The reactants were t-butylbenzene, diisobutylene, butenes, and t-butanol. It was concluded that skeletal transformations require the presence of very strong acid sites, that double-bond isomerization occurs over moderately strong acid sites, and that alcohol dehydration can occur on weak acid sites.

F. Recommendations

Of the surface acidity measurements discussed in this critique, the one we would recommend depends on the objectives of the investigator. If he wishes to screen a large number of solid samples for potential acid-catalyzed activity, the most appropriate method would be the rapid estimation of acid strength from colors of adsorbed indicators (12, 18, 21). However, such acid strength values should be treated as a necessary, but not sufficient, condition for catalytic activity. It has been our experience, for example, that every catalyst that is active for cracking gas–oil at 500°C will give acid colors for Hammett indicators having pK_a values of +1.5 or higher. However, every sample that gives such acid colors is not necessarily active for catalytic cracking.

If the objective is to "fingerprint" surface acidity (so that one acidic solid can be distinguished more clearly from another), the nonaqueous titration method (26) would be appropriate. The original method could be made more selective for titration of Brønsted acids by choosing recently recommended indicators (21) and using a sterically hindered amine (54) as the basic reagent. However, even an improved indicator method may have its drawbacks: a sterically hindered amine is not perfectly selective, and adsorption equilibrium in the amine-indicator-catalyst system may be difficult to establish (31). As a result, the amine titration method yields acidity values that are generally higher than the true number of strong Brønsted acid sites.

When the objective is complete identification of the number, strength and type of acid sties on a catalyst surface, the infrared spectroscopic method is unsurpassed. It has its drawbacks: it is time consuming and requires the use of sophisticated equipment and sample preparation techniques. But once these obstacles are overcome, complete acidity characterization is possible. The total number of acid sites can be obtained from the amount of amine adsorbed by the catalyst sample; the relative amounts of Brønsted and Lewis acid sites can be determined from the relative intensities of absorption bands arising from the protonated and coordinated forms of the adsorbed amine; and the relative strengths of Brønsted acid sites can be deduced by using a series of amines that differ in basicity. In addition, relative strengths of Lewis acid sites can be estimated from frequency shifts in the absorption band arising from the coordinate bond formed when the amine reacts with type of acid. One must keep in mind, however, that the strength of a coordinate bond depends on steric factors as well as on the intrinsic acceptor ability of the Lewis acid.

If the objective is a pragmatic study of the effects of variables such as composition, preparation, and pretreatment on the acidic properties of a solid catalyst, the catalytic titration method would be the most relevant way to determine number and strength of the acid sites that form the seat of catalytic activity. Further understanding could then be gained by complementing the catalytic method with an independent study using a tool such as infrared spectroscopy.

As already indicated, catalytic titrations are carried out by monitoring conversion of a model reactant while the surface concentration of adsorbed reagent (usually an amine) is increased in measured steps. Conversion of the reactant is determined by means of a pulse reactor or a conventional flow reactor, depending on the needs of the investigator. In order to obtain the most meaningful titration results, we make the following recommendations.

1. Uncomplicated model reactions should be chosen in which the rate-determining step is an acid-catalyzed transformation. Reactants should therefore be selected that do not undergo extensive side reactions and that undergo kinetically simple reactions such as unimolecular cracking or isomerization.

2. Reaction conditions (temperature, flow rate, choice of reactant) should be adjusted to give relatively low conversions (20% or less; 2% or

less would be even better) of the reactant so that conversion values closely approximate a direct index of catalytic activity.

3. To provide additional information concerning relative activities of acid sites, two or more reactions could be used that belong to a single class in which the reactant changes systematically in its acid strength requirement. Good examples are cracking of *tert*-butyl-, *sec*-butyl-, and *n*-butylbenzenes (*56, 57*); cracking of isopropyl- and *n*-propylbenzenes; or isomerization of dimethyl- and diethylbenzenes.

4. Poisoning effectiveness of the basic reagent should be optimized by choosing an appropriate amine (e.g., 2,6-lutidine) and by using a titration technique that will enable distribution of amine among active centers within a convenient time interval.

III. Review of Acidity and Catalytic Activity of Specific Solids

In these sections of our chapter, we emphasize research advances in the area of surface acidity of specific solids that have occurred during the period from 1970 to the fall of 1976. As stated earlier, the class of solids with which we are chiefly concerned are metal oxides that catalyze skeletal rearrangements of hydrocarbons via carbonium ion intermediates. However, we have included reviews of silica gel and alumina, which are relatively inactive, because the properties of these solids form a useful frame of reference. The initial sections (Sections III.A–III.D) deal predominantly with amorphous catalysts; the final sections (Sections III.E and III.F), with crystalline catalysts.

Before discussing individual solids, we call attention to the summary (Table V) we have provided of recently reported methods for measurement of surface acidity other than those already discussed in the critique (*44, 47, 53, 54*).

A. Silica Gel

Pure silica gel is inactive for demanding acid-catalyzed reactions, evidently because surface SiOH groups have only a feeble acid strength. The pK_a values of such groups fall somewhere in the range of 4 to 7, depending on the mode of measurement (*18, 63, 64*). Infrared studies of bases chemisorbed on silica gel also show that the SiOH groups are only weakly acidic. Morimoto *et al.* (*59*) recently carried out an infrared study showing that silica gel has weak Brønsted acid sites which react with *n*-butylamine but are inert towards pyridine under the same conditions. In addition, we

TABLE V
Recently Reported Methods for Determination of Surface Acidity

Type of solid	Basis of method	Refs.
Silica, alumina, silica–alumina	IR and UV spectra of chemisorbed diazines	58
Silica, alumina, silica–alumina	IR spectra of chemisorbed *n*-butylamine	59
Alumina, chlorided alumina, silica–alumina	Laser Raman spectra of chemisorbed pyridine	60
Metal oxides	IR spectra of surface alcoholates; shift in CH stretching band	61
Silica–alumina series	IR spectra of surface OH groups; shift in OH stretching frequency upon adsorption	62
Metal oxides	IR spectra of surface OH groups; shift in OH stretching frequency upon adsorption	63
Silica gel	IR spectra of surface OH groups; shift in OH stretching frequency upon adsorption	64
Silica, alumina, NaX	UV spectra of chemisorbed pyridine	65
Silica–alumina series	Determination of SiOH and AlOH concentrations by nuclear magnetic resonance	66
Alumina	EPR spectrum of adsorbed nitric oxide; H_2S poisoning	67
Alumina	Temperature-programmed desorption of amines	68
Silica–alumina	Amine titration with Hammett indicators; extension of the acid strength scale	69
Silica–alumina	Calorimetric determination of heat of adsorption of butylamine	70
Alumina, silica–alumina, clays	Calorimetric titration with *n*-butylamine and trichloroacetic acid	71
Silica, alumina, silica–alumina, zeolites	Oxidation of iodide ion in aqueous solution in presence of oxygen.	72
Alumina, silica–alumina	Thermal desorpiton of butylamine and pyridine	73, 73a
Alumina	Adsorption of aliphatic amines by means of pulse chromatography and temperature-programmed desorption.	74
Alumina, silica–alumina	Amine titration before and after chemisorption of water, alcohols, and amines.	75
Silica–alumina	IR spectra of chemisorbed pyridine; modified procedure.	75a
Silica, alumina, silica–alumina	^{13}C-NMR spectra of adsorbed amines	75b

have observed that NH_4^+ absorption bands are formed when pure silica gel is exposed to ammonia but that the NH_4^+ is completely removed when the sample is outgassed at room temperature.

When silica gel is outgassed at temperatures higher than 500°C, most surface SiOH groups are removed as well as the weak Brønsted acidity associated with such groups (59). Reactive sites are formed when silica gel is dehydroxylated, as confirmed in a recent infrared study by Morrow and Cody (76). They find that the new sites chemisorb H_2O, NH_3, and CH_3OH to form SiOH, $SiNH_2$, and $SiOCH_3$ groups, respectively, and suggest that such reactive sites consist of strained Si—O—Si bridges. Earlier work by Low and Morterra (77, 78) showed that an even more reactive form of silica gel can be prepared by a three-step process consisting of methoxylation of silica gel, pyrolysis of the product, and removal of surface silanes and SiOH groups formed during the pyrolysis. They reported that the resulting silica will not only chemisorb the previously mentioned molecules but more inert molecules such as H_2, O_2, and CH_4 as well. They postulate that the reactive sites consist of biradical centers.

$$Si\overset{O}{\underset{O}{|}}Si$$

Despite the presence of sites that strongly chemisorb a variety of molecules, pure silica gel is catalytically inactive for skeletal transformations of hydrocarbons. However, as has recently been emphasized by West et al. (79), only trace amounts of acid-producing impurities such as aluminum need be present in pure silica gel to provide catalytic activity— especially when a facile reaction such as olefin isomerization is used as a test reaction. They found that addition of 0.012% Al to silica gel resulted in a 10,000-fold increase in the rate of hexene-1 isomerization at 100°C over the pure gel. An earlier study by Tamele et al. (22) showed that introduction of 0.01% wt Al in silica gel produces a 40-fold increase in cumene conversion when this hydrocarbon is cracked at 500°C. The more highly acidic solids that are formed when substantial concentrations of metal oxides are incorporated with silica are discussed in following sections.

Recent studies verify earlier conclusions (39) that surface acidity and catalytic activity can also be generated when nonmetals interact with silica gel. Taniguchi et al. (80) report that Brønsted acidity and activity for cumene cracking at 450°C are created when silica gel is halogenated. Both cracking activity and amount of acids stronger than a pK_a of −5.6 decrease in the order Cl > Br > I > F. Majewski et al. (81) find that phosphoric acid mounted on silica gel is also active for cumene cracking. Maximum activity for the P_2O_5/SiO_2 system is obtained at a P_2O_5 content of 2% when catalysts are calcined at 600°C.

B. ALUMINA

Alumina is one of the most widely used catalyst supports in the petroleum industry because it is robust, porous, relatively inexpensive, and—what is especially important—it is capable of contributing acid-catalyzed activity that can be "tailored" to suit the requirements of a diverse array of catalytic processes. These include reforming (82, 83), hydrotreating (84, 85), and paraffin isomerization (86–88). Since pure alumina is relatively inactive for the skeletal isomerization reactions that are necessary in such processes, its "acid" activity is promoted through the addition of catalyst components such as fluoride, chloride, phosphate, silica, or boria. After a discussion of pure alumina itself, we will review pertinent studies of surface acidity and catalytic activity of the promoted aluminas.

Aluminas having high surface areas (100–500 m²/g) can be prepared either by thermal decomposition of well-crystallized hydroxides or by precipitation of colloidal gels. The colloidal gels are often preferred as supports in commercial reforming and hydrotreating catalysts because they can be prepared in a pure state and their porosities and surface areas are readily adjusted by appropriate choice of gelation conditions. In most cases, the dehydrated products of alumina gels have a disordered structure resembling that of γ-alumina (which is formed by thermal decomposition of crystalline boehmite). Further details regarding structural properties of aluminas are available in the informative review by Lippens (89).

The surface of uncalcined aluminas consists of amphoteric hydroxyl groups that are even less acidic than those in silica gel. Electrophoretic measurements by Stigter *et al*. (90) show that the isoelectric point for alumina is attained at a pH of 9, much higher than the value of 2 obtained in the case of silica gel. Thus, hydroxylated aluminas are relatively nonacidic, unless they are "promoted" with acid-producing impurities.

A complex variety of surface groups are formed when hydroxylated aluminas are "activated" by outgassing or calcination above 500°C (38, 91). The primary change that occurs is the removal of most hydroxyl groups. Those that remain are still nonacidic (23) but now exist in a variety of coordination states (92). One type of hydroxyl group, namely that exhibiting the highest stretching frequency at 3800 cm⁻¹, is sufficiently basic to react with CO_2 to form bicarbonate ion (93, 94). However, the major portion of an activated-alumina surface consists of several types of oxide ions (93) that apparently differ from one another in nature or number of nearest neighbors. Since each new oxide ion is formed by condensation of two hydroxyl ions, oxide-rich portions of the alumina surface must contain the incompletely coordinated aluminum ions that are the seat of Lewis acidity (23, 95). An acid site created in this way is unusual in two respects: it becomes accessible through the creation of oxide ions that are strongly basic; and both its acid strength and accessibility increase as the proportion of nearest-neighbor oxide ions increases.

As has been emphasized in Knözinger's recent review (91), the exact identification of active sites in alumina for "acid-catalyzed" reactions can be a challenging problem. The complexity of the problem depends on the model reaction that is chosen. An example of an especially difficult case is the double-bond isomerization in olefins. This is a reaction which at one time was postulated to occur over metal oxides by an acid-catalyzed "hydrogen-switch" mechanism (96, 97). However, Peri (95) found that isomerization of butene-1 to cis- and trans-butene-2 did not involve surface hydroxyl groups in the case of γ-alumina; he thus proposed that isomerization occurred over active sites (Lewis acids) consisting of incompletely coordinated aluminum ions. He later reported (98) that CO_2 was strongly adsorbed on the active sites (called α sites) in γ-alumina but that the CO_2 could be displaced by butene-1. Peri then concluded that an active site consists of an oxide ion (basic site) in close proximity to an exposed aluminum ion. Medema and Houtman (99) found that isomerization of n-butenes over γ-alumina was strongly poisoned by triethylamine and therefore used this reagent to carry out catalytic titrations by means of a pulse reactor. Clark and Finch (100) reported that isomerization of butene-1 over activated alumina was poisoned by ammonia. Since butene isomerization occurred without an initial buildup of polymeric complex, they concluded that the reaction proceeds by a different mechanism over alumina than it does over silica–alumina. Gati and Knözinger (101) measured isomerization rates of 14 terminal olefins over η-alumina and concluded that isomerization takes place by intramolecular proton transfer on aluminum ions with the assistance of neighboring oxide ions. Ghorbel et al. (102) found that isomerization of butene-1 over an activated form of amorphous alumina was poisoned by ammonia, acetic acid, phenothiazine, and tetracyanoethylene. They concluded that isomerization activity arises both from acid sites having oxidizing character and basic sites having reducing character. Lunsford et al. (67) recently found that hydrogen sulfide is selectively adsorbed on exposed aluminum ions and that this adsorption process poisons double-bond isomerization of butene-1 over γ-alumina. However, previous work by Rosynek et al. (103) indicated that the presence of SO_2, HCl, NH_3, or pyridine had no effect on isomerization rates over γ-alumina. In summary, it is evident that agreement on the exact identity of active sites for double-bond isomerization has not yet been reached. However, the overall results do suggest that aluminum ion is the primary seat of this type of activity in pure alumina.

Pure activated aluminas are also capable of catalyzing the skeletal isomerization of olefins (*104, 105*), but at considerably higher temperatures (350°–400°C) than those required for double-bond isomerization. The results obtained by Pines and Haag (*105, 106*) leave little doubt that this type of isomerization is acid catalyzed. They found that (a) skeletal isomerization of cyclohexane or 3,3-dimethylbutene-1 over pure alumina was poisoned upon ammonia addition and (b) the order of appearance of products from 3,3-dimethylbutene-1 isomerization as contact time is increased was that predicted from carbonium ion theory. They also used indicator tests to show that the seat of acid activity in γ-alumina consists of Lewis, not Brønsted, acidity. Independent infrared studies of pyridine chemisorbed on pure alumina have verified the existence of Lewis acidity and the absence of Brønsted acidity in pure alumina (*23, 107*).

1. *Fluorided Alumina*

Fluoride addition promotes the cracking (*108–110*) and isomerization (*108, 111*) activity of alumina, presumably, because of the formation of Brønsted acid sites. In a comprehensive study of fluorided aluminas, Antipina *et al.* (*110*) demonstrated that there is a close parallelism between generation of cumene cracking activity at 430°C and surface acidity when fluoride content is increased from 0 to 7% wt (see Fig. 15). Acidity was measured by n-butylamine titration; endpoints were determined by means of an arylcarbinol indicator that detected Brønsted acids stronger than those corresponding to a pK_a of -13.3. In a separate article (*112*), Anti-

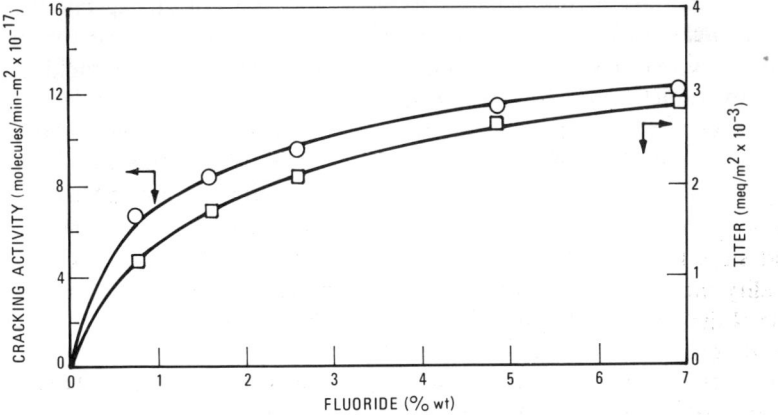

FIG. 15. Dependence of cumene cracking activity and Brønsted acidity on fluorine content in samples of fluorided alumina (*110*).

pina reported infrared measurements of chemisorbed pyridine which showed that Brønsted acid sites were indeed formed in increasing concentrations as fluoride content was increased. On the other hand, concentration of nonacidic OH groups decreased upon fluoride addition and were absent at a level of 6% F. Hughes *et al.* *(113)* carried out an independent infrared study of pyridine chemisorbed on fluorided alumina preparations. They also found that Lewis and Brønsted acids are both present on the surface of fluorided aluminas that had been dehydrated. Exposure of the dehydrated product to water vapor increased the concentration of Brønsted sites and decreased that of Lewis sites. The effectiveness of fluoride addition for the production of Brønsted acid sites was low: a maximum of about 0.01 such sites per fluoride ion was formed at a level of 7% F. The latter result can be reconciled with the much higher titration value (0.15 acid sites per F) reported by Antipina *et al.* (Fig. 15) when one takes into account the indiscriminate chemisorption of the basic reagent by Brønsted *and* Lewis sites during most amine titrations (see pg. 117). The infrared method probably gives a more accurate measure of Brønsted acidity.

2. *Chlorided Alumina*

As in the case of fluoride promotion, addition of chloride to alumina enhances its activity for skeletal transformations of hydrocarbons. At low chloride levels (0.1 to 1% wt), chlorided alumina is widely used as the acidic component of bifunctional reforming catalysts because of its high selectivity for the skeletal isomerization of olefin intermediates produced under reforming conditions *(82, 83)*. The high selectivity of chlorided alumina apparently arises because its acid sites are not strong enough to crack such intermediates extensively, as would be the case for most strongly acidic oxides at reforming conditions. The cumene cracking results by Tamele *et al.* *(22)*, which are plotted in Fig. 16, illustrate the relatively low enhancement of alumina cracking activity at low chloride levels. Comparison of the promotional effects of chloride and fluoride (Figs. 15 and 16) is also pertinent. Fluoride is the stronger promoter of cracking activity, presumably because introduction of fluoride produces acid sites that are stronger than those created in the case of chloride. Acidity measurements with arylcarbinol indicators *(24)* show that this is indeed the case. Addition of fluoride to a chlorided alumina (0.5% wt Cl) greatly increases its acid strength.

Though chlorided alumina is not strongly acidic at low chloride levels, an infrared study by Tanaka and Ogasawara *(114)* demonstrates that chemisorption of HCl on γ-alumina does form Brønsted acid sites as well as

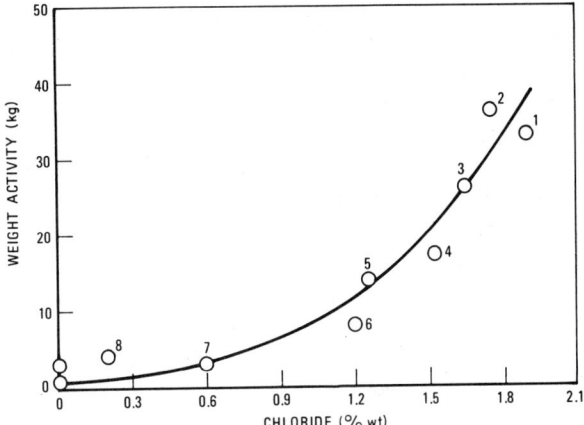

FIG. 16. Cracking of cumene at 500°C over samples of chlorided alumina (22). (Reprinted with permission of N. V. Boekhandel & Drukkerij, Leiden.)

nonacidic OH groups. Spectra of chemisorbed pyridine were examined before and after the alumina had been exposed to gaseous HCl. The formation of the 1545-cm^{-1} absorption band due to pyridinium ion is direct evidence for the creation of Brønsted acid sites in the chlorided product (1.6% wt Cl). Comparison of absorption intensities of the 1545-cm^{-1} and 1450-cm^{-1} bands also shows that the ratio of Brønsted and Lewis acid site concentrations is about 0.3, much lower than the 3.2 ratio that would have been obtained if each HCl molecule had produced one Brønsted acid site. In the same study, measurements of butene-1 isomerization over deuterated alumina samples with varying chloride contents showed that initial rates of double-bond isomerization and hydrogen–deuterium exchange increased as the chloride level increased.

Strongly acidic chlorided aluminas that contain unusually high chloride contents have been prepared by reaction of alumina with chlorinating agents such as aluminum chloride (86), carbon tetrachloride (86), thionyl chloride (87), and anhydrous hydrogen chloride (115, 116). Such chlorided aluminas, which can contain as much as 15% wt chloride, are several orders of magnitude more active for skeletal isomerization and cracking of hydrocarbons than the promoted aluminas discussed in the preceding paragraphs. Though surface acidity measurements of highly chlorinated aluminas have not yet been published, their high "acid" activities for paraffin isomerization suggest that these materials contain Brønsted acid sites having acid strengths approaching those of Friedel–Crafts catalysts such as HF/SbF$_5$ (117) and HCl/AlCl$_3$ (118).

Work by Goble and Lawrence (86) illustrates the preparation and catalytic properties of a representative member of this remarkable class of acidic solids. They reported that chlorinated alumina can be prepared by passing carbon tetrachloride vapor in an inert carrier gas over predried alumina at temperatures above 150°C. The carbon tetrachloride reacts with oxide and hydroxyl ions on the alumina surface, as evidenced by the appearance of carbon dioxide, phosgene, hydrogen chloride, and water in the exit gases. Initial buildup of chloride on alumina (see Fig. 17) is not accompanied by hydrogen removal from the catalyst; thus only oxide ions are replaced by chloride during the first stage of chlorination. After the chloride content increases beyond 7%(wt) (stage II), the hydrogen content of the catalyst decreases in amounts roughly equivalent, on an atomic basis, to the increase in chloride content. Most of the hydroxyl groups in alumina appear to be replaced by chloride ions during stage II. The effect of this type of chlorination procedure on activity for n-hexane isomerization of a sample of mildly acidic platinum–alumina catalyst is illustrated in Fig. 18, which is taken from the same study by Goble and Lawrence. In the case of each of the two catalysts, raising the reaction temperature results in increased production of isohexanes which then falls off as cracking sets in. However, the maximum in the production of isohexanes is about 250°C lower in the case of the CCl_4-treated catalyst, showing that

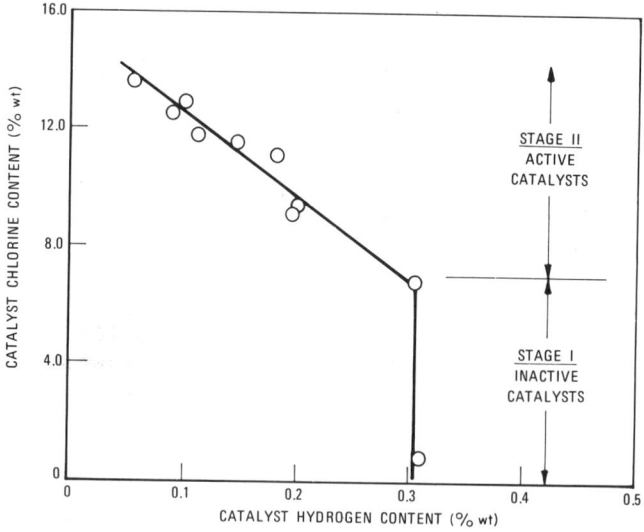

Fig. 17. Variation of catalyst chlorine and hydrogen contents during CCl_4 treatment (86). (Reprinted with permission of North-Holland Publishing Company).

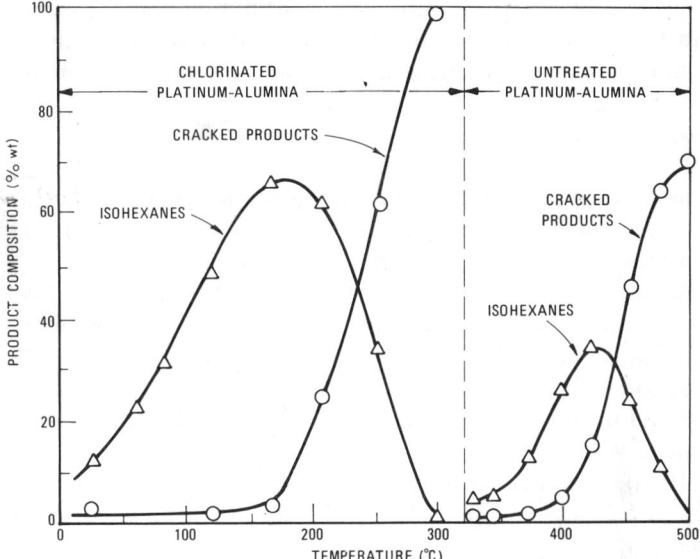

FIG. 18. Conversion of n-hexane over CCl_4 chlorinated platinum–alumina catalyst; effect of temperature on reaction products (86). (Reprinted with permission of North-Holland Publishing Company.)

chloride treatment brings about an enormous increase in isomerization activity. Isohexanes production is considerably greater over the CCl_4-treated catalyst, both because of its higher selectivity and because formation of isohexanes is thermodynamically favored to an increasing extent as temperature is lowered.

Goble and Lawrence attributed the high isomerization activity of chlorinated platinum–alumina catalyst to the creation of a "localized dual site" comprising a Lewis acid site and an adjacent platinum site. However, as has since been pointed out by Asselin et al. (88), carbonium ion intermediates over "low-temperature" isomerization catalysts are probably created by the same process as that observed for Friedel–Crafts catalyst: abstraction of hydride ion from the paraffin by a strong Brønsted acid according to the equation

$$H^+A^- + n\text{-}C_mH_{2m+2} = [n\text{-}C_mH_{2m+1}]^+A^- + H_2 \qquad (7)$$

where A^- denotes a surface anion. The main evidence against the necessity of postulating an active site that contains platinum is the high initial activity of low-temperature catalysts when platinum is completely absent.

The chief role of platinum (with hydrogen) during paraffin isomerization appears to be the *reduction* of excessive carbonium ion concentrations that lead to enhanced cracking and unstable catalyst performance (*119*).

3. *Other Promoted Aluminas*

Alumina acidity can also be promoted by addition of other anions as well as by combination of alumina with other metal oxides (*2*). Recent work in this area of research is reviewed below.

Aluminum phosphate, which can be considered a mixed oxide of Al and P, is a strongly acidic solid having high catalytic activity for cracking West Texas gas–oil (*120*). An infrared study of pure $AlPO_4$ by Peri (*121*) confirmed that this type of promoted alumina is highly acidic. He found that spectra of dry $AlPO_4$ contain two OH-stretching bands at 3800 and 3680 cm^{-1}, which he assigned to Al—OH and P—OH groups, respectively. IR spectra of chemisorbed ammonia and pyridine showed the presence of both Lewis and Brønsted acid sites. Adsorption of CO_2 and HCl revealed very few of the reactive oxide ions (α sites) found in the case of pure alumina. Peri concluded that the inactivity of $AlPO_4$ for butene isomerization probably reflects the inadequate basicity of P=O groups.

Kiviat and Petrakis (*122*) investigated the surface acidity of Co-, Ni-, and Mo-promoted aluminas by means of IR and NMR spectroscopy of chemisorbed pyridine. They observed that alumina and both Co- and Ni-impregnated alumina contained only Lewis acid sites whereas alumina inpregnated with Mo, in either the presence or the absence of Co or Ni, contained both Lewis and Brønsted acid sites. In a later publication, Ratnasamy *et al.* (*123*) reported surface acidity measurements of the Co–Mo–alumina system that were obtained by *n*-butylamine titration with Hammett indicators. These authors emphasized that the sodium content of the alumina support and the order of impregnation of cobalt and molybdenum strongly influenced the total acidity and acid strength distribution of the final product. In agreement with the results of Kivat and Petrakis, they found that strong acids ($pK_a \leq -5.6$) are generated when MoO_3 is deposited on alumina and that such strong acids are absent when CoO is deposited on alumina. However, Co gives a synergistic effect; the Co–Mo–alumina catalyst is more acidic than Mo–alumina catalyst.

Walvekar and Halgeri (*124*) reported *n*-butylamine titration measurements with Hammett indicators for a variety of metal oxides supported on alumina. They observed that amine titers and acid strengths of such binary oxides decrease in the following order:

$$MoO_3/Al_2O_3 > WO_3/Al_2O_3 > V_2O_5/Al_2O_3 > MgO/Al_2O_3 > BeO/Al_2O_3 > ThO_2/Al_2O_3$$

Of the preceding oxides, only those having an acid strength equal to or greater than that corresponding to an H_0 of -3.0 were catalytically active for the depolymerization of paraldehyde.

Since addition of boria or fluorine strongly promotes the surface acidity of alumina, it is not surprising that BF_3 is a good promoting agent. In a continuation of the study of BF_3-treated aluminas, Matsuura et al. (125) investigated the nature of the acid centers responsible for catalytic activity in a series of samples prepared by adsorption of BF_3 on γ-alumina that had been calcined at temperatures ranging from 300° to 1000°C. Surface acidities were measured by n-butylamine titration; Hammett and arylcarbinol indicators were used to determine endpoints. They concluded that strong Brønsted acid sites ($H_0 \leq -8.2$) were created when BF_3 reacted with dehydrated alumina. A good correlation was also found between n-butylamine titers ($H_0 \leq -5.6$) and catalytic activity of BF_3-treated alumina samples for cumene cracking, o-xylene isomerization, toluene disproportionation, and propylene polymerization.

A recent study by Yamamura and Nakatomi (126) shows that phosphotungstic acid impregnated on alumina is more active than alumina for codimerization of ethylene with propylene. The authors concluded that the newly formed acid sites ($H_0 \leq -5.6$) increased the yield of pentenes.

C. Silica–Alumina

The surface acidity of silica–alumina has been studied more extensively than that of any other amorphous solid primarily because of the widespread utilization of this highly acidic oxide as a cracking catalyst in the petroleum industry. Although amorphous silica–alumina has been replaced in most petroleum refineries by the more active and selective zeolite catalysts of today, it remains a high-volume product because of its use as the major matrix component in zeolite-containing cracking catalysts (127). The recent literature, reviewed in the following discussion, reflects the continuing interest in the acidic properties of silica–alumina. The earlier work in this area is covered in the excellent reviews by Tanabe (2) and Ryland et al. (11).

A unique way of identifying acid sites in amorphous silica–alumina was tried by Bourne et al. (128). These authors decided to synthesize, then characterize, two extreme types of acid site structures that they felt existed in commercial silica–aluminas. The two catalyst types consisted of low concentrations (<1.4% wt) of aluminum atoms incorporated (a) on the surface of silica gel (termed *aluminum-on-silica*) and (b) within the silica lattice (termed *aluminum-in-silica*). From infrared measurements of pyridine chemisorbed on the two materials, they conclude that dehydrated aluminum-on-silica contains only Lewis acid sites and that dehydrated alumina-in-silica contains only Brønsted acid sites. Addition of water to dehydrated aluminum-on-silica converted part of the Lewis acid sites to Brønsted acid sites.

Ballivet et al. (129) carried out infrared and thermogravimetric studies of pyridine chemisorbed on a silica–alumina series of varying alumina contents. The series was prepared by leaching aluminum from an amorphous silica–alumina containing 14% wt alumina with HCl solutions until contents of 9.4, 2.4, and 0.1% wt alumina were obtained. Neither Brønsted nor Lewis sites could be detected in the sample with 0.1% wt alumina. The other acid-extracted samples contained less acid sites than the starting material, though the acid sites that remained were stronger than those removed by HCl extraction.

An extensive study of the effect of sodium poisoning on the surface acidity of silica–alumina was carried out by Bremer et al. (130). These investigators added varying amounts of sodium ethylate from nonaqueous solution to members of an amorphous silica–alumina series containing 5 to 35% wt Al_2O_3. The products were characterized by titration with potassium methylate, chemisorption of ammonia, D_2O exchange, desorption studies, infrared spectroscopy, and ESR spectroscopy. The authors concluded that silica–alumina surfaces contain a variety of strong Brønsted and Lewis acid sites as well as weakly acidic OH groups. As sodium is added to silica–alumina in increasing amounts, the strongest Brønsted acids are poisoned first, then the Lewis and remaining Brønsted acids, and finally the weakly acidic OH groups.

As a continuation of their former studies of the states of aluminum and silicon ions in amorphous silica–aluminas, Leonard et al. (131) reported on the use of radial electron distribution and X-ray fluorescence spectroscopy to determine structure defects in alumina and in two series of cogelled silica–aluminas. The three main kinds of structure defects appear to be (1) aluminum ions tetrahedrally coordinated to oxide ions, (2) aluminum cations in a perturbed tetrahedral arrangement, and (3) silicon cations in a perturbed tetrahedral arrangement. In these perturbed tetrahedral arrangements, one of the four oxide ions may be displaced and thereby partially expose the cations. The authors conclude that in alumina, type (2) defects are the most probable and that they are related to Lewis acidity. In silica–aluminas, type (1) defects constitute the Brønsted acid sites over the entire composition range from 0 to 100% Al_2O_3. At Al_2O_3 contents greater than 50% (m), type (3) defects can provide a new source of Lewis acidity. This picture is consistent with the patterns of cumene cracking activity versus catalyst composition that these investigators obtained at several reaction temperatures for an appropriate series of silica–alumina preparations.

Parera et al. (*132*) investigated the relation between acidity and the catalytic activity for a silica–alumina catalyst that was poisoned to varying degrees by impregnation with NaOH solution. Catalytic activity was measured by means of a flow reactor for three separate reactions: cracking of cumene at 350°C, dehydration of methanol at 230°C, and methylation of methyl aniline with methanol at 230°C. Surface acidity of each catalyst was measured by n-butylamine titration with Hammett indicators at acid strengths corresponding to an H_0 of $+3.3$, -5.6, and -8.2. Plots of relative activities and acidities versus the degree of sodium poisoning were in good agreement if the proper acid strength range was chosen. Dehydration and cracking activities correlated best with the relative amounts of strong acid sites ($H_0 < -8.2$). Methylation activity closely paralleled the relative amounts of total acid sites ($H_0 < +3.3$). In another set of tests, dehydration and cracking activity of the original unpoisoned silica–alumina catalyst was measured as pulses of n-butylamine were added *in situ* at the respective reaction temperatures. In terms of millimoles per gram of catalyst, n-butylamine was 10 times as effective a catalyst poison as sodium.

Sugioka and Aomura (*133*) provide kinetic evidence indicating that the rate-determining step in the hydrocracking of aliphatic sulfur compounds over silica–alumina is catalyzed by Brønsted acid sites. Conversions of reactants were measured by use of a pulse reactor; hydrogen was used as the carrier gas. They found that reactivities of mercaptans are in the following order:

$$C_2H_5SH < n\text{-}C_3H_7SH < n\text{-}C_4H_9SH < i\text{-}C_3H_7SH < sec\text{-}C_4H_9SH$$

and those of sulfides are

$$(CH_3)_2S < (C_2H_5)_2S < (n\text{-}C_3H_7)_2S < (n\text{-}C_4H_9)_2S$$

These orders are consistent with the order of stability of the alkyl carbonium ions produced from the corresponding alkyl groups. A linear relationship was obtained between the logarithms of apparent rate constants for cracking and the heats of formation of the corresponding carbonium ions. Sugioka and Aomura therefore conclude that catalytic cracking of sulfur compounds over silica–alumina takes place by a carbonium ion mechanism.

A recent infrared study of the interrelationship of silica–alumina acidity and activity for olefin polymerization was carried out by Mizuno *et al.* (*134*). They showed that Lewis acid sites on silica–alumina could be selectively poisoned with pyridine in an infrared cell by a three-step process: (1) outgassing the silica–alumina wafer at 110°C after it had been exposed to pyridine (which produces the 1460-cm^{-1} and 1540-cm^{-1} bands arising from coordinately bound pyridine and pyridinium ion, respectively), (2) raising the outgassing temperature of the pyridine-covered wafer to 350°C (which completely removes the 1540-cm^{-1} band but does not completely remove the 1460-cm^{-1} band), and (3) allowing the outgassed wafer to stand overnight at 110°C (which regenerates the fraction of Brønsted acid sites lost during step (2). The final product thus consisted of silica–alumina in which the strong Lewis acid sites were selectively poisoned by pyridine without loss of the original Brønsted acid sites. Polymerization of propylene or *cis*-2-butene at 30°C over such silica–alumina wafers was also followed by infrared spectroscopy. By the measurements of peak intensities at 2970–2960 cm^{-1}, the formation of polymeric species on the silica–alumina surface could be monitored before and after the surface was selectively poisoned by pyridine. From the results of such experiments, the authors conclude that strong Lewis acid sites (25–30% of the total Lewis acid content) are active for olefin polymerization over silica–alumina.

The other studies of silica–alumina acidity reviewed in this section fulfill a twofold purpose: the characterization of surface acidity, and an advance in the method of measuring acidity. References to these studies are also collected in Table V, which contains a general compilation of recently reported methods for acidity measurements.

Andreu *et al.* (57) investigated the cracking of alkylbenzenes over several compositions of silica–alumina by means of the pulse technique. The principle silica–alumina series they examined (which contained 5, 10, 18, 25, and 50% wt alumina) was prepared by impregnating silica with varying amounts of aluminum nitrate solution and calcining the dried products at 700°C. Surface acidities of these materials were also measured by *n*-butylamine titration in benzene using *p*-dimethylaminoazobenzene as the indicator. Maximum acidity was obtained at an alumina content of approximately 25% wt, although only minor differences in the conversions of *tert*-butylbenzene (at 280°) and *sec*-butylbenzene (at 380°) were observed over the entire silica–alumina series. The lack of agreement between acidity and activity measurements probably arises because the titration method does not enable differentiation between Lewis and Brønsted acid sites. The most noteworthy aspect of the above study was the use of isomeric butylbenzenes as model reactants to differentiate between catalytic activities of silica–alumina preparations. The authors demonstrated that the order of reactivity for the reactants is *tert*-butylbenzene > *sec*-butylbenzene > *n*-butylbenzene, which is the order expected when cracking occurs via carbonium ion intermediates. Also, the products are those expected from such intermediates. As a consequence, this class of reactants appears to provide a broad index of "acid" activity and thereby increases the experimental capability of comparing widely differing catalyst activities at a variety of reaction temperatures.

Rouxhet and Sempels (62) characterized the hydroxyl groups in a series of silica–alumina gels (0 to 50% wt Al_2O_3) from the shifts in the OH-stretching frequencies when a variety of weakly basic molecules were adsorbed on the gels. These authors made use of previously established correlations showing that such frequency shift measurements provide a reliable estimate of acidities of OH groups. Their results for silica–alumina show the existence of two types of hydroxyl groups. Hydroxyl groups of the first type appear to have an acid strength identical with those on weakly acidic silica gel. Hydroxyl groups of the second type are more strongly acidic (pK_a of -4 to -8); their proportion increases with increasing alumina content.

Take et al. (69) extended the acid strength range of the n-butylamine titration method. They employed indicators such as 4-nitrotoluene ($pK_a = -10.5$) and 2,4-dinitrotoluene ($pK_a = -12.8$), which are considerably less basic than the other Hammett indicators used to measure surface acidity. Endpoints were determined spectrophotometrically. These authors found that the acid sites on silica–alumina catalyst had an acid strength corresponding to an H_0 between -10.5 and -12.8; a few sites had even higher acid strengths ($H_0 < 12.8$). Strong acid sites were eliminated when silica–alumina was poisoned with sodium ions.

Yoshizumi et al. (70) determined acid strength distributions on silica–alumina catalyst calorimetrically by measuring the heat adsorption of n-butylamine from benzene solution. They found that the differential heat of adsorption of n-butylamine ranged from 3.7 kcal/mole (weak acid sites) to 11.2 kcal/mole (strongest acid sites).

Tomida et al. (73) investigated the temperature-programmed desorption of n-butylamine from silica–alumina and alumina. The desorbed amine products were different in the two cases. n-Butylamine and n-butene were obtained from silica–alumina; dibutylamine and n-butene were obtained from alumina. In a subsequent paper by Takahashi et al. (73a), the authors conclude that two types of adsorption sites on silica–alumina account for the desorption behavior of n-butylamine. One type chemisorbs the amine and the other catalyzes the decomposition of the amine to lower olefins at temperatures above 300°C. On the other hand, amine decomposition was not observed when pyridine was desorbed from silica–alumina. The effects of sodium poisoning on desorption behavior of n-butylamine and pyridine were also examined.

In an infrared study of pyridine chemisorbed on silica–alumina catalysts of varying silica content, Schwarz (75a) describes a modified procedure for the measurement of Lewis and Brønsted acid sites. Quantitative determination of the relative proportions of Lewis and Brønsted acid sites is made possible because of the stoichiometric conversion of part of the coordinately bound pyridine to pyridinium ion when pyridine-covered silica–alumina is exposed to water. Brønsted acid site concentrations measured in this way are in good agreement with previously reported activities for the isomerization of o-xylene over similar silica–alumina catalysts.

Gay and Liang (75b) have recently investigated the surface acidities of silca, alumina, and silica–alumina by examining ^{13}C-NMR spectra of a variety of aliphatic and aromatic amines adsorbed on these solids. In the case of silica, only weak interactions of amines with surface hydroxyl groups are observed. Much stronger interactions are obtained in the case of alumina, although large chemical shifts that would occur on protonation are not observed. The strong acid sites on alumina appear to be sterically hindered. In the case of silica–alumina, the chemical shifts due to protonation are observed. The authors also show that the ^{13}C-NMR technique can be used to estimate Brønsted acid site concentrations.

D. Other Binary Oxides

In addition to the aluminosilicate systems, a wide variety of other binary metal oxides exhibit surface acidities and acid-catalyzed activities that are greater when the oxides are chemically combined than would be the case for a physical mixture of the same oxides. The existence of this type of synergism is most easily explained in terms of Pauling's electrostatic valence rules (135). In a recent paper, Tanabe et al. (136) have extended such rules to account for the generation of acidity in 26 binary oxides. They make two postulates:

1. The coordination number (and valence) for each metal ion in its own oxide is retained in the binary oxide.
2. The coordination number (and valence) for the oxygen ion in the oxide of the major metal component is retained for all oxygen ions in the binary oxide.

If these postulates hold, excess electrostatic charges will be created in most cases when metal oxides form chemical bonds with one another. Excess negative charges would produce Brønsted acids; positive charges would produce Lewis acids.

The majority of the experimental evidence for the previously mentioned model was obtained from acidity measurements reported by Shibata et al. (137) from the same laboratories. Eighteen binary metal oxides were prepared that consisted of coprecipitated gels of TiO_2 with oxides of Al, Si, Zr, Mg, Bi, Cd, and Sn; of ZnO with oxides of Al, Si, Zr, Mg, Sb, Bi, and Pb; of Al_2O_3 with oxides of Zr, Sb, and Bi; and of ZrO_2 with CdO. Surface acidities of the calcined products were measured by n-butylamine titration using Hammett indicators to determine acid strengths in an H_0 range of $+4.8$ to -8.2. The acid strengths of 14 of the tested binary oxides (1:1 molar ratio) were found to be much higher than those of the component single oxides. Acid strengths of the most acidic binary oxides decreased in the order: TiO_2–SiO_2 > TiO_2–Al_2O_3, Al_2O_3–ZrO_2 > TiO_2–CdO, TiO_2–SnO_2, ZnO–SiO_2. The effect of composition on acidity was also examined for the TiO_2–Al_2O_3, ZnO–Al_2O_3, and Al_2O_3–ZrO_2 systems. A fairly good correlation was obtained between the highest acid strengths observed and the average electronegativities of the two metal ions in each oxide system.

Silica–magnesia is a binary oxide of special interest because of its potential utility as a matrix in zeolite-containing cracking catalysts (127). A recent infrared study by Kermarec et al. (138) shows that the surface acidity of silica–magnesia appears to be more complex than that of silica–alumina. From infrared spectra of pyridine and 2,6-dimethylpyridine chemisorbed on silica–magnesia (18% MgO), they deduce that Brønsted and Lewis acid sites are present. Unlike the case of silica–alumina, rehydration of silica–magnesia does not convert Lewis to Brønsted acids. What is especially surprising is that the spectrum of ammonia chemisorbed on the same silica–magnesia catalyst does *not* contain the 1450-cm^{-1} band for NH_4^+ that should have been produced if Brønsted acid sites were indeed available. In separate tests, the cationic exchange capacity of the same silica–magnesia gel was also determined.

Moffat and Neeleman (139) investigated the adsorption of ammonia on boron phosphate by means of infrared spectroscopy. Ammonia appear to dissociate on this solid. Although absorption bands arising from NH_4^+ and coordinated ammonia were obtained, the authors feel that the presence of NH_4^+ does not necessarily indicate that Brønsted sites were initially present on the BPO_4 surface. Hydroxyl groups that might be formed when ammonia dissociates could react with dry ammonia to form NH_4^+.

The remaining studies reviewed in this section deal with binary oxides that contain titania. Shibata and Kiyoura (140) measured surface acidities by the n-butylamine titration method of the TiO_2–ZrO_2 system as a function of composition and method of preparation. Samples were prepared by calcination of coprecipitated mixtures of titanium and zirconium hydroxides that were made by ammonia or urea addition. The products had high acid strengths in the composition range, 40 to 75% m ZrO_2. The highest titer of strong acid sites ($H_o < -5.6$) was obtained at a composition of 50% m ZrO_2. An unexpectedly high surface area (250 m²/g) was also obtained at this composition. Pure TiO_2 and ZrO_2 prepared by precipitation of the respective gels had relatively low surface areas (40 and 70 m²/g, respectively).

Surface acidities of Al_2O_3–TiO_2, SiO_2–TiO_2, and TiO_2–ZrO_2 preparations heated at various temperatures have been reported by Walvekar and Halgeri (141). Acidities were measured by the n-butylamine titration method and by aqueous titration methods. Butylamine titers of each binary oxide increase as calcination temperature is increased, then attain maximum values, and finally decrease. The calcination temperature at which maximum acidity is obtained is 420°C for Al_2O_3–TiO_2, 550°C for SiO_2–TiO_2, and 440°C for TiO_2–ZrO_2. All three binary oxides have high acid strengths ($H_0 < -3.0$).

Tanabe et al. (142, 143) find that silica–titania is highly acidic and has high catalytic activity for phenol amination with ammonia and for double-bond isomerization in butenes. Its acidity determined by n-butylamine titration varies with pretreating temperature and catalyst composition. The highest acidity per unit weight of catalyst was obtained when TiO_2–SiO_2 (1:1 molar ratio) was heated at 500°C.

E. Zeolites

Zeolites are crystalline aluminosilicates that have exhibited catalytic activities ranging from one to four orders of magnitude greater than amorphous aluminosilicates for reactions involving carbonium ion mechanisms such as catalytic cracking (144). As a result extensive efforts have been undertaken to understand the nature of the catalytic sites that are responsible for the observed high activity. The crystalline nature of zeolites permits more definite characterization of the catalyst than is possible for amorphous acidic supports such as alumina and silica–alumina. Spectral techniques, in conjunction with structural information derived from X-ray diffraction studies, have led to at least a partial understanding of the nature of the acidic sites in the zeolite framework.

Most of the published information regarding surface acidity and its relation to catalytic activity has involved zeolites of the faujasite structure as found in zeolites X and Y. A smaller number of investigations of mordenite have been reported. This discussion will concentrate on studies of these two types of zeolites because their acidic and catalytic properties have been most widely investigated, and because they are both of significant industrial importance.

The structures of zeolites important to this review have been thoroughly studied (*145*) and will not be discussed in detail. Zeolites are composed of silcon and aluminum atoms joined by bridging oxygen atoms to form an open three-dimensional framework. The silicon and aluminum atoms are each tetrahedrally bonded to four oxygen atoms. Tetrahedral coordination of aluminum provides one negative charge to the framework for each aluminum atom. The charge on the framework is balanced by exchangeable cations, the nature of which directly affect the acidity of the zeolite. Synthetic zeolites are typically obtained in the sodium form, in which the ion-exchange sites on the framework are occupied by sodium ions.

The open three-dimensional nature of zeolite structures permits diffusion of reactant molecules into the interior voids in the crystal and accounts for the high effective surface area of these materials. Faujasitic zeolites have channels of about 8-Å diameter connecting cavities of 13-Å diameter (supercages) in a three-dimensional network. The zeolite mordenite has parallel channels with a diameter of about 7-Å. The intracrystalline surface of the zeolite is, therefore, accessible to molecules with kinetic diameters equal to or smaller than the channel diameters.

1. *Faujasitic Zeolites*

a. *Hydrogen Forms. Infrared studies.* Detailed infrared studies of the nature of the hydroxyl groups on H—X and H—Y zeolites have been reported by Uytterhoeven and co-workers (*146*), by Angell and Schaffer (*147*), and were extended in a later study by Ward (*148*). Ammonia is evolved upon heating ammonium ion-exchanged Y zeolite, and typically three distinct hydroxyl frequencies are observed (*149*). The most intense band appears at 3650 cm^{-1}. Samples which have been extensively exchanged with ammonium ion so that only a small fraction of the ion-exchange sites are occupied by residual sodium cations give rise to another hydroxyl group near 3550 cm^{-1}. A third band occurs near 3745 cm^{-1}.

Uytterhoeven et al. (*146*) proposed that the protons liberated by decomposition of the ammonium ions attacked the oxygen atoms of the zeolite framework to form structural hydroxyl groups. The infrared absorption bands at 3650 and 3550 cm^{-1} were ascribed to hydroxyl groups of this type. The mechanism of formation of the hydroxyl groups is shown in the following two equations.

$$\text{O-Al(-O)(-O)-O-Si(-O)(-O)-O with H}^{\oplus} \rightleftharpoons \text{O-Al(-O)(-O)-O-Si(-O)(-O)-O with H}$$

The band observed at 3745 cm^{-1} is similar in frequency to that of silanol groups in silica, and Angell and Schaffer (147) attributed it to a Si–OH-stretching vibration. No structural position was assigned with certainty, although it possibly arises from siliceous material occluded within the zeolite structure. It has also been ascribed to Si–OH groups terminating the zeolite framework.

The appearance of the hydroxyl bands at 3650 and 3550 cm^{-1} upon heating the ammonium form accompanies the decrease and disappearance of the NH-stretching bands as ammonia is evolved. The rate of decomposition of the ammonium ions appears to be influenced by the calcination conditions. Ward observed that most of the ammonium ions decomposed between 200° and 350°C, and at 420° only discreet hydroxyl bands were present (148). With extensively exchanged samples (>90% of the exchange sites occupied by ammonium ions), the 3550-cm^{-1} band was more intense than that at 3650 cm^{-1}, in contrast to the intensity relationship observed at lower ammonium-exchange levels. Angell and Schaffer also noted the variability of the relative intensities of the two bands with different extents of ammonium ion exchange.

Ward showed that the maximum hydroxyl group intensity was reached at 350°C and remained constant to 500°C. (148). As the temperature was raised above 500°C, the 3650- and 3550-cm^{-1} bands decreased until at 800°C, very few hydroxyl groups were observed on the surface. The loss of hydroxyl group absorption intensity upon heating above 500°C was accompanied by marked weight loss of the sample. This phenomenon was attributed to the loss of water by dehydroxylation by a mechanism resulting in formation of tricoordinated aluminum atoms (Lewis acid sites) and tricoordinated silicon atoms (146), as represented by structure I. This

$$\text{Si-O-Al}^{\ominus}\text{-O-Si-O-Si}^{\oplus}\text{-O-Al}$$

(I)

interpretation was substantiated by the observation that addition of water to samples calcined at 600°C resulted in only partial restoration of the hydroxyl groups.

The acidic nature of the hydroxyl groups corresponding to the absorption bands at 3650 and 3550 cm^{-1} was demonstrated by Uytterhoeven et al. (146) by addition of ammonia to H—Y zeolite that had been evacuated at 415°C. The intensity of the absorption bands in the hydroxyl stretching region decreased and eventually disappeared with increasing amounts of adsorbed ammonia. Simultaneously, NH_4^+ absorption bands, corresponding to NH_3 bound at Brønsted acid sites, appeared in both the stretching and bending regions. Infrared absorption bands arising from NH_3 coordinated to Lewis acid sites were not observed with samples preheated at temperatures lower than 300°C, but were evident with samples heated above 500°C (which had undergone dehydroxylation).

Infrared spectral studies of adsorbed pyridine provided further insight into the nature of the acidic hydroxyl groups. Liengme and Hall (150) found that chemisorption of pyridine on samples of Na,H—Y zeolite preheated at 480°C resulted in complete elimination of the band at 3650 cm^{-1} and weakening of the 3550-cm^{-1} band. Most of the chemisorbed pyridine was in the form of pyridinium ions, which have a characteristic absorption band at 1550 cm^{-1}. Further studies were carried out by Hughes and White (41) and by Ward (148), who examined samples which were more extensively exchanged with ammonium ions. They observed that pyridine was predominantly bound at Brønsted sites in samples that had been pretreated in the temperature range of 325° to 500°C. Adsorption of pyridine at room temperature eliminated the higher frequency band and reduced the intensity of the 3550-cm^{-1} band. Hughes and White reported no interaction with the 3550-cm^{-1} band upon adsorption at 150°C, and Eberly (151) observed the same result at 260°C. In both cases the hydroxyls at 3650 cm^{-1} were eliminated.

Hughes and White observed that adsorption of the stronger base, piperidine, resulted in the disappearance of both the 3650- and 3550-cm^{-1} bands. Partial restoration of the 3550-cm^{-1} band was achieved only after heating to 200°C, whereas heating to 400°C was necessary to restore the 3650-cm^{-1} band. The reaction of the 3650- and 3550-cm^{-1} OH groups with piperidine indicated that both were accessible to strongly basic molecules. The apparent difference in acid strength of the two hydroxyl groups suggested that they were in different structural locations in the faujasite framework.

Both hydroxyl groups were shown to be stronger acids than the hydroxyl groups in silica and alumina (41). The zeolitic hydroxyl groups were capable of protonating piperidine [(pK_a (aq) = 11.22 at 25°C (152)] and ammonia (146) [(pK_a (aq) = 9.27 at 25°C (153)], whereas the hydroxyl groups of silica (154) and alumina (155) cannot under the same conditions. The 3650-cm^{-1} hydroxyl group appeared to be the stronger Brønsted acid, since it interacted more strongly with pyridine and piperidine than did the

3550-cm^{-1} hydroxyl group. Higher temperatures were required for removal of the bases bound at the 3650-cm^{-1} hydroxyl group.

The nature of the surface acidity is dependent on the temperature of activation of the NH$_4$-faujasite. With a series of samples of NH$_4$—Y zeolite calcined at temperatures in the range of 200° to 800°C, Ward (148) observed that pyridine-exposed samples calcined below 450°C displayed a strong infrared band at 1545 cm^{-1}, corresponding to pyridine bound at Brønsted (protonic) sites. As the temperature of calcination was increased, the intensity of the 1545-cm^{-1} band decreased and a band appeared at 1450 cm^{-1}, resulting from pyridine adsorbed at Lewis (dehydroxylated) sites. The Brønsted acidity increased with calcination temperature up to about 325°C. It then remained constant to 500°C, after which it declined to about 1/10 of its maximum value (Fig. 19). The Lewis acidity was virtually nil until a calcination temperature of 450°C was reached, after which it increased slowly and then rapidly at calcination temperatures above 550°C. This behavior was considered to be a result of the combination of two adjacent hydroxyl groups followed by loss of water to form tricoordinate aluminum atoms (structure I) as suggested by Uytterhoeven et al. (146). Support for the proposed dehydroxylation mechanism was provided by Ward's observations of the relationship of Brønsted site concentration with respect to Lewis site concentration over a range of calcination tem-

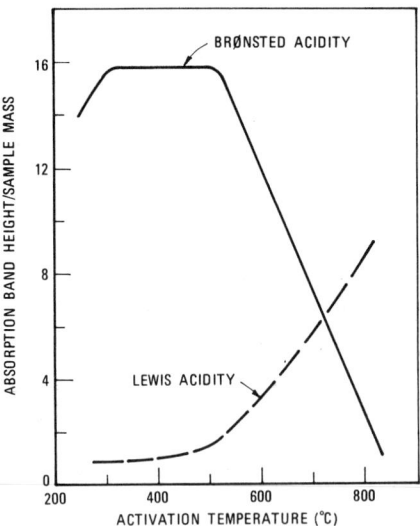

FIG. 19. Intensity of absorption bands of pyridine chemisorbed on protonic and Lewis acid sites (148).

peratures (*148*). The elimination of water by dehydroxylation should result in the formation of one Lewis acid site from two Brønsted sites; consequently, the concentration of Brønsted acid sites plus twice the concentration of Lewis sites should be constant. Such behavior was, in fact, observed by Ward (*148*) and by Hughes and White (*41*) in the temperature range of 300° to 600°C. Above 600°C, the total acidity decreased sharply, which was attributed to structural degradation.

The relationship of Lewis and Brønsted acid site concentrations on H—Y zeolite was explored further in a study by Ward (*156*) of the effect of added water. At low calcination temperatures (<500°C) only a small increase in the Brønsted acid site concentration occurred upon addition of water to the sample. Rehydration of samples dehydroxylated by calcination above 600°C resulted in a threefold increase in the amount of Brønsted-bound pyridine. However, no discreet hydroxyl bands were present in the infrared spectrum after rehydration. Thus, the hydroxyl groups reformed upon hydration must be in locations different from those present in the original H—Y zeolite, which gave rise to discreet OH bands at 3650 and 3550 cm^{-1}.

Surface acidity and catalytic activity. Numerous investigations have been made relating the nature and concentration of zeolite acid sites to catalytic activity for reactions involving carbonium ion mechanisms. One approach is to measure the catalytic activity of NH_4—Y zeolite samples that have been preheated or activated at various temperatures. As discussed previously, decomposition of the ammonium ions results in evolution of ammonia, and the residual protons form structural hydroxyl groups by combining with oxygen atoms of the zeolite framework. By preheating samples at various temperatures, the degree of ammonium ion decomposition can be controlled with resultant variation in type and concentration of surface acid sites.

Thermogravimetric investigations by Benesi (*157*), Hickson and Csicsery (*158*), and Chu (*159*) demonstrated that physically sorbed water, ammonia, and water of constitution (structural hydroxyl groups) are evolved in successive stages as the zeolite is heated. Desorption of physically adsorbed water occurs in the range of 100° to 200°C, followed by evolution of ammonia from ammonium cations at 200°–400°C (*159*). Dehydroxylation occurred at temperatures above 600°C. These thermoanalytical data are consistent with infrared spectral evidence (*41, 146–148, 151*) indicating the presence of structural hydroxyl groups in the temperature range of 300° to 500°C, which disappeared upon calcination above 500°C.

A maximum in catalytic activity occurred in samples which had been calcined at temperatures above that at which ammonia evolution had ceased. Venuto *et al.* (*160*) found that the maximum activity for alkylation of benzene with ethylene occurred upon calcination at 550°–600°C (Fig. 20). Similar activation temperatures were reported to result in maximum activity for toluene disproportionation (*157*) and *n*-hexane cracking (*161*), also shown in Fig. 20. The activity in each case increased from a very low value after activation at 300°C to the observed maximum and then declined rapidly as the activation temperature exceeded 600°C. The zeolites were essentially inactive after calcination at 700°C. The maximum activity for isomerization of 1-methyl-2-ethylbenzene observed by Hickson and Csicsery was reached at higher activation temperatures (650°C), and the activity did not diminish so precipitously. Substantial activity was observed with calcination temperatures as high as 750°C.

These results strongly pointed toward the involvement of the acidic hydroxyl groups in the catalytic reaction as suggested by Benesi (*157*), since the maximum activity was obtained when the zeolite was completely deammoniated. In addition, catalysts which had been dehydroxylated by high-temperature calcination demonstrated low activity. Thus, Benesi proposed that the Brønsted acid sites rather than the Lewis acids were the seat of activity for toluene disproportionation. This conclusion was supported by the enhancement in toluene disproportionation activity observed when the dehydroxylated (Lewis acid) Y zeolite was exposed to small quantities of water. As discussed previously, Ward's IR studies (*156*) indicated a substantial increase in Brønsted acidity upon rehydration of dehydroxylated Y sieve.

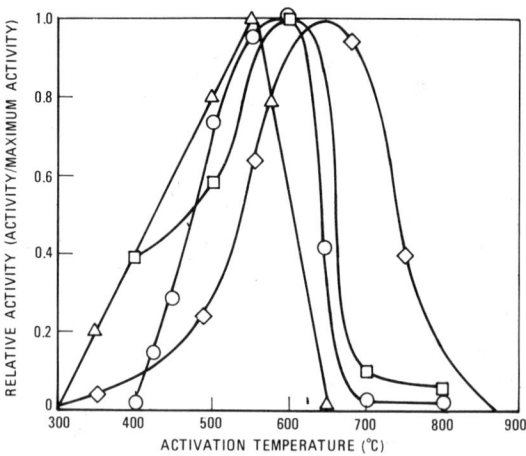

FIG. 20. Effect of activation temperature on catalytic activity. ○, ethylene-benzene alkylation (*160*); □, toluene disproportionation (*157*); △, *n*-hexane cracking (*161*); ◇, 1-methyl-2-ethylbenzene isomerization (*158*).

The observed catalytic behavior in the case of 1-methyl-2-ethylbenzene isomerization (158) was not so straightforwardly related to the Brønsted site concentration. The maximum activity was observed with samples activated at temperatures where significant dehydroxylation had occurred. This reaction occurs more readily over acid catalysts than toluene disproportionation or xylene isomerization and may require fewer Brønsted acid sites, or the reaction mechanism may involve Lewis sites.

Amine titration measurements of surface acidity. Amine titration measurements of acid site concentrations employing Hammett or arylmethanol indicators have been obtained by several investigators. Although the findings are in general agreement with the results of infrared spectral investigations, there are some dissimilarities among the titration results.

Otouma et al. (162) determined the acid site strength distribution of H—Y sieve (4.73 SiO_2/Al_2O_3, 87.5% m H form) by n-butylamine titration with Hammett indicators. The samples are presumed to have been completely deammoniated, although the activation temperature was not given. The acid strength distribution is shown in curve 1 of Fig. 21. The curves show the concentration of sites (in mmole/g) with acid strengths equal to or greater than the H_o value at a given point on the abscissa. The distribution indicates that most of the acid sites have H_o values in the range of -6 to -8. About 20% of the sites are almost uniformly distributed in the range $+3.3 \geq H_o \geq -6$. The remaining 20% have the highest acid strength ($H_o \leq -8.2$).

In a similar study, Ikemoto et al. (163) determined the influence of silica-to-alumina mole ratio on the number and strength of acid sites on H—Y zeolite. By means of ultraviolet spectroscopy, indicators with values of pK_a to -12.8 could be applied to characterize the strongest acid sites. The acid strength distributions of two samples of H—Y sieve activated at 500°C are shown in Fig. 21. Curve 2 represents a sample with a $SiO_2:Al_2O_3$ molar ratio of 4.6 (89% m H), and curve 3 shows a sample with $SiO_2:Al_2O_3$ of 5.0 (94% m H). The higher $SiO_2:Al_2O_3$ ratio sample has almost 50% more acid sites, and in both cases the vast majority of sites have H_o values in the range of -8.2 to -10.8.

Further acid site strength and concentration measurements were reported by Morita et al. (164), who related the acidity measurements to various catalytic reactions. Using Y zeolite (Linde SK-40, 90% H form) activated at 450°C, they observed no acid sites stronger than an H_o of -8.2, although the total acid site concentration was almost twice that of the former investigations (Fig. 21, curve 4). They also measured acid site concentration as a function of decomposition temperature for NH_4Y, and found that n-butylamine titration values paralleled results obtained from pyridine adsorption studies (41, 151). The maximum total acidity occurred

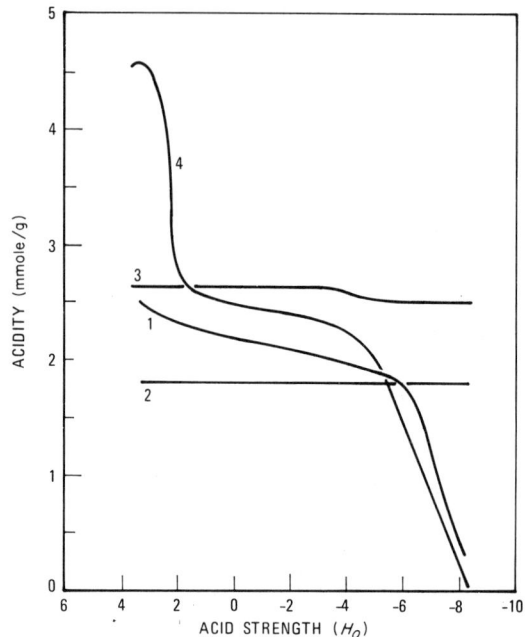

Sample	Ion exchange level (% mole H)	SiO$_2$: Al$_2$O$_3$	Activation temperature (°C)	Refs.
1	87.5	4.7	—	*162*
2	89	4.6	500	*163*
3	94	5.0	500	*163*
4	90	5.0	450	*164*

FIG. 21. Acid strength distributions of H—Y zeolites.

at 450°C, declining to less than half the maximum after calcination at 700°C, as shown in Fig. 22. A dependence on the reaction type was observed when a correspondence was sought between the maximum observed catalytic activity and the concentrations of sites of various acid strengths (*164*). Cumene cracking activity was linearly related to the concentration of sites of $H_0 \leqq -3.0$, as was the activity for o-xylene isomerization. Disproportionation of o-xylene was most closely related to the total acid site concentration ($H_0 \leqq +3.3$).

Figure 21 illustrates some of the difficulties in comparing the results of different investigations. Although the four zeolite samples shown fall within a relatively narrow range of compositions, the measured total acid site concentrations differ by as much as 250%. In addition the acid sites of

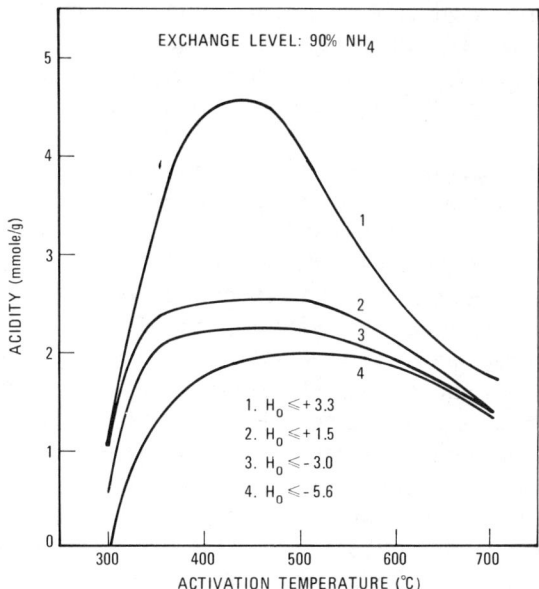

FIG. 22. Effect of activation temperature on acidity of NH_4—Y zeolite (*164*). (Copyright by the Chemical Society of Japan. Reprinted with permission.)

two of the samples are all strong ($H_0 \leq -8.2$), whereas no sites stronger than $H_0 \leq -8.2$ were observed with the other samples. It is possible that small compositional differences could result in the observed behavior; however, sample preparation procedures can also affect the nature of the zeolite. The determination of the relative contributions of these two factors requires further investigation.

Topchieva and co-workers (*165, 166*) also demonstrated that the acid sites of H—Y zeolite encompass a spectrum of acid strengths by measuring the irreversible adsorption of ammonia on the zeolite. The acid strength distribution was indicated by the change in activation energy of ammonia desorption versus surface coverage and additionally by measuring the amount of irreversibly adsorbed ammonia at various temperatures. In agreement with the results of Morita *et al.*, it was found that cumene cracking activity was related to the concentration of sites of medium strength, whereas the strongest sites were not involved (*167*). Iso-octane cracking also was related to the concentration of centers of medium strength.

Poisoning experiments. Poisoning of acidic sites by adsorption of bases is another technique that has been applied to identify the centers of catalytic activity for various reactions. The results of many of the earlier studies were interpreted in terms of interaction of the adsorbed bases with

Lewis acid centers, which were thought to be the primary catalytic sites. Boreskova et al. (51) studied the poisoning effect of quinoline on the cracking of cumene over Na, H—Y zeolite and observed a linear decrease in activity with the amount of quinoline added until a constant level of activity was reached. The catalytic activity was attributed to trivalent aluminum centers (Lewis acids), which were poisoned by coordinately bound quinoline. In a similar study of cumene cracking, Turkevich et al. (50) also concluded on the basis of magnetic resonance experiments that Lewis centers were the active sites.

More recent experiments by Turkevich and Ono (168), however, led to the conclusion that cumene cracking over Na, H—Y zeolite at 325°C was catalyzed by Brønsted acid sites. The acid site concentration determined by quinoline poisoning decreased rapidly upon pretreatment of the catalyst at temperatures above 500°C. The activity for cumene cracking displayed similar behavior. As discussed previously, Brønsted acid sites predominate in samples calcined at 500°C, whereas higher activation temperatures result in dehydroxylation with formation of Lewis acid sites. Thus, the acid sites at which quinoline was adsorbed were presumed to have been Brønsted acid sites, rather than Lewis sites. The quantitative determination of acid site concentrations by quinoline poisoning was shown by Romanovskii et al. (52) to be subject to complications resulting from adsorption–desorption occurring at sites other than those responsible for catalytic activity. They observed that the cumene cracking activity of H—Y zeolite upon exposure to quinoline declined slowly for several hours after which the activity slowly increased. The behavior was ascribed to gradual adsorption–desorption of quinoline on the zeolite surface. Pulsed poisoning methods, therefore, result in quinoline concentrations on the catalyst surface that are not necessarily directly related to the number of acid sites active for catalytic cracking (see p. 116).

That adsorption of quinoline can occur at sites other than those which are catalytically active was further shown by Goldstein and Morgan (53). Using continuous rather than pulsed introduction of quinoline, they determined the poisoning effect of quinoline on cumene cracking activity. They measured the equilibrium adsorption of quinoline on H—Y and Na,H—Y zeolite. The results with samples of Na,H—Y exchanged to various extents showed that the quinoline titer of active sites was almost constant when the acid site concentration was greater than one site per supercage, as calculated from the number of exchange sites occupied by hydroxyl groups. At active site concentrations less than one site per supercage, the quinoline titer decreased. Of all the catalysts measured, included Ce-, Ca-, and NH_4-exchanged forms, the maximum titers were nearly equal to the supercage densities, corresponding to the adsorption of one quinoline molecule per supercage. Thus, the observed value was significantly lower than the potential maximum of seven sites per supercage, which would result if every univalent exchange site were an acidic site. The results were interpreted in terms of physical blockage of the cumene from the active sites, rather than poisoning of the individual acid sites. Apparently, one molecule of quinoline per supercage was sufficient to prevent access of cumene to the active sites.

Jacobs et al. (169) investigated the poisoning of cumene cracking activity of H—Y zeolite with pyridine. The concentrations of Brønsted and Lewis acid sites were estimated from the observed intensities of infrared absorption bands of adsorbed ammonia and pyridine, respectively. The number of molecules of pyridine required at 400°C to completely poison the initial activity was equivalent to the number of Brønsted sites on catalysts that had been preheated at temperatures in the range of 450° to 650°C. At pretreatment temperatures above 650°C, the amount of pyridine required was greater than the number of Brønsted sites. Although there was a parallel relationship between hydroxyl concentration and cumene cracking activity, it was observed that the concentration of Brønsted sites and the sum of the concentrations of Brønsted and Lewis sites displayed similar behaviors at activation temperatures below 650°C. As a result the Brønsted sites appeared to be the primary source of catalytic activity, but participation by the Lewis sites could not be excluded.

It was recognized that pyridine molecules do not react selectively with Brønsted sites at 400°C (169), but also interact with Lewis acid sites formed during dehydroxylation. Thus, the concentration of acid sites measured by the pyridine-poisoning technique was considered to be an upper limit. Evidence was presented that only a fraction of the hydroxyl groups were active in cumene cracking, whereas the adsorbed base reacted with all the hydroxyls. With increasing degrees of dehydroxylation the OH concentration decreased much more rapidly than the decrease in initial catalytic activity (Fig. 23). Considerable activity was present after heat treatment at 750°C, whereas the hydroxyl band intensity was practically nil. Thus, it appeared that only a small number of OH groups was responsible for the catalytic activity. Poisoning experiments with ammonia supported this interpretation, since only 1.5×10^{19} NH_3 molecules per gram of zeolite (0.6% of the structural aluminum atoms) reduced the catalytic activity by 50%. In addition, the initial activity per active site increased with increasing extent of dehydroxylation as shown in Table VI, which further indicates that the OH groups are heterogeneous in nature and display a range of catalytic activities.

In order to distinguish clearly between the contributions of Lewis and Brønsted acid sites to the cumene cracking activity of H—Y zeolite, a method of selective poisoning was applied by Jacobs and Heylen (44). As was shown previously (54), 2,6-dimethylpyridine (DMPy) selectively ad-

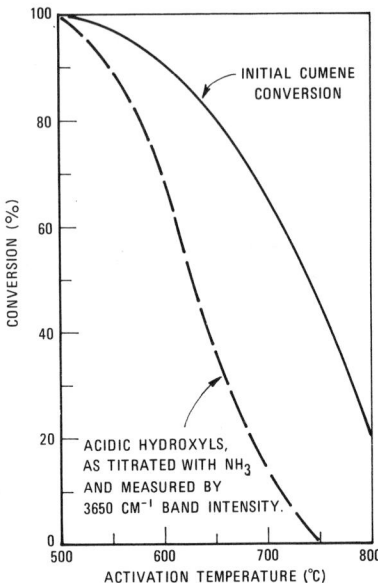

FIG. 23. Relation between cumene cracking activity and acidic hydroxyl concentration (169).

sorbs on protonic acid sites at 400°C. This effect is a result of the steric hindrance of the methyl groups to strong interactions of the basic nitrogen with Lewis acid sites such as tricoordinate aluminum atoms. By comparison of the infrared spectra of adsorbed 2,6-DMPy with that of adsorbed pyridine, and in conjunction with adsorption measurements, they were

TABLE VI

Initial Cumene Cracking Activity per Active Site of H—Y Zeolite (169) (% Conversion per OH Group)

Pretreatment temperature (°C)	Activity	
500	0.65[a]	0.66[b]
550	0.71	0.68
600	0.85	0.88
700	1.42	2.80
750	1.50	

[a] OH concentration determined by pyridine poisoning at 400°C.
[b] OH concentration from ammonia poisoning at room temperature.

able to obtain quantitative information on the amount of base adsorbed at the two types of acid sites. Room temperature adsorption of 2,6-DMPy on H—Y sieve preheated at 400°C (HY-400) resulted in selective disappearance of the 3650-cm^{-1} hydroxyl band. Only upon adsorption at 400°C did the 3550-cm^{-1} band decrease in intensity, which was ascribed to migration of protons (which give rise to the 3550-cm^{-1} absorption) into the supercages with consequent interaction with the base molecules. Formation of tricoordinate aluminum Lewis acid sites by dehydroxylation at 600°C resulted in interaction of the 2,6-DMPy first with the hydroxyl groups and only thereafter with the Lewis acid sites. The same selectivity of adsorption was observed in desorption experiments. Upon heating to 400°C, bands due to Lewis-bound 2,6-DMPy disappeared, whereas the 2,6-DMPyH$^+$ (Brønsted-bound molecules) frequencies were undisturbed.

Jacobs and Heylen also placed an upper limit on the number of active (Brønsted) sites in H—Y zeolite by poisoning the cumene cracking activity with 2,6-DMPy and pyridine. As shown in Table VII, samples treated at 400° and 600°C required equal amounts of 2,6-DMPy to reduce the cumene dealkylation activity to zero. The amount required was two orders lower than the number of aluminum atoms in the structure. The required amount of pyridine was less at the higher pretreatment temperature and was considerably higher than the amount of 2,6-DMPy. Thus, selective adsorption of 2,6-DMPy on Brønsted sites resulted in total poisoning of the cumene cracking activity at much lower levels of adsorption than did adsorption of pyridine, which interacts with both Brønsted and Lewis acidic sites. The results indicated that Brønsted sites act as the primary source of cumene cracking activity.

Nature of acidic sites. The location of the acidic hydroxyl groups in the faujasite structure has been the subject of numerous investigations and much discussion. The results of adsorption experiments with several molecules led Eberly (*170*) to conclude that the 3550-cm^{-1} hydroxyl absorption band represented hydroxyl groups located in the hexagonal prisms of the faujasite framework [(S_I sites (*171*)], where they were relatively inac-

TABLE VII
Active Sites of H—Y Zeolite for Cumene Cracking as Titrated with Pyridine and 2,6-Dimethylpyridine (Sites g^{-1} × 10^{-20})^a

Pretreatment temperature (°C)	2,6-Dimethylpyridine	Pyridine
400	1.86	10.0
600	1.80	6.3

^{a} From Jacobs and Heylen (*44*).

cessible to large molecules. The hydroxyls associated with the 3650-cm^{-1} band were ascribed to positions in the supercages (near S_{II} sites). White et al. (172) and Hall et al. (146, 150) concluded that the 3650-cm^{-1} hydroxyls were in the supercages and the 3550-cm^{-1} hydroxyls were in the cubooctahedral cages of the zeolite framework.

Arguing on the basis of Si(Al)—O bond distances obtained from a single-crystal X-ray diffraction structural determination of hydrogen faujausite, Olson and Dempsey (173) proposed that the 3650 cm^{-1} band resulted from hydrogen atoms bonded to the oxygen atoms O(1) [Broussard and Shoemaker (171)] bridging the double six-membered rings (hexagonal prisms). These positions are in the supercages and are accessible to large molecules. The 3550 cm^{-1} hydroxyl groups were attributed to hydrogen atoms within the hexagonal prisms, in agreement with Eberly's interpretation. Ion-exchange investigations by Uytterhoeven et al. (146, 174), White et al. (172), and Ward (175) supported the latter assignments. Comparison of Na, H—Y zeolite samples containing various amounts of sodium indicated that the 3650 cm^{-1} hydroxyl band was formed by replacement of the most readily exchanged sodium ions with protons. Ward (175, 176) showed that the intensity of the 3650 cm^{-1} band increases linearly with decreasing sodium content, whereas the 3550 cm^{-1} band intensity increases very little until about 60% of the sodium ions have been removed. The lower frequency band intensity then increases rapidly with decreasing sodium content until it is comparable to that of the 3650 cm^{-1} band at very low sodium levels. The assignment of the 3550 cm^{-1} hydroxyl band to a position within the hexagonal prisms is reasonable since it is formed by replacement of the most difficultly exchangeable cations. Further evidence was provided by Ward (175) when he treated ammonium Y zeolite with cesium ion, which is too large to enter the hexagonal prisms and sodalite cages of the structure (177). The 3650 cm^{-1} absorption was eliminated, whereas the 3550 cm^{-1} band was essentially unchanged in the cesium–ammonium sample.

Beaumont et al. (178, 179) employed n-butylamine titrations with Hammett and arylmethanol indicators to measure the acid strength distribution of samples of Na,H—Y zeolite exchanged with various amounts of sodium. During exchange of sodium ions with protons, the cations associated with the weaker sites were exchanged first. Sodium ions at the strongest sites were exchanged only when no more sodium ions were linked to the weaker acid centers. Thus, the weaker sites are related to the 3650 cm^{-1} band observed in IR studies, and the formation of the strongest acid sites to the appearance of the IR band at 3550 cm^{-1}. The concentration of the strongest sites was related to iso-octane cracking activity (180), which was low until 30–35% of the sodium ions, i.e., those located at the weakest sites, were replaced by protons.

The assignment of the 3650 cm^{-1} hydroxyl groups to the weakest acid sites results in an apparent disagreement between the n-butylamine titration results (179) and previous pyridine desorption experiments of Hughes and White (41), which indicated that the 3650 cm^{-1} hydroxyls retained pyridine more strongly than did the 3550 cm^{-1} hydroxyls. Dempsey and Olson (181) and Beaumont and Barthomeuf (179) suggested that sodium ions in the supercages act as partial poisons of the acidity of the hydroxyl groups. On the basis of pyridine desorption experiments and observed shifts in the IR frequencies of the acidic hydroxyl groups, Bielanski and Datka (182) suggested that the acid strength of the 3650 cm^{-1} hydroxyl groups may be increased by increasing the degree of exchange of sodium cations with protons. Using infrared spectra of adsorbed pyridine, it was shown that at temperatures above 500°C, samples with the highest sodium ion content desorbed pyridine most easily. As the sodium ion content decreased pyridine was held more strongly, indicating an increase in the strength of the protonic acid centers. Over the same range of sodium ion contents the positions of the 3650 and 3560 cm^{-1} hydroxyl bands shifted to lower frequency with decreasing sodium content. The changes in acidity and OH band position were attributed to increasing polarization of the OH bond.

Interaction of large basic molecules such as piperidine and pyridine with the seemingly inaccessible 3550 cm^{-1} hydroxyl groups (41, 170, 183) has been explained on the basis of mobility of the protons, as suggested by Olson and Dempsey (173). Evidence for the occurrence of such a phenomenon has been presented on the basis of temperature dependent IR and NMR measurements. The infrared spectral studies have been the source of conflicting interpretations. Ward (184) suggested that the reversible 30% decrease in the intensities of the 3650 and 3550 cm^{-1} hydroxyl bands upon heating from 150° to 450°C and the observed frequency shifts could be due to delocalization of the protons. Cant and Hall (185) observed smaller decreases in intensity and concluded that they were insufficient to require postulation of the dissociation of hydroxyl groups. On the basis of further work, Schoonheydt and Uytterhoeven (186) concluded that the reversible decrease in hydroxyl band intensity could not be explained entirely by mobility of the protons, but that the increased thermal motion and slightly greater distances between the framework constituents were sufficient to cause the observed frequency shifts and changes in bandwidth. They proposed that the variation in the relative intensities of the higher and lower frequency hydroxyl bands was evidence for proton migration. Uytterhoeven et al. (174) showed previously that the higher frequency hydroxyls are more thermally stable than the lower frequency band, and it was proposed that, upon heating, a given fraction of the protons become delocalized and depopulate the least stable sites in favor of the most stable. The distribution of protons over the surface sites was proposed to correspond to an equilibrium state. The position of the equilibrium depends on external factors such as temperature and residual cation content. In addition, the presence of the basic molecules was also thought to influence the proton distribution.

Nuclear magnetic resonance techniques afford a more direct method of detecting proton motions. The technique is capable of detecting jump frequencies of considerably lower magnitude than the 10^9–10^{11} sec^{-1} estimated to be required for detection by IR spectral techniques (187). A short review of NMR applications to surface chemistry has been given by Fripiat (188). Mestdagh et al. (189) presented evidence for proton delocalization, although the conclusions drawn required some extrapolation of the data and assumptions relating to the mechanism for the longitutional (T_1) and transverse relaxation (T_2) times. They estimated jump frequencies of 3.2×10^6 sec^{-1} at 300°C and 1.96×10^7 sec^{-1} at 450°C.

Freude et al. (190), also found NMR evidence for proton migration and estimated the activation energy of the proton jumps to be 5–10 kcal/mole. Proton jump frequencies at 200°C were about 5×10^4 sec^{-1} for H—Y zeolite. Adsorption of basic molecules such as pyridine was found to decrease the activation energy necessary for proton motion. Adsorption of two pyridine molecules per supercage at 200°C resulted in a 60-fold increase in the jump frequency. The acidity of H—Y zeolite was considered to be directly connected to the proton jump frequency in the presence of adsorbed molecules. Assuming that acidity corresponds to free protons and that the mean lifetime of the protons at framework oxygen atoms is much greater than the lifetime of the "free" state, *acidity* was defined to be proportional to the product of the population of acidic structural hydroxyl groups and the proton jump frequency.

b. *Ultrastable and Aluminum-Deficient Forms.* The term *ultrastable faujasite* was introduced by McDaniel and Maher in 1967, when they reported the preparation of a form of H—Y zeolite that had unusually high thermal stability (191). This type of zeolite required specific high-temperature calcination conditions for its formation and had a lower ion-exchange capacity and a smaller unit cell than the original material. Investigations by Kerr demonstrated that NH$_4$—Y zeolite could be transformed into either normal H—Y zeolite or the ultrastable form depending on the calcination environment (192). Calcination at 500°C under conditions permitting rapid removal of the evolved water and ammonia (shallow-bed conditions) resulted in H—Y zeolite, whereas conditions impeding removal of the evolved gases (deep-bed conditions) gave rise to the ultrastable form. Ward reported that ultrastable Y samples also could be prepared by calcination in flowing steam (193).

Although the structural composition of ultrastable Y zeolite has not been firmly established, a considerable amount of information has been reported. Kerr recently presented a review of its chemical, physical, and catalytic properties (*194*) which gives a broader background than can be given here. It has been shown (*192, 195*) that the ultrastabilization procedure results in the loss of aluminum from the zeolite framework. Aluminum atoms migrate to cation positions in the zeolite structure and can be removed by treatment with NaOH solution. The removal of aluminum from the zeolite framework apparently occurs by a hydrolysis mechanism, which is promoted by the presence of significant amounts of water vapor during calcination. The nature of the sites vacated by aluminum atoms remains unclear, although it has been suggested (*196–198*) that silicon atoms migrate into the vacant aluminum sites. In the following paragraphs of this section, the effect of ultrastabilization and dealumination of the faujasite framework on the acidity and related catalytic activity of the resulting zeolites will be discussed.

Infrared spectral studies. Several investigations (*193, 198–200*) of the effect of ultrastabilization on the hydroxyl stretching region of the infrared spectrum have been reported. Ward (*193*) prepared ultrastable Y samples by several different procedures and observed a significant decrease in hydroxyl absorption intensities in comparison with H—Y samples. In addition, significant changes in the hydroxyl stretching frequencies and relative intensities were apparent. The intensity of the band at 3740 cm^{-1}, which is usually attributed to amorphous silica impurities or to silanol groups on the crystal exterior, was much increased. The most intense band occurred at 3590–3570 cm^{-1}, which is not present in H—Y zeolite. The acidic hydroxyl groups found at 3650 and 3550 cm^{-1} in H—Y were greatly reduced in intensity in the ultrastable samples.

Jacobs and Uytterhoeven (*199, 200*) observed a band in the 3700 to 3675 cm^{-1} region in addition to the bands reported by Ward. The intensities of the acidic bands at 3650 and 3550 cm^{-1} were greater than those observed by Ward, which probably resulted from a lesser degree of aluminum removal. The new bands at 3700 and 3600 cm^{-1} arose from hydroxyls that were nonacidic to ammonia (*199, 200*) and pyridine (*198, 199*), although bands from pyridinium ions were observed in the IR spectrum. The latter bands were attributed to interaction of pyridine with the 3650 cm^{-1} hydroxyls (*200*). Jacobs and Uytterhoeven (*199*) and Scherzer and Bass (*198*) attributed the 3700 and 3600 cm^{-1} bands to structural hydroxyl groups associated with removal of aluminum from the zeolite framework. The 3600 cm^{-1} band arose from weakly acidic hydroxyls (*200*) since the band was removed by treatment with 0.1 N NaOH solution. The 3700 cm^{-1} band was unaffected by a similar treatment.

Scherzer and Bass (*198*) found that calcination at 820°C during the ultrastabilization procedure resulted in substantial dehydroxylation. Only weak shoulders were observed at 3680 and 3600 cm^{-1}. Adsorption of pyridine showed the presence of both Lewis and Brønsted acid sites after the high-temperature calcination. Thus, Brønsted acidity was present even in the absence of the absorption bands at 3650 and 3550 cm^{-1} observed on H—Y zeolite. Rehydration of the dehydroxylated sample did not result in a significant degree of regeneration of Brønsted sites.

Surface acidity and catalytic activity. In a series of papers (*179, 180, 201*) Beaumont and Barthomeuf investigated the dependence of the acidity of X, Y, dealuminated X and Y, and ultrastable Y zeolites on the aluminum content of the zeolite framework. They employed *n*-butylamine titrations using Hammett and arylmethanol indicators to measure the total acidity (sites with $H_0 \leq +3.3$) of samples of Na,H—Y, K,—Y, La,H—Y, and Na,H—X zeolite exchanged with various amounts of the metal ions. Linear decreases in the number of equivalents of acidity per unit cell were observed with increasing cation content of Y zeolite (*201*). A structural acidity parameter α_0 was defined, which was the slope of the line relating acid site concentration (in equivalents per unit cell) to the number of cations per unit cell. An α_0 value of 0.6 was found for all the Y zeolite samples, irrespective of the alkali cation type. The value of α_0 obtained with X zeolite was 0.16. In separate tests, samples of Y zeolite with various degrees of aluminum deficiency were prepared by extraction of framework aluminum atoms using EDTA or acetylacetone. The value of α_0 increased with decreasing aluminum content.

The parameter α_0 was interpreted as a measure of the increase in acid site concentration corresponding to the removal of one metal cation. Since a given aluminum content resulted in a constant value of α_0 irrespective of cation type or content, Beaumont and Barthomeuf proposed that the result could be explained in terms of a migration of cations toward the supercages during the *n*-butylamine titrations, resulting in a constant metal ion population in the large cages. At lower metal ion contents than those measured, a scarcity of cations would be expected, and the acidity as reflected by α_0 would be modified.

The decrease in α_0 with increasing aluminum content is consistent with the lower acidity of X zeolite in comparison with the lower aluminum content Y zeolite. The observed acidity decrease with increasing aluminum content was interpreted as a "self-inhibition" of the acidic sites as their number increases. Such a phenomenon is also consistent with the electrostatic field calculations of Pickert *et al.* (*202*) and Dempsey (*203*), who found that for a cation at a given distance from a cation site the electric field is smaller for X zeolite than for Y.

In a further study, Beaumont and Barthomeuf (*179*) related the change in the concentration of acidic sites of various strengths to the framework aluminum content. Aluminum was removed by extraction with organic complexing agents, and preparation conditions were chosen to avoid the deep-bed stabilization effect. The concentration of the various acid sites depended upon the aluminum content as shown in Fig. 24. The sites of weak acidity ($3 \times 10^{-4}\%$ H_2SO_4) declined from the start of aluminum removal, whereas the strongly acidic (88% H_2SO_4) site concentration remained constant until removal of about 21 of the original 56 aluminum atoms had occurred. For samples with fewer than 35 Al atoms per unit cell, the three curves superimposed, and the acid site concentration decreased with decreasing aluminum content. Extraction initially removes selectively those aluminum atoms giving rise to the sites of weakest acidity. The remaining aluminum atoms, which are associated with strong acid sites, are extracted only after the first type have been removed. The results suggested that the 30–40 aluminum atoms associated with strong acid sites are strongly bound to the zeolite framework; the others (from a total of 56 per unit cell) are more weakly bound. The numbers of strongly bound aluminum atoms, 35, agreed reasonably well with the results of Kerr (192), who showed that 40 aluminum atoms remain within the framework in deep-bed calcined (ultrastable) samples. Measurements by

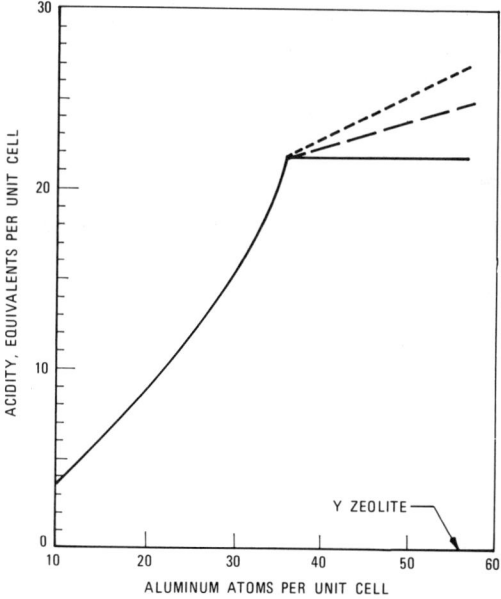

FIG. 24. The dependence of acidity on aluminum content of Y zeolites (179). (---) total acidity ($3 \times 10^{-4}\%$ H_2SO_4); (--) intermediate acidity (77% H_2SO_4); (-) strong acidity (88% H_2SO_4).

Jacobs and Uytterhoeven (199) and Maher et al. (195) indicated that 38 aluminum atoms remained in framework positions in ultrastable samples. It was suggested that each aluminum site in ultrastable zeolites gives rise to one strong acid site (179).

The change in the acid strength distribution with various framework aluminum contents also was related to iso-octane cracking activity by the same investigators (180). The observed dependence of cracking activity upon framework aluminum content is shown in Table VIII. Samples with 37 or more framework aluminum atoms per unit cell had similar activities, whereas lower aluminum contents resulted in decreasing catalytic activity with progressive aluminum deficiency. These results also indicate a relationship between the strong acid site concentration and iso-octane cracking activity, since the change in strong acid site concentration in Fig. 24 parallels the change in activity.

Topchieva and T'Huoang (204) reported a similar dependence of cumene cracking activity on the framework aluminum content. They observed a maximum in catalytic activity in dealuminated samples with $SiO_2:Al_2O_3$ molar ratio of 8, which corresponds to about 33 aluminum atoms per unit cell. Further removal of aluminum resulted in a decrease in activity.

Examination of the effect of various pretreatment conditions on the catalytic activity of aluminum-deficient and ultrastable samples of Y zeolite led Jacobs et al. (205) to propose that different hydroxyl groups were responsible for cumene cracking than were involved in toluene disproportionation. Toluene disproportionation appeared to require stronger acid sites, since the last ammonia molecules evolved during formation of the

TABLE VIII

Iso-octane Cracking Activity of Aluminum-Deficient Na, H—Y Zeolites (300°C, H_2 Atmosphere)[a]

Aluminum content (Al atoms per u.c.)	Initial rate of isobutene formation, $r_a \times 10^6$ (mole sec^{-1} g^{-1})
56	1.96
46.2	2.18
42.2	2.00
39.3	2.00
37.5	2.06
30.5	1.27
26.5	1.06
17.4	0.97
15	0.68

[a] From Beaumont and Barthomeuf (180).

hydrogen zeolite from the ammonium form resulted in enhancement of toluene disproportionation activity relative to cumene cracking. Other factors which appeared to affect the catalytic activity were the formation of Lewis acid sites, the extent of dealumination, and the presence of nonframework (cationic) aluminum atoms.

c. *Group Ia Forms.* The surface acidity of zeolites is strongly dependent upon the cation occupying the exchange sites on the structural framework. The previous discussion of hydrogen zeolites, those containing protons at the ion-exchange sites, demonstrated a relationship between surface acidity and the extent of occupancy of the sites by protons. Similar results have been shown for other cations. In the latter case, however, the acidity of the zeolite also depends upon the type of cation present as well as the extent of site occupancy. Unless severe dehydration has occurred, Brønsted acidity is generated by the presence of most multivalent metal ions. This acidity apparently arises from water molecules that react with the multivalent ions. The following discussion will be concerned with the Group Ia, alkaline and rare earth, and transition metal forms of X and Y zeolites.

Infrared spectral studies. Studies by several investigators have shown that infrared spectra of X and Y zeolite with complete occupancy of the exchange sites by Group Ia metal ions do not contain absorption bands in the OH stretching region that would arise from structural hydroxyl groups. Bertsch and Habgood (206) examined Li—, Na—, and K—X zeolites containing small amounts of water. Their spectra were consistent with isolated water molecules adsorbed at the exchangeable cations. In those cases where structural hydroxyl groups have been reported (147, 207), their presence can be explained by a partial deficiency of cations, resulting in the presence of protons at the cation-exchange sites (149).

Infrared spectral studies of pyridine adsorbed on alkali metal ion-exchanged faujasites have demonstrated the absence of Brønsted acidity, as reported by Eberly (151), Ignat'eva et al. (208), and Ward (156, 209–211). Pyridine is adsorbed weakly by coordination to the alkali metal ions (151, 156). Addition of small amounts of water does not result in formation of Brønsted acid sites, indicating that the coordinately bound pyridine is not associated with Lewis acid sites in the zeolite framework (210).

Catalytic activity measurements. Alkali metal ion X and Y zeolites show no activity for hydrocarbon conversions involving carbonium ion intermediates. Ward (210) showed that the decomposition of cumene to benzene and propylene did not occur over Na—Y zeolite at 260°C, whereas extensive conversion took place with H—Y and alkaline earth ion-exchanged forms. Similarly, isomerization of *o*-xylene at 250°C did not take place over Na—Y zeolite (176), whereas increasing the extent of ion exchange of sodium ions by protons resulted in increasing conversion. A sample exchanged to 99.8% H form gave 46% conversion. Substantial conversion of o-xylene by Na—Y zeolite occurred only at temperatures above 500°C, where free-radical mechanisms are involved. The nonacidic catalytic nature of alkali metal ion-exchanged faujasite is consistent with the observation of the lack of structural hydroxyl groups and the absence of formation of pyridinium ion or strongly Lewis-bound pyridine upon adsorption of pyridine.

d. *Alkaline and Rare Earth Forms. Infrared spectral studies.* The presence of acidic hydroxyl groups in multivalent metal ion-exchanged faujasite-type zeolites has been shown by several studies. Carter *et al.* (*149*) observed residual OH groups present on the zeolite surface even after evacuation at 450°C. Subsequent investigations by Angell and Schaffer (*147*), Eberly (*151*), Ignat'eva *et al.* (*208*), and Ward (*156, 210–213*) have led to the assignment of the IR bands to acidic hydroxyl groups and have defined the effect of cation type and content on surface acidity.

Bands at 3745 and 3690 cm^{-1} are also present. The 3745 cm^{-1} band has usually been attributed to silica impurities occluded within the zeolite framework or to Si—OH groups terminating the lattice. The 3690 cm^{-1} band is due to undissociated water since its intensity is dependent upon the degree of hydration.

The spectra of alkaline earth ion-exchanged samples, with the exception of the barium form (*211*), have hydroxyl absorption bands at 3645 and 3540 cm^{-1}, similar to those found in H—Y zeolite. The barium form behaves like the alkali-exchanged zeolites. The similarity of the spectra of the alkaline earth forms with that of the hydrogen form suggests that the acidic hydroxyls are associated with the same structural features (*151*). Band frequencies in the region of 3600 to 3560 cm^{-1} vary with the cations and are thought to result from hydroxyl groups associated with the divalent cations (*211*). They are weakly acidic or inaccessible to adsorbate molecules since the band intensity is not affected by adsorption of pyridine (*209*).

Infrared spectra of alkaline earth-exchanged Y sieve containing adsorbed pyridine showed that the bands at 3650 and 3550 cm^{-1} arise from hydroxyl groups which function as Brønsted acids (*151, 209, 211*). The zeolites undergo dehydroxylation upon treatment at temperatures in the range of 400° to 600°C. Experiments with Ca—Y zeolite by Eberly (*151*) indicated that extensive dehydroxylation occurred upon heating to 427°C. There was little evidence of hydroxyl groups in the 3800 to 3500 cm^{-1}

region, and exposure to pyridine resulted in a strong band at 1440 cm^1, which was attributed to pyridine bound to tricoordinate aluminum atoms (Lewis centers).

Alkaline earth-exchanged samples examined by Ward (211) were more resistant to thermal dehydroxylation; hydroxyl bands were present in the spectra after dehydration at 500°C. The concentration of OH groups was, however, much smaller than found in H—Y zeolite, and was dependent on the cation type. An almost linear inverse relationship was found between the alkaline earth cation radius and the concentration of acidic hydroxyl groups (210).

Eberly (151) found that exposure of dehydroxylated alkaline earth samples to water vapor resulted in reappearance of the acidic hydroxyl bands at 3650 and 3550 cm^{-1}. Subsequent exposure to pyridine resulted in interaction with the 3650 cm^{-1} groups and the formation of pyridinium ions. A band at 3585 cm^{-1} has also been reported (156), which appears simultaneously with the reformation of the acidic hydroxyl groups upon exposure to water. This band does not react with pyridine and is thought to arise from hydroxyl groups associated with the cations that result from dissociation of the added water.

The formation of structural hydroxyl groups in the presence of divalent cations has been explained on the basis of a hydrolysis mechanism (148) involving water initially coordinated to the metal ions (210, 214–216). The formation of a nonacidic hydroxyl group on the metal ion and an acidic hydroxyl on the zeolite framework by dissociation of the water molecule is consistent with the observed IR spectra and pyridine adsorption experiments. Further calcination at higher temperatures results in dehydroxylation and formation of Lewis acid sites at tricoordinate aluminum atoms in the zeolite framework (149).

Infrared spectral studies of rare earth (RE) ion-exchanged faujasites have been reported by Rabo et al. (214), Christner et al. (217), Ward (211, 212), and Bolton (218). Distinct hydroxyl absorption bands are observed at 3740, 3640, and 3522 cm^{-1} after calcination at temperatures in the range of 340° to 450°C. As previously discussed, the hydroxyl groups at 3740 cm^{-1} are attributed to silanol groups either located at lattice termination sites or arising from amorphous silica associated with the structure. The hydroxyl groups that form the 3522 cm^{-1} band are nonacidic to pyridine or piperidine and are thought to be associated with the rare earth cations.

Pyridine and piperidine interact with the 3640 cm^{-1} hydroxyls (212), which resulted in their assignment to acidic structural hydroxyl groups similar to those found in the alkaline earth and hydrogen forms. The formation of the structural hydroxyls has been attributed by most investigators (219, 220) to the hydrolysis of the rare earth cations by adsorbed water molecules, similar to the mechanism proposed for alkaline earth cations. The mechanism has been the subject of some controversy, centering on the extent of hydrolysis of the rare earth ions. Venuto et al. (219) and Plank (220) proposed the following single-step mechanism, as described for alkaline earth-exchanged faujasite:

$$[RE(H_2O)]^{3+} + O^--Zeol \rightarrow [RE(OH)]^{2+} + HO-Zeol$$

The intensity of the 3640 cm^{-1} band is several times greater in rare earth Y zeolite than in the corresponding alkaline earth forms, which led Ward (212) to propose a further step in the hydrolysis mechanism, resulting in formation of two structural hydroxyl groups per cation:

$$RE(H_2O)_2^{3+} \rightleftarrows RE(OH)(H_2O)^{2+} + H^+ \rightleftarrows RE(OH)_2^+ + 2H^+$$

Bolton (218) investigated the mechanism of hydroxyl formation by measuring the hydroxyl content of Na,RE—Y samples using a combination of thermogravimetric and IR methods. He concluded that only one water molecule was associated with each rare earth ion after removal of physically adsorbed water. With increasing extent of exchange of sodium by RE cations, the 3640 cm^{-1} (structural OH) band intensity decreased relative to the 3520 cm^{-1} (RE—OH) band. This observation was inconsistent with the proposed hydrolysis mechanisms, which would predict a constant relative intensity of the two bands at all ion-exchange levels. Bolton proposed that the band at 3640 cm^{-1} resulted from a combination of hydrolysis of the RE ion and exchange of sodium ions by protons present in the acidic rare earth salt solution that is used to prepare the RE samples. His measurements of the hydroxyl group contents of such samples were confirmed by OH population determinations by Oehme et al. (216). Using NMR pulse techniques and wide-line measurements, they found that the hydroxyl concentration was proportional to the multivalent cation content, and the proportionality was dependent on the cation content and the pretreatment temperature. Rare earth-exchanged Y zeolite contained 2 to 3 OH groups per trivalent ion after pretreatment at 300°C and less than 2 at 400°C, consistent with Bolton's determination. After pretreatment at 500°C, approximately 0.5 OH groups per RE cation were present as a result of the onset of dehydroxylation at that temperature.

The IR spectra of pyridine adsorbed on RE—Y zeolite calcined at 480°C indicate that its acidity is predominantly of the Brønsted type (212). Only a weak absorption at 1450 cm^{-1} due to Lewis-bound pyridine is observed, whereas the pyridinium ion band at 1550 cm^{-1} is quite intense. Such measurements show that calcination at higher temperatures decreases the concentration of Brønsted acid sites with a simultaneous increase in Lewis acid sites. The behavior is similar to that of H—Y zeolite upon high-temperature calcination.

In contrast with the observed behavior of the alkaline earth forms, addition of water to dehydroxylated rare earth ion-exchanged Y zeolite did not result in formation of new structural hydroxyl groups (*211*). This is consistent with Bolton's interpretation that a significant fraction of the structural hydroxyl groups are formed by exchange with protons in solution rather than by hydrolysis of the rare earth ion.

Amine titration measurements. Kladnig measured the surface acidities of alkaline and rare earth-exchanged Y zeolites by butylamine titration using Hammett indicators (*221*). With K-, Na-, Sr-, Ca-, La-, and Gd-exchanged Y zeolite, an essentially linear dependence was found of both total acidity ($H_0 < +6.8$) and strong acidity ($H_0 < -5.6$) upon the charge to radius ratio for the preceding cations. The total surface acidity (meq acid sites/g zeolite) of the rare earth-exchanged forms was approximately twice that of the alkaline earth forms. Hydration–dehydration studies indicated that rare earth ions bind one water molecule, in agreement with Bolton's proposal (*218*). Roughly 30% of the acid sites of the alkaline and rare earth exchanged forms were strongly acidic ($H_o < -5.6$).

Surface acidity and catalytic activity. Faujasitic zeolites exchanged with multivalent ions demonstrate significant catalytic activity for reactions involving carbonium ion mechanisms, in contrast to the inactivity of the alkali metal ion-exchanged forms. Several possible sources of the observed activity were proposed initially. Rabo *et al.* (*202, 214*) suggested that electrostatic fields associated with the multivalent ions were responsible for the catalytic activity. Lewis acid sites were proposed as the seat of catalytic activity by Turkevich *et al.* (*50*) and by Boreskova *et al.* (*222*). Brønsted acid sites formed by hydrolysis of the multivalent metal ions were proposed as the catalytic centers by Venuto *et al.* (*219*) and by Plank (*220*).

Catalytic activity measurements and correlations with surface acidity have been obtained by numerous investigators. The reactions studied most frequently are cracking of cumene or normal paraffins and isomerization reactions; both types of reactions proceed by carbonium ion mechanisms. Venuto *et al.* (*219*) investigated alkylation reactions over rare earth ion-exchanged X zeolite catalysts (REX). On the basis of product distributions, patterns of substrate reactivity, and deuterium tracer experiments, they concluded that zeolite-catalyzed alkylation proceeded via carbonium ion mechanisms. The reactions that occurred over REX catalysts such as alkylation of benzene/phenol with ethylene, isomerization of *o*-xylene, and isomerization of paraffins, resulted in product distributions typical for reactions involving very strong protonic acids such as sulfuric, hydrofluoric, and promoted Lewis acids (223). Enhancement of the ethylene–benzene alkylation rate occurred upon introduction of proton donors such as hydrogen chloride and water, which strongly supports the concept of proton involvement in the reaction.

Ward (210) demonstrated that Brønsted acid sites play an important role in the catalytic activity of alkaline earth ion-exchanged Y zeolite for cumene and hexane cracking. Zeolite catalysts calcined below 500°C (which contain only Brønsted acid sites) cracked cumene at 260°C, indicating that Lewis sites were not involved in the reaction. The catalytic activity increased as the cation radius decreased, indicating an effect of the electrostatic field generated by the cation. The Brønsted acidity as indicated by IR band intensities also increased with decreasing cation radius. Thus, the activity and acidity of the zeolites decreased in the order Mg—Y > Ca—Y > Sr—Y > Ba—Y. The maximum activity observed was less than that of H—Y. The observed activity relationship is therefore consistent with Brønsted sites providing the active catalytic centers.

Hopkins (161) found that a steady decrease in n-heptane cracking activity occurred over La- and Ca-exchanged Y zeolites as the catalyst calcination temperature was increased from 350° to 650°C. The lanthanum form was about twice as active as the calcium form. Reduction in activity with increasing activation temperature was attributed to removal of acidic framework hydroxyl sites as dehydration becomes more extensive. The greater activity of La—Y with respect to the calcium form was thought to result from the greater hydrolysis tendency of lanthanum ion, which would require more extensive dehydration to result in the same concentration of acidic OH groups as found on Ca—Y.

Ward measured the o-xylene isomerization activities of Na, Mg, RE, and H—Y zeolites and found the rare earth form to be intermediate in activity between the magnesium and hydrogen forms as shown in Table IX (212). The sodium form was essentially inactive. He interpreted the activity relationship RE—Y > Mg—Y to result from the formation of two acidic structural hydroxyl groups per trivalent rare earth cation. The formation of acidic structure hydroxyl groups by exchange of sodium ions with protons in the rare earth solution, as proposed by Bolton (218), may also account for the greater activity of the rare earth-exchanged zeolite.

Moscou and Mone measured the acidity distribution of various rare earth-exchanged Y zeolites and correlated the results with the performance of the zeolite in cracking of gas–oil (224). They found that steaming the zeolite at 675°–750°C resulted in a significant decrease in the concentration of strongly acidic sites ($H_o < -8.2$), while the intermediate

SURFACE ACIDITY OF SOLID CATALYSTS 165

TABLE IX
Brønsted Acidity and Cumene Cracking Activity of Y Zeolite[a]

Zeolite	Brønsted acidity[b]	Temperature for 25% conversion of o-xylene (°C)
Na—Y	0	>500
Mg—Y	6.9	284
RE—Y	11.3	260
H—Y	15.8	230

[a] From Ward (212).
[b] Determined from peak height/sample mass of the 1545-cm^{-1} band that arises from pyridinium ion.

($-8.2 < H_0 < -3.0$) and weakly acidic ($-3.0 < H_0 < +3.3$) site concentrations decreased only slightly. Steaming had only a minor effect on conversion of the gas–oil, but increasing the steaming temperature from 600° to 750°C resulted in increased amounts of gasoline (C_5 and higher molecular weight hydrocarbons) and decreased coke and gas (C_1 through C_4 hydrocarbons) formation. Y zeolite containing 1% w Na_2O, which also had a lower concentration of strongly acidic sites, similarly showed an increase in gasoline formation.

e. *Transition Metal Forms.* Several studies have shown that acidic sites are present on transition metal-exchanged faujasitic zeolites. Angell and Schaffer (147) and Christner et al. (217) discussed the infrared spectra of various molecules adsorbed by such solids, and Hattori and Shiba (225) and Ward (213) described the acidic properties of X zeolites that contained transition metals.

A study of the acidic properties of various transition metal ion-exchanged Y zeolites was also made by Ward (226). Protonic acidity was determined by infrared spectral measurements of adsorbed pyridine, and the o-xylene isomerization activity was compared with the observed acidity. Y zeolite containing the transition metal ions Cr, Mn, Fe, Co, Ni, Cu, Zn, and Cd displayed Brønsted acidity as shown by the formation of pyridinium ions upon adsorption of pyridine. The Brønsted sites were attributed to structural hydroxyl groups (3640 cm^{-1}) formed upon hydrolysis of the transition metal ion by the mechanism proposed for alkaline and rare earth ion-exchanged zeolites. A relationship was observed between acid site concentration and the electrostatic field calculated for each ion. The data were consistent with the interpretation that the cations generating the strongest fields would result in the highest concentration of structural hydroxyl groups, because of stronger interaction of water with the cations. No Lewis acidity was observed with samples calcined at 480°C, although higher activation temperatures might be expected to generate Lewis sites by dehydroxylation. All of the transition metal ion-exchanged forms were catalytically active for isomerization of o-xylene; however, there was no simple relationship between activity and acid site concentration. It was suggested that acid site strength might be affected by the transition metal ion. As discussed previously with H—Y zeolites, acid-site strength distribution is an important parameter for reactions proceeding by carbonium ion mechanisms. It is also possible that the hydrocarbon interacts directly with the transition metal ion since these ions demonstrate catalytic behavior of themselves in numerous reactions.

2. Mordenite

The nature of the acidity of mordenite and its relation to catalytic activity have been investigated by Benesi (157), Lefrancois and Malbois (227) and Eberly et al. (228). Eberly et al. observed two absorption bands in the hydroxyl region of the infrared spectrum of H-mordenite. A band at 3740 cm^{-1} was attributed to silica-type hydroxyl groups, and a lower frequency band, 3590 cm^{-1}, was thought to arise from hydroxyl groups associated with aluminum atoms in the structure. Acid extraction of the aluminum atoms from the framework, although leaving the structure intact resulted in a loss of the lower frequency hydroxyl band.

Benesi (157) suggested that toluene disproportionation and cracking of n-butane and n-pentane occur at Brønsted acid sites generated upon decomposition of the ammonium form, similar to the mechanism proposed for NH$_4$—Y zeolite. The maximum rate of disproportionation of toluene occurred at activation temperatures of 600° to 700°C. Cracking of n-butane and n-pentane displayed the same behavior. In all cases the activity declined as the catalyst was preheated at temperatures above 700°. It was concluded that the Brønsted sites rather than the Lewis sites were the primary source of catalytic activity. Substantiating evidence was provided by the observation that addition of water to a catalyst deactivated by heating to 800° resulted in restoration of the original n-butane cracking activity. Evidence was also presented which suggested that the sites generated upon ammonia desorption were not of equal catalytic activity. Readdition of less than 20% of the amount of ammonia originally evolved resulted in essentially complete poisoning of the cracking activity. Thus, the most active sites were preferentially neutralized by adsorbed ammonia.

Results obtained by Hopkins (161) substantiated Benesi's ammonia poisoning observations. The activities for n-hexane and n-heptane cracking of NH_4-mordenite and H-mordenite were measured after exposure of the zeolites to different activation temperatures. Samples of H-mordenite demonstrated equal activity after activation at 350° and 500°, whereas NH_4-mordenite was more active after activation at the higher temperature. At 350°, the NH_4-mordenite was extensively deamminated, but the strongest acid sites were still occupied by ammonium ions since they retain ammonia most strongly. The results suggested that the strongest acid sites of mordenite are most active for paraffin cracking reactions.

Lefrancois and Malbois (227) determined the types of acidity present on H-mordenite and various cationic forms by obtaining infrared spectra of pyridine adsorbed on the zeolite. H-mordenite activated at 400° contained both Brønsted and Lewis acid sites. Upon addition of water, the band due to Lewis-bound pyridine disappeared and the Brønsted site concentration increased. Removal of the added water by evacuation restored some of the Lewis acid sites.

H-mordenite was shown to be more active than H—Y zeolite for acid-catalyzed reactions such as toluene disproportionation and n-paraffin cracking (157). A similar relationship has been shown for the corresponding zeolites containing Pd or Pt in paraffin isomerization and hydrocracking (230). The greater activity of mordenite could possibly result from an interaction of Lewis sites with the acidic hydroxyl groups, which increases the Brønsted acid strength by an inductive mechanism, according to the scheme proposed by Lunsford (231). This possibility remains to be confirmed, however, since data obtained by Kouwenhoven (119) indicated that low sodium Pd,H-mordenite and low sodium Pd,H—Y zeolite have the same activity for isomerization of n-hexane. It was concluded that optimum performance for either zeolite could be obtained only after extensive removal of sodium. Low levels of alkali cations thus appear to have a marked effect on the acidic and catalytic properties of the two zeolites. Kouwenhoven's results are consistent with earlier findings by Turkevich et al. (50) and Ward (176), who reported a rapid increase in the catalytic activity of Y zeolite with increasing sodium removal for cumene dealkylation and o-xylene isomerization, respectively. The results indicated an increasing catalytic effectiveness per site with increasing extent of ion exchange. Measurement of the effect of small amounts of alkali cations on the acid strength distribution and on the rates of various acid catalyzed reactions would perhaps provide relevant information on a question which has important catalytic consequences.

Weeks et al. investigated the changes which occur in NH_4-mordenite by thermal analysis and IR spectroscopy (229). Three distinct steps were observed to occur upon heating in air. Adsorbed water is evolved from room temperature to 250°C. Decomposition of the ammonium cations occurs in the temperature range of 350° to 550°C, and the zeolite undergoes dehydroxylation at 700°–900°C. Upon deammoniation an intense IR band develops in the hydroxyl region. The frequency of the band decreases with increasing activation temperature. After activation at 300°C the band is located at 3621 cm^{-1}, and with increasing activation temperature the band frequency shifts to 3612 cm^{-1}. The maximum intensity occurred after 500°C activation. No hydroxyl bands are observed after calcination at 800°C, as a result of dehydroxylation of the surface. Addition of water to a dehydroxylated sample does not result in reappearance of hydroxyl group bands.

F. Clays

Clays have been used to promote acid-catalyzed reactions in the petroleum industry for several decades. Their use was of considerable economic importance during the early years of catalytic cracking, which achieved commercial status in 1936. Acid-treated montmorillonite and halloysite clays (232) accounted for most of the early cracking catalyst production until they were largely supplanted by synthetic silica–alumina gel catalysts by 1950. Although zeolite-containing cracking catalysts now have replaced silica–alumina catalysts, primarily because of their greater gasoline yield, substantial amounts of clays are consumed in production of the matrix for the zeolite component.

1. Structure

Attention will be focused on two clay mineral groups: kaolin and montmorillonite. Both groups have layered structures consisting of parallel sheets of oxide and hydroxide ions coordinated to Si^{4+} and Al^{3+} ions (233). Differences in the arrangement of the sheets and in their chemical composition give rise to the division into two groups. Within each group are several minerals of similar structure and composition.

The two most important minerals in the kaolin group are kaolinite and halloysite. Their chemical compositions are close to the ideal oxide ratio $Al_2O_3:2SiO_2:2H_2O$. Each layer is 7.15 Å in thickness and is composed of three parallel sheets of oxide and hydroxide ions. Silicon ions in tetrahedral coordination are sandwiched between a sheet of oxide ions and a mixed sheet of oxide and hydroxide ions. Octahedrally coordinated aluminum ions lie between the mixed oxide–hydroxide ion sheet and a third sheet of hydroxyl ions. The clay crystallites are composed of stacks of these layers, which are thought to be held together in a regular manner by hydrogen bonding between the oxide sheet of one layer and the hydroxide sheet of another layer.

Halloysite has a chemical composition similar to kaolinite, but with a higher water content. The layers of halloysite are like those in kaolinite, but they are stacked with highly random displacements parallel to the layers, as opposed to the regular stacking found in kaolinite. The interlayer distance is greater in halloysite, allowing for the presence of a sheet of water molecules. A small ion-exchange capacity is measurable in kaolinite and halloysite minerals, which arises from a small amount of isomorphous replacement of Si^{4+} or Al^{3+} in the framework (234).

The montmorillonite group is made up of several mineral subgroups, which differ in Si:Al ratio and coordination environments and have various degrees of substitution of other cations for aluminum. Within the montmorillonite group is the mineral, montmorillonite, which is composed of alumino–silicate layers stacked in a parallel arrangement. Each layer is composed of four parallel oxide or mixed oxide–hydroxide sheets. The two external sheets in each layer contain oxide ions, which form part of the environment around tetrahedrally coordinated silicon ions. The two internal oxide–hydroxide sheets complete the coordination environment around the silicon ions and form an octahedral coordination environment around aluminum and/or magnesium ions, which lie midway between the two oxide–hydroxide layers. Substitution of magnesium for aluminum contributes to the layer a net negative charge, which is balanced by cations present between the three-layer "sandwiches." The cations are exchangeable and can be coordinated by water molecules, which penetrate between the structural layers. In other montmorillonite minerals, such as beidellite, nontronite, and saponite, charge is contributed by substitution of aluminum for silicon in the tetrahedral environment.

2. Surface Acidity

The first measurement of clay acidity was reported by Walling (12), who found that an acid-washed bentonite clay of the montmorillonite group imparted the acid color to *p*-nitrobenzene-azo(*p'*-nitro) diphenylamine, which has a pK_a of 0.43. At least part of this acidity, however, was thought to result from residual acid from the acid wash.

Benesi (18) employed a broader range of indicators with pK_a values ranging from +6.8 to −8.2 to measure the acidity of clays in various cation-exchanged forms. The sodium, ammonium, and hydrogen forms of kaolinite were prepared by passing a suspension of the clay through a cation-exchange column in the appropriate cation form. The measured acid strengths of the dried clays are shown in Table X. The hydrogen forms of the clays were the strongest acids, $-5.6 \leq pK_a \leq -8.2$, and the sodium and ammonium forms of kaolinite were stronger than the corresponding forms of montmorillonite. The strongly acidic surfaces found for

TABLE X
Acid Strengths of Clay Surfaces in H_0 Units[a]

Original kaolinite	−3.0 to −5.6
Sodium kaolinite	−3.0 to −5.6
Ammonium kaolinite	−3.0 to −5.6
Hydrogen kaolinite	−5.6 to −8.2
Original montmorillonite	+1.5 to −3.0
Sodium montmorillonite	+1.5 to −3.0
Ammonium montmorillonite	+1.5 to −3.0
Hydrogen montmorillonite	+5.6 to −8.2

[a] From Benesi (18). (Reprinted with permission of the American Chemical Society.)

the dried sodium and ammonium forms of both clays was unexpected, because these solids had been completely neutralized in their aqueous suspensions. The appearance of surface acidity upon drying was explained in terms of acid sites that were exposed when water was removed from highly solvated cations. A natural clay-based cracking catalyst was more strongly acidic than the clay samples; the measured pK_a was less than −8.2, similar to synthetic silica–alumina. Since it has been shown that the clay structure is partially destroyed during the acid activation step of catalyst preparation (232), the noncrystalline portion of the resulting catalyst is probably quite similar to silica–alumina.

Benesi (26) later expanded the study of clay surface acidity to include determination of the number of acid sites as well as acid strength. The observed acid strength distributions of natural kaolinite and montmorillonite are shown in Table XI. Most of the acid sites on kaolinite are strong; however, the total number of acid sites is small. Montmorillonite contains a large number of acid sites falling into a single acid strength range. Natural montmorillonite and the hydrogen- and sodium-exchanged forms were also compared (Table XI). The butylamine titer of the hydrogen form is much higher than the sodium form or natural montmorillonite. The number of protonic sites in the hydrogen form was less than the total ion-exchange capacity of the clay (0.98 meq/g), suggesting that some sites are not accessible or that the hydrogen form of the clay is unstable and readily converts to the aluminum form by removal of structural aluminum ions from the octahedral layer. The latter explanation is more probable.

The effect of water content on the acidity of kaolinite was examined by Solomon and co-workers (235, 236). Kaolinite dried at 110°C (0% water) had strongly acidic sites, comparable to 90% sulfuric acid ($pK_a < -8.2$), as indicated by the Hammett indicator method. The strongest sites were readily poisoned by water. At 1% wt water content the strongest acid sites were equivalent in strength to 48% sulfuric acid ($pK_a = 3.0$). With

TABLE XI
Acidity of Clays Dried at 120°C[a]

	Butylamine titer in the H_0 range			
	+3.3 to 1.5	+1.5 to −3.0	<−3.0	Total
Kaolinite	0.004	0.004	0.014	0.022
Montmorillonite				
Natural form	0.00	0.09	0.00	0.09
Sodium form	0.03	0.01	0.00	0.04
Hydrogen form	0.10	0.00	0.55	0.65

[a] From Benesi (26). (Reprinted with permission of the American Chemical Society.)

increasing water content the acid site strength decreased, so that at 10% wt water the strongest acid sites had pK_a values in the range of +3 to +4.

The change in surface acidity with water content was also demonstrated by the ability of kaolinite to promote acid-catalyzed polymerization (236). Styrene, p-methylstyrene, and p-methoxymethylstyrene polymerized vigorously on kaolinite that was dried at 110°C. At 0.2% wt water content, p-methylstyrene and p-methoxystyrene polymerized; and at 0.6% wt water content, only p-methoxystyrene polymerized. The polymerization results are consistent with lower acidity at higher water contents since the susceptibility of these monomers to acid-catalyzed polymerization is in the order p-methoxystyrene > p-methylstyrene > styrene.

Infrared spectra of pyridine adsorbed on kaolinite indicated that the dry clay (110°C) contained both Brønsted and Lewis acid sites (235). At 1% water content only protonic acid sites were observed. It was not possible to assign the polymerization activity to either type of acid site, since both were present on samples which were catalytically active.

Recently, Frenkel (237) measured the surface acidity of various cation-exchanged samples of montmorillonite by the Hammett indicator method, using the spectrophotometric method of Drushel and Sommers (21) to monitor the formation of the acid forms of the organic indicators. No acid sites stronger than $pK_a = -3.7$ were observed on any of the samples, even with the protonic or aluminum forms. He attributed the stronger acidity ($pK_a \leq -8.2$) of similar clays observed by previous workers to over titration of the samples, resulting from mistaking the slightly colored physisorbed forms of benzalacetophenone ($pK = -5.6$) and anthraquinone ($pK = -8.2$) for the true acid form. Although the two forms are difficult to distinguish by eye, the spectra obtained by instrumental means are distinctive.

The source of the proton-donating ability of clays has been investigated by several workers (238–243). Mortland and co-workers (238) used a variety of techniques to study the interactions of ammonia with montmorillonites. Infrared spectra of ammonia adsorbed on hydrogen- and calcium-exchanged montmorillonites showed the presence of ammonium ions on both samples. In the hydrogen-exchanged clays the ammonium ions formed either from the exchangeable protons or from protons provided by hydrated aluminum ions, which were present at exchange positions due to gradual decomposition of the H-montmorillonite. In the calcium-exchanged sample the protons must have been provided either by constitutional hydroxyl groups or from adsorbed water. Deuterium-exchange studies showed that reaction between ND_3 and structural hydroxyl groups became appreciable only above 200°C, suggesting that the protons arise from adsorbed water (which undergoes H—D exchange with ND_3 at room temperature). It was proposed that strongly polarized water molecules in the vicinity of exchangeable cations dissociate to provide protons and hydroxyl ions according to the following reaction:

$$M^{z+}(H_2O)_{ads} + NH_3 \rightarrow (M^{z+})_{ads} + (OH^-)_{ads} + (NH_4^+)_{ads}$$

Calorimetric data were consistent with the chemisorbed water being in a dissociated state before exposure to ammonia. Additional support for this view comes from NMR studies (244, 245), which showed that water adsorbed by montmorillonite is more highly dissociated than in the normal state. The average lifetime of a proton in a particular molecule is sufficiently short to suggest that the adsorbed water is 1000 times more highly dissociated than in the liquid state.

The relationship of surface acidity of montmorillonites to the state of hydration and nature of the exchangeable cation was investigated by Mortland and Raman (243). By a combination of infrared spectroscopy and gravimetric and elemental analyses the direction of the equilibrium in the following reaction:

$$M(H_2O)_x^{n+} + NH_3 \rightleftarrows [M(H_2O)_{x-1}(OH)]^{(n-1)+} + NH_4^+$$

was determined for samples of bentonite and nontronite (a member of the montmorillonite group with partial substitution of Fe^{3+} for Al^{3+} in the octahedral environment). Samples containing different cations were examined at various levels of hydration. The ability of the homoionic clays to protonate ammonia decreased roughly in the order of polarizing power of the exchangeable cation. Thus, at low humidity ($p/p_0 = 0.20$) the order of proton donating ability of bentonite was Al = Mg > Ca > Li > Na > K, and at high humidity ($p/p_0 = 0.98$) the order was Al > Mg > Ca = Li > Na = K. A similar trend was apparent with homoionic nontronite samples.

The amount of NH_4^+ ion formed was found to increase with decreasing water content as shown for Ca-bentonite at various levels of hydration in Table XII. The increasing proton donating ability of the exchangeable cation with increasing dehydration was explained on the basis of sharing of the polarizing forces of the cation among the coordinated water molecules. As the number of water molecules coordinated to the cation decreases, the polarization forces are concentrated on the remaining water molecules with a subsequent increase in the extent of hydrolysis and, therefore, in the protonating ability.

Frenkel (237) also observed a correlation between the polarizing ability of the exchangeable cation in montmorillonite and the number and strength of acid sites. The greatest surface acidity was observed with protons as the exchangeable cation. The hydrogen form of the clay is transformed upon heating into the aluminum form by movement of aluminum ions from the framework into exchangeable cation positions, and the acidity of the resulting clay is nearly equal to that of Al-montmorillonite.

By measuring the rate of t-butyl alcohol dehydration catalyzed by the hydrogen forms of various montmorillonite-type clays, Davidtz (246) found that the acid activity of the clay was related to the extent of isomorphous substitution of Al^{3+} ions for Si^{4+} ions in the tetrahedral silicate layer. Specifically, a linear relationship was observed between the rate of isobutylene formation and the tetrahedral-layer cation-exchange site density. The exchange sites arising from isomorphous substitution of divalent ions in the octahedral Al^{3+} positions had no effect on the acid activity. Two possible causes of the observed behavior were proposed. The protons located at the octahedral layer may be physically shielded from interaction with reactant molecules due to their relatively inaccessible positions within the aluminosilicate layer. Alternatively, it was suggested that the acidic hydroxyls in the octahedral layer are lost during decomposition of the ammonium form. The latter suggestion was consistent with the observed

TABLE XII
Formation of NH_4^+ on Ca-Bentonite at Various Levels of Hydration[a]

Percent H_2O on Clay[b]	Equilibrium Constant, K_{eq}[c]
40.7	4.4×10^{-1}
39.0	9.4×10^{-1}
31.1	13.9×10^{-1}
5.9	120×10^{-1}

[a] From Pickett and Lemcoe (244).
[b] Oven dry basis (105°C)
[c] $K_{eq} = [NH_4^+][OH^-]/[H_2O][NH_3]$

behavior of clays having isomorphous substitution only at the octahedral layer.

3. *Synthetic Mica-Montmorillonite*

Recently, interest in clays as acidic catalysts has been quickened by the reported high catalytic activity of a synthetic mica-montmorillonite clay and its nickel-containing analogs. Wright *et al.* *(247)* have described the structure, thermal modification and surface acidity of the clay, which they designated SMM for *synthetic mica-montmorillonite*.

The clay is related structurally to muscovite mica, a clay of composition and structure similar to the montmorillonite previously described. It is composed of parallel layers with an interlayer spacing of 9.4 Å. As in montmorillonite each layer contains four parallel oxide or mixed oxide–hydroxide ion sheets. The two external oxide ion sheets in each layer form half of the tetrahedral coordination environment around silicon atoms. The remaining half of the silicon environment is composed of the two internal sheets of mixed oxide and hydroxide ions. The internal oxide–hydroxide sheets also form an octahedral environment around aluminum ions, which are sandwiched between the two silicate lamellae. The oxide ions of the internal sheets form bridges between the silicon and aluminum atoms. The structural hydroxyl groups are substituted to various degrees by fluoride ions. The framework is negatively charged, resulting from substitution of four-coordinate aluminum atoms in the tetrahedral silicate layer. The charge is largely balanced by ammonium ions and to some extent by aluminum cations or partially hydroxylated aluminum cations, $[Al(OH)_x(H_2O)_y]^{3-x}$.

Thermal activation of the clay up to 600°C results in a simultaneous loss of ammonia and structural hydroxyl groups. During decomposition, the ammonium ions release protons to the clay framework. Infrared spectra of dehydrated SMM samples exposed to ammonia vapor showed small amounts of ammonium ions and Lewis-bound ammonia. The ammonium ion was thought to result from interaction with silanol groups at crystal edges. Partially dehydrated samples adsorbed larger amounts of ammonia, and a greater proportion was present as ammonium ion, probably because of the lesser extent of dehydroxylation.

Adsorption of pyridine on dehydroxylated SMM resulted in formation of a small amount of pyridinium ions and at least two Lewis-bound species. Rehydration occurred upon addition of water to the sample, and the Lewis-bound species were converted to pyridinium ions. The relative amounts of pyridine bound at Lewis and Brønsted sites was found to be a strong function of the residual water content. Quantitative infrared data obtained from partially dehydroxylated SMM (evacuated at 150°C for 1 hr) gave values of 2.5 meq/100 g pyridine bound at Brønsted sites and 1.5 meq/100 g bound at Lewis sites. The acid site density of 2.4×10^{19}/g measured by pyridine adsorption was much lower than that found in the zeolite faujasite, 32×10^{19}/g (41), and on the basis of site density alone one would predict a much lower acidic catalytic activity. On the contrary, the acidic catalytic activity of the SMM clays is comparable to that of zeolites, and was explained on the basis of greater accessibility of the layered clay surface in comparison with the channel structure of zeolites.

Hattori and co-workers (248) investigated the mechanism of isomerization of cyclopropane, methylcyclopropane and n-butenes over SMM clay to characterize the active sites and to determine the source of the high activity of the clay, which is much greater than that of amorphous silica–alumina. By a combination of deuterium-exchange and microwave spectroscopic techniques, it was demonstrated that hydrogen atoms from the catalyst participate in the isomerization reactions. The results were consistent with only a small fraction (less than 10%) of the catalyst hydrogen atoms serving as active sites. The deuterium-exchange data indicated, however, that the hydrogen atoms on the surface are quite labile and can undergo exchange with hydrogens at inactive sites.

The distribution of deuterated products resulting from isomerization of methylcyclopropane over SMM was consistent with carbonium ion formation by interaction with Brønsted acid sites. There was no evidence for hydride ion abstraction occurring over Lewis sites. The rate-limiting step of the reaction appeared to be the transfer of a proton from the surface to form the carbonium ions. By comparison, amorphous silica–alumina Brønsted sites undergo exchange with hydrocarbon molecules much less readily than SMM clay. The greater lability of the protons on the clay surface was thought to be a factor in the higher cracking activity of the layered silicate in relation to that of silica–alumina.

Rather remarkable increases in the catalytic activity of the SMM clays by introduction of nickel or cobalt ions into the structure during synthesis were reported by Swift and Black (249). Although the structure of the nickel- or cobalt-substituted clays has not been reported in detail, incorporation of the metal ions into the structure apparently occurs by substitution into octahedral aluminum sites. With an acid-catalyzed reaction, such as n-hexane isomerization, a palladium-containing nickel-substituted clay was found to be 20 times more active than a Pd-SMM clay without nickel and 1.7 times more active than Pd-mordenite on a weight basis. The importance of incorporating the metal ions into the clay structure was demonstrated by the much lower activity observed with SMM clays impregnated with nickel ions in comparison with clays containing nickel ions in the octahedral layer. The maximum activity was obtained with catalysts containing five nickel atoms per unit cell.

Hexane hydrocracking experiments demonstrated that the specific activity (activity per unit surface area) of the P-nickel clay was approximately 15 times greater than Pd—H—Y zeolite and 1.2 times greater than Pd—H—mordenite. An increase in hexane hydrocracking activity was also observed with clays incorporating cobalt ions, although the effect was not so large as with catalysts containing an equivalent number of nickel ions per unit cell.

The reason why incorporation of nickel or cobalt ions into the structure results in such substantial increases in activity is not clear. An increase in surface area occurs upon incorporation of nickel into the structures, but the activity increases are greater than can be attributed to surface area alone.

REFERENCES

1. Goldstein, M. S., *in* "Experimental Methods in Catalytic Research" (R. B. Anderson, ed.), p. 361. Academic Press, New York, 1968.
2. Tanabe, K., "Solid Acids and Bases." Academic Press, New York, 1970.
3. Forni, F., *Catal. Rev.* **8,** 69 (1973).
4. Voge, H. H., *Catalysis* **6,** 407 (1958).
5. Condon, F. E., *Catalysis* **6,** 43 (1958).
6. Kennedy, R. M., *Catalysis* **6,** 1 (1958).
7. Olah, G. A., "Friedel-Crafts Chemistry," p. 43. Wiley, New York, 1973.
8. Oblad, A. G., Mills, G. A., and Heinemann, H., *Catalysis* **6,** 341 (1958).
9. Whitmore, F. C., *Ind. Eng. Chem.* **26,** 94 (1934).
10. Flockhart, B. D., and Pink, R. C., *J. Catal.* **8,** 293 (1967).
11. Ryland, L. B., Tamele, W. W., and Wilson, J. N., *Catalysis* **7,** (1960).
12. Walling, C., *J. Am. Chem. Soc.* **72,** 1164 (1950).
13. Hammett, L. P., and Deyrup, A. J., *J. Am. Chem. Soc.* **54,** 2721 (1932).
14. Hammett, L. P., "Physical Organic Chemistry," p. 251. McGraw-Hill, New York, 1940.
15. Brown, H. C., and Johanneson, R. B., *J. Am. Chem. Soc.* **75,** 16 (1953).
16. Brown, H. C., and Kanner, B., *J. Am. Chem. Soc.* **75,** 3865 (1953).
17. Weil-Malherbe, H., and Weiss, J., *J. Chem. Soc.* p. 2164 (1948).
18. Benesi, H. A., *J. Am. Chem. Soc.* **78,** 5490 (1956).
19. Leftin, H. P., and Hobson, M. C., Jr., *Adv. Catal.* **14,** 115 (1963).
20. Kotsarenko, N. S., Karakchiev, L. G., and Dzisko, V. A., *Kinet. Katal.* **9,** 158 (1968).
21. Drushel, H. V., and Sommers, A. L., *Anal. Chem.* **38,** 1723 (1966).
22. Tamele, M. W., Ryland, L. B., Rampino, L., and Schlaffer, W. G., *World Pet. Congr., Proc. 3rd* **4,** 98 (1951).
23. Parry, E. P., *J. Catal.* **2,** 371 (1963).
24. Hirschler, A. E., *J. Catal.* **2,** 428 (1963).
25. Johnson, O., *J. Phys. Chem.* **59,** 827 (1955).
26. Benesi, H. A., *J. Phys. Chem.* **61,** 970 (1957).
27. Hantzsch, A., *Ber. Deut. Chem. Ges.* **54,** 2573 (1921).
28. Gold, V., and Hawes, B. W. V., *J. Chem. Soc.* p. 2102 (1951).

29. Newman, M. S., and Deno, N. C., *J. Am. Chem. Soc.* **73,** 3644 (1951).
30. Deno, N. C., Berkheimer, H. E., Evans, W. L., and Peterson, H. J., *J. Am. Chem. Soc.* **81,** 2344 (1959).
31. Take, J., Nomizo, Y., and Yoneda, Y., *Bull. Chem. Soc. Jpn.* **46,** (11), 3568 (1973).
32. Mills, G. A., Boedeker, E. R., and Oblad, A. G., *J. Am. Chem. Soc.* **72,** 1554 (1950).
33. Clark, A., Holm, V. C. F., and Blackburn, D. M., *J. Catal.* **1,** 244 (1962).
34. Holm, V. C. F., and Clark, A., *J. Catal.* **2,** 16 (1963).
35. Mapes, J. E., and Eischens, R. P., *J. Phys. Chem.* **58,** 1059 (1954).
36. Basila, M. R., and Kantner, T. R., *J. Phys. Chem.* **71,** 467 (1967).
37. Eischens, R. P., and Pliskin, W. A., *Adv. Catal.* **10,** 1 (1958).
38. Little, L. H., "Infrared Spectra of Adsorbed Species." Academic Press, New York, 1966.
39. Hair, M. L., "Infrared Spectroscopy in Surface Chemistry." Dekker, New York, 1967.
40. Basila, M. R., and Kantner, T. R., *J. Phys. Chem.* **70,** 1681 (1966).
41. Hughes, T. R., and White, H. M., *J. Phys. Chem.* **71,** 2192 (1967).
42. Ward, J. W., *J. Catal.* **11,** 271 (1968).
43. Ward, J. W., and Hansford, R. C., *J. Catal.* **13,** 154 (1969).
44. Jacobs, P. A., and Heylen, C. P., *J. Catal.* **34,** 267 (1974).
45. Glasstone, S., Laidler, K. J., and Eyring, H., "Theory of Rate Processes," p. 376. McGraw-Hill, New York, 1941.
46. Prater, C. C., and Lago, R. M., *Adv. Catal.* **8,** 293 (1956).
47. Maatman, R. W., *J. Catal.* **19,** 64 (1970).
48. Stright, P., and Danforth, J. D., *J. Phys. Chem.* **57,** 448 (1953).
49. Kokes, R. J., Tobin, H., and Emmett, P. H., *J. Am. Chem. Soc.* **77,** 5860 (1955).
50. Turkevich, J., Nozaki, F., and Stamires, D., *Proc. Int. Congr. Catal., 3rd, Amsterdam, 1964* **1,** 586 (1965).
51. Boreskova, E. G., Topchieva, K. V., and Piguzova, L. I., *Kinet. Katal.* **5,** 903 (1964).
52. Romanovskii, V. B., Ho Shi Thong, and Topchieva, K. V., *Kinet. Katal.* **7,** 179 (1966).
53. Goldstein, M. S., and Morgan, T. R., *J. Catal.* **16,** 232 (1970).
54. Benesi, H. A., *J. Catal.* **28,** 176 (1973).
55. Misono, M., Saito, Y., and Yoneda, Y., *Proc. Int. Congr. Catal., 3rd, Amsterdam, 1964* **1,** 408 (1965).
56. Greensfelder, B. S., Voge, H. H., and Good, G. M., *Ind. Eng. Chem.* **37,** 1168 (1945).
57. Andreu, P., Martin, G., and Noller, H., *J. Catal.* **21,** 255 (1971).
58. Pichat, P., *J. Phys. Chem.* **78,** 2376 (1974).
59. Morimoto, T., Imai, J., and Nagao, M., *J. Phys. Chem.* **78,** 704 (1974).
60. Hendra, P. J., Turner, I. D. M., Loader, E. J., and Stacey, M., *J. Phys. Chem.* **78,** 300 (1974).
61. Takezawa, N., and Kobayashi, H., *J. Catal.* **25,** 179 (1972).
62. Rouxhet, P. G., and Sempels, R. E., *J. Chem. Soc., Faraday Trans.* **70,** 2021 (1974).
63. Hair, M. L., and Hertl, W., *J. Phys. Chem.* **74,** 91 (1970).
64. Marshall, K., Ridgewill, G. L., Rochester, C. H., and Simpson, J., *Chem. Ind. (London)* (1974).
65. Kageyama, Y., Yotsuyanagi, T., and Aomura, K., *J. Catal.* 36, 1 (1975).
66. Schreiber, L. B., and Vaughan, R. W., *J. Catal.* **40,** 226 (1975).
67. Lunsford, J. H., Zingery, L. W., and Rosynek, M. P., *J. Catal.* **38,** 179 (1975).
68. Koubek, J., Volf, J., and Pasek, J., *J. Catal.* **38,** 385 (1975).
69. Take, J., Tsuruya, T., Sato, T., and Yoneda, Y., *Bull. Chem. Soc. Jpn.* **45,**(11), 3409 (1972).
70. Yoshizumi, H., Shimada, Y., and Shirasaki, T., *Chem. Lett.* p. 107 (1973).

71. Bakshi, K. R., and Gavalas, G. R., *J. Catal.* **28**, 312 (1975).
72. Flockhart, B. D., Liew, K. Y., and Pink, R. C., *J. Catal.* **32**, 10 (1974).
73. Tomida, M., Tanaka, M., Iwasawa, Y., and Ogasawara, S., *Chem. Lett.* p. 375 (1973).
73a. Takahashi, M., Iwasawa, Y., and Ogasawara, S., *J. Catal.* **45**, 15 (1976).
74. Koubek, J., Volf, J., and Pasek, J., *J. Catal.* **38**, 385 (1975).
75. Figoli, N. S., and Parera, J. M., *J. Res. Inst. Catal., Hokkaido Univ.* **18**, 142 (1970).
75a. Schwarz, J. A., *J. Vac. Sci. Technol.* **12**, 32 (1975).
75b. Gay, I. D., and Liang, S., *J. Catal.* **44**, 306 (1976).
76. Morrow, B. A., and Cody, I. A., *J. Phys. Chem.* **79**, 761 (1975).
77. Low, M. J. D., *J. Catal.* **32**, 103 (1974).
78. Morterra, C., and Low, M. J. D., *J. Catal.* **28**, 265 (1973).
79. West, P. B., Haller, G. L., and Burwell, R. L., *J. Catal.* **29**, 489 (1973).
80. Taniguchi, K., Toshida, S., and Tarama, K., *Bull. Jpn. Pet. Inst.* **12**, 136 (May, 1970).
81. Majewski, W., Malinowski, S., and Jaczewska, A., *Rocz. Chem.* **49**(6), 1119 (1975).
82. Sinfelt, J. H., *Adv. Chem. Eng.* **5**, 37 (1964).
83. Ciapetta, F. G., and Wallace, D. N., *Catal. Rev.* **5**, 67 (1971).
84. Ahuja, S. P., Derrien, M. L., and Le Page, J. F., *Ind. Eng. Chem., Prod. Res. Dev.* **9**, 272 (1970).
85. Ripperger, W., and Saum, W., *Int. Conf. Chem. Uses Molybdenum, 2nd, 1976.*
86. Goble, A. G., and Lawrence, P. A., *Proc. Int. Congr. Catal., 3rd, Amsterdam, 1964* p. 320 (1965).
87. Giannetti, J. P., and Sebalsky, R. T., *I & EC Res. Dev.* **8**, 356 (1969).
88. Asselin, G. F., Bloch, H. S., Donaldson, G. R., Haensel, V., and Pollitzer, E. L., *Am. Chem. Soc., Div. Pet. Chem., Prepr.* **17**, B4 (1972).
89. Lippens, B. C., Doctoral thesis, Delft Tech. Hogesch., Delft, The Netherlands, 1961.
90. Stigter, D., Bosman, J., and Ditmarch, R., *Rec. Trav. Chim. Pays-Bas* **77**, 430 (1958).
91. Knözinger, H., *Adv. Catal.* **25**, (1976).
92. Peri, J. B., *J. Phys. Chem.* **69**, 220 (1965).
93. Parkyns, N. D., *J. Phys. Chem.* **75**, 526 (1971).
94. Fink, P., *Z. Chem.* **7**, 324 (1967).
95. Peri, J. B., *Int. Congr. Catal., 2nd, Paris, 1960* p. 1333 (1961).
96. Brouwer, D. M., *J. Catal.* **1**, 22 (1962).
97. Turkevich, J., and Smith, R. K., *J. Chem. Phys.* **16**, 466 (1948).
98. Peri, J. B., *J. Phys. Chem.* **70**, 3168 (1966).
99. Medema, J., and Houtman, J. P. W., *J. Catal.* **6**, 322 (1966).
100. Clark, A., and Finch, J. N., *Int. Congr. Catal., 4th, Moscow* Paper no. 75 (1968).
101. Gati, G., and Knözinger, H., *Int. Congr. Catal., 5th, Miami Beach, Fla., 1972* **1**, 819 (1973).
102. Ghorbel, A., Hoang-Van, C., and Teichner, S. J., *J. Catal.* **33**, 123 (1974).
103. Rosynek, M. P., Smith, W. D., and Hightower, J. W., *J. Catal.* **23**, 204 (1971).
104. Oblad, A. G., Messenger, J. V., and Brown, H. T., *Ind. Eng. Chem.* **39**, 1462 (1947).
105. Pines, H., and Haag, W. O., *J. Am. Chem. Soc.* **82**, 2471 (1960).
106. Haag, W. O., and Pines, H., *J. Am. Chem. Soc.* **82**, 2488 (1960).
107. Knözinger, H., and Kaerlein, C., *J. Catal.* **25**, 436 (1972).
108. Holm, V. C. F., and Clark, A., *Ind. Eng. Chem., Prod. Res. Dev.* **2**, 38 (1963).
109. Covini, R., Fattore, V., and Giordano, N., *J. Catal.* **7**, 126 (1967).
110. Antipina, T. V., Bulgakov, O. V., and Uvarov, A. V., *Int. Congr. Catal., 4th Moscow Prepr.,* Pap. No. 77 (1968).
111. Gerberich, H. R., Lutinski, F. E., and Hall, W. K., *J. Catal.* **6**, 209 (1966).
112. Antipina, T. V., *J. Catal.* **12**, 108 (1968).
113. Hughes, T. R., White, H. M., and White, R. J., *J. Catal.* **13**, 58 (1969).

114. Tanaka, M., and Ogasawara, S., *J. Catal.* **16,** 157 (1970).
115. Myers, J. W., *Ind. Eng. Chem., Prod. Res. Dev.* **10,** 200 (1971).
116. Massoth, F. E., and Kivat, F. E., *Proc. Int. Congr. Catal., 5th, Miami Beach, Fla., 1972* p. 807 (1973).
117. Brouwer, D. M., and Oelderik, J. M., *Am. Chem. Soc., Div. Pet. Chem., Prepr.* p. 184 (1968).
118. Olah, G. A., "Friedel-Crafts Chemistry," p. 420. Wiley, New York, 1973.
119. Kouwenhoven, H. W., *Adv. Chem. Ser.* No. 121, p. 529 (1973).
120. Kearby, K. K., *Int. Congr. Catal., 2nd, Paris, 1960* p. 2567 (1961).
121. Peri, J. B., *Discuss. Faraday Soc.* **52,** 55 (1971).
122. Kiviat, F. E., and Petrakis, L., *J. Phys. Chem.* **77,** 1232 (1973).
123. Ratnasamy, P., Sharma, D. K., and Sharma, L. D., *J. Phys. Chem.* **78,** 2069 (1974).
124. Walvekar, S. P., and Halgeri, A. B., *J. Indian Chem. Soc.* **50,** 387 (1973).
125. Matsuura, K., Suzuki, A., and Itoh, M., *Bull. Chem. Soc. Jpn.* **45,** 2079 (1972).
126. Yamamura, T. and Nakatomi, S., *J. Catal.* **37,** 142 (1975).
127. Magee, J. S., and Blazek, J. J., *ACS Monogr.* No. 171, p. 615 (1976).
128. Bourne, K. H., Cannings, F. R., and Pitkethyly, R. C., *J. Phys. Chem.* **74,** 2197 (1970).
129. Ballivet, D., Barthomeuf, D., and Pichat, P., *J. Chem. Soc., Faraday Trans. 1* **68,** 1712 (1972).
130. Bremer, H., Steinberg, K. H., and Chuong, T., *Z. Anorg. Allg. Chem.* **400,** 115 (1973).
131. Leonard, A. J., Ratnasamy, P., Declerek, F. D., and Fripiat, J. J., *Discuss. Faraday Soc.* **52,** 98 (1971).
132. Parera, J. M., Hillar, S. A., Vincenzini, J. C., and Figoli, N. S., *J. Catal.* **21,** 70 (1971).
133. Sugioka, M., and Aomura, K., *Bull. Jpn. Pet. Inst.* **15,** 136 (1973).
134. Mizuno, K., Ikeda, M., Imokawa, T., Take, J., and Yoneda, Y., *Bull. Chem. Soc. Jpn.* **49,** 1788 (1976).
135. Pauling, L., "The Nature of the Chemical Bond," p. 547. Cornell Univ. Press, Ithaca, New York, 1960.
136. Tanabe, K., Sumiyoski, T., Shibata, K., Kiyoura, T., and Kitagawa, J., *Bull. Chem. Soc. Jpn.* **47,** 1064 (1974).
137. Shibata, K., Kiyoura, T., Kitagawa, J., Sumiyoshi, T., and Tanabe, K., *Bull. Chem. Soc. Jpn.* **46,** 2985 (1973).
138. Kermarec, M., Briend-Faure, M., and Delafosse, D., *J. Chem. Soc., Faraday Trans. 1* **70,** 2180 (1974).
139. Moffat, J. B., and Neeleman, J. F., *J. Catal.* **39,** 419 (1975).
140. Shibata, K., and Kiyoura, T., *J. Res. Inst. Catal., Hokkaido Univ.* **19,** 35 (1971).
141. Walvekar, S. P., and Halgeri, A. B., *Indian J. Chem.* **11,** 662 (1973).
142. Tanabe, K., Ito, M., and Sato, M., *J. Chem. Soc. Chem. Commun.* p. 676 (1973).
143. Itoh, M., Hattori, H., and Tanabe, K., *J. Catal.* **35,** 225 (1974).
144. Miale, J. N., Chen, N. Y., and Weisz, P. B., *J. Catal.* **6,** 278 (1966).
145. Breck, D., "Zeolite Molecular Sieves." Wiley, New York, 1974.
146. Uytterhoeven, J. B., Christner, L. G., and Hall, W. K., *J. Phys. Chem.* **69,** 2117 (1965).
147. Angell, C. L., and Schaffer, P. C., *J. Phys. Chem.* **69,** 3463 (1965).
148. Ward, J. W., *J. Catal.* **9,** 225 (1967).
149. Carter, J. L., Lucchesi, P. J., and Yates, D. J. C., *J. Phys. Chem.* **68,** 1385 (1964).
150. Liengme, B. V., and Hall, W. K., *Trans. Faraday Soc.* **62,** 3229 (1966).
151. Eberly, P. E., Jr., *J. Phys. Chem.* **72,** 1042 (1968).
152. Searles, S., Tamres, M., Block, F., and Quarterman, L. A., *J. Am. Chem. Soc.* **78,** 4917 (1956).
153. Bissot, T. C., Parry, R. W., and Campbell, D. H., *J. Am. Chem. Soc.* **79,** 796 (1957).

154. Cant, N. W., and Little, L. H., *Can. J. Chem.* **43,** 1252 (1964).
155. Peri, J. B., *J. Phys. Chem.* **69,** 231 (1965).
156. Ward, J. W., *J. Catal.* **11,** 238 (1968).
157. Benesi, H. A., *J. Catal.* **8,** 368 (1967).
158. Hickson, D. A., and Csicsery, S. M., *J. Catal.* **10,** 27 (1968).
159. Chu, P., *J. Catal.* **43,** 346 (1976).
160. Venuto, P. B., Hamilton, L. A., Landis, P. S., and Wise; J. J., *J. Catal.* **4,** 81 (1966).
161. Hopkins, P. D., *J. Catal.* **12,** 325 (1968).
162. Otouma, H., Arai, Y., and Ukihashi, H., *Bull. Chem. Soc. Jpn.* **42,** 2449 (1969).
163. Ikemoto, M., Tsutsumi, K., and Takahashi, H., *Bull. Chem. Soc. Jpn.* **45,** 1330 (1972).
164. Morita, Y., Kimura, T., Kato, F., and Tamagawa, M., *Bull. Jpn. Pet. Inst.* **14,** 192 (1972).
165. Topchieva, K. V., and Tkhoang, K. S., *Russ. J. Phys. Chem.* **47,** 1185 (1973).
166. Topchieva, K. V., Tkhoang, K. S., Romanovskii, B. V., Ivanova, T. M., and Vishnevskaya, L. M., *Kinet. Katal.* **14,** 1355 (1973).
167. Huo Shih Thoah and Topchieva, K. V., *Dokl. Akad. Nauk SSSR* **211,** 870 (1973).
168. Turkevich, J., and Ono, Y., *Adv. Chem. Ser.* No. 102, p. 315 (1971).
169. Jacobs, P. A., Leeman, H. E., and Uytterhoeven, J. B., *J. Catal.* **33,** 17 (1974).
170. Eberly, P. E., Jr., *J. Phys. Chem.* **71,** 1717 (1967).
171. Broussard, L., and Shoemaker, D. P., *J. Am. Chem. Soc.* **82,** 1041 (1960).
172. White, J. L., Jelli, A. N., Andre, J. M., and Fripiat, J. J., *Trans. Faraday Soc.* **63,** 461 (1967).
173. Olson, D. H., and Dempsey, E., *J. Catal.* **13,** 221 (1969).
174. Uytterhoeven, J. B., Jacobs, P., Makay, K., and Schoonheydt, R., *J. Phys. Chem.* **72,** 1768 (1968).
175. Ward, J. W., *J. Phys. Chem.* **73,** 2088 (1969).
176. Ward, J. W., *J. Catal.* **13,** 364 (1969).
177. Sherry, H. S., *J. Phys. Chem.* **70,** 1158 (1966).
178. Beaumont, R., Barthomeuf, D., and Trambouze, Y., *Adv. Chem. Ser.* No. 102, p. 327 (1971).
179. Beaumont, R., and Barthomeuf, D., *J. Catal.* **27,** 45 (1972).
180. Beaumont, R., and Barthomeuf, D., *J. Catal.* **30,** 288 (1973).
181. Dempsey, E., and Olson, D. H., *J. Catal.* **15,** 309 (1969).
182. Bielanski, A., and Datka, J., *J. Catal.* **37,** 383 (1975).
183. Ward, J. W., *J. Phys. Chem.* **71,** 3106 (1967).
184. Ward, J. W., *J. Catal.* **9,** 396 (1967).
185. Cant, N. W., and Hall, W. K., *Trans. Faraday Soc.* **64,** 1093 (1968).
186. Schoonheydt, R. A., and Uytterhoeven, J. B., *J. Catal.* **19,** 55 (1970).
187. Rouxhet, P. G., Touillaux, R., Mestdagh, M., and Fripiat, J. J., *Proc. Int. Clay Conf.* **1,** 109 (1969).
188. Fripiat, J. J., *Catal. Rev.* **5,** 269 (1971).
189. Mestdagh, M. M., Stone, W. E., and Fripiat, J. J., *J. Phys. Chem.* **76,** 1220 (1972).
190. Freude, D., Oehme, W., Schmiedel, H., and Staudte, B., *J. Catal.* **32,** 137 (1974).
191. McDaniel, C. V., and Maher, P. K., "Molecular Sieves," p. 186. Soc. Chem. Ind., London, 1967.
192. Kerr, G. T., *J. Catal.* **15,** 200 (1969).
193. Ward, J. W., *J. Catal.* **18,** 348 (1970).
194. Kerr, G. T., *Adv. Chem. Ser.* No. 121, p. 219 (1973).
195. Maher, P. K., Hunter, F. D., and Scherzer, J., *Adv. Chem. Ser.* No. 101, p. 266 (1971).
196. Kerr, G. T., *J. Phys. Chem.* **71,** 4155 (1967).
197. Peri, J. B. *Int. Congr. Catal., 5th, Miami Beach, Fla., 1972,* p. 329.

198. Scherzer, J., and Bass, J. L., *J. Catal.* **28,** 101 (1973).
199. Jacobs, P. A., and Uytterhoeven, J. B., *J. Catal.* **22,** 193 (1971).
200. Jacobs, P. A., and Uytterhoeven, J. B., *J. Chem. Soc., Faraday Trans.* **69,** 373 (1973).
201. Beaumont, R., and Barthomeuf, D., *J. Catal.* **26,** 218 (1972).
202. Pickert, P. E., Rabo, J. A., Dempsey, E., and Schomaker, V., *Proc. Int. Congr. Catal., 3rd, Amsterdam* **1,** 714 (1964).
203. Dempsey, E., "Molecular Sieves," p. 293. Soc. Chem. Ind., London, 1967.
204. Topchieva, K. V., and Huo Shih T'Huoang, *Kinet. Katal.* **11,** 490 (1970).
205. Jacobs, P. A., Leeman, H. E., and Uytterhoeven, J. B., *J. Catal.* **33,** 31, (1974).
206. Bertsch, L., and Habgood, H. W., *J. Phys. Chem.* **67,** 1621 (1963).
207. Watanabe, Y., and Habgood, H. W., *J. Phys. Chem.* **72,** 3066 (1968).
208. Ignat'eva, L. A., Moskovskaya, I. F., Oppengeim, V. D., Spozhakina, A. A., and Topchieva, K. V., *Kinet. Katal.* **9,** 135 (1968).
209. Ward, J. W., *J. Phys. Chem.* **72,** 2689 (1968).
210. Ward, J. W., *J. Catal.* **10,** 34 (1968).
211. Ward, J. W., *J. Phys. Chem.* **72,** 4211 (1968).
212. Ward, J. W., *J. Catal.* **13,** 321 (1969).
213. Ward, J. W., *J. Catal.* **14,** 365 (1969).
214. Rabo, J. A., Angell, C. L., Kasai, P. H., and Schomaker, V., *Discuss. Faraday Soc.* **41,** 328 (1966).
215. Hall, W. K., *Chem. Eng. Prog., Symp. Ser.* **63,** 68 (1967).
216. Oehme, W., Freude, D., and Schmiedel, H., *Z. Phys. Chem. (Leipzig)* **255,** 566 (1974).
217. Christner, L. A., Liengme, B. V., and Hall, W. K., *Trans. Faraday Soc.* **64,** 1679 (1968).
218. Bolton, A. P., *J. Catal.* **22,** 9 (1971).
219. Venuto, P. B., Hamilton, L. A., and Landis, P. S., *J. Catal.* **5,** 484 (1966).
220. Plank, C. J., *Proc. Int. Congr. Catal., 3rd, Amsterdam, 1964* **1,** 727 (1965).
221. Kladnig, W., *J. Phys. Chem.* **80,** 262 (1976).
222. Boreskova, E. G., Lygin, V. I., and Topchieva, K. V., *Kinet. Katal.* **5,** 1115 (1964).
223. Patinken, S. H., and Friedman, B. S., in "Friedel-Crafts and Related Reactions" (G. A. Olah, ed.), Vol. 2, Part 1, p. 31. Wiley (Interscience), New York, 1964.
224. Moscou, L., and Mone, R., *J. Catal.* **30,** 417 (1973).
225. Hattori, H., and Shiba, T., *J. Catal.* **12,** 111 (1968).
226. Ward, J. W., *J. Catal.* **22,** 237 (1971).
227. Lefrancois, M., and Malbois, G., *J. Catal.* **20,** 350 (1971).
228. Eberly, P. E., Jr., Kimberlin, C. N., Jr., and Voorhies, A., Jr., *J. Catal.* **22,** 419 (1971).
229. Weeks, T. J., Jr., Hillery, H. F., and Bolton, A. P., *J. Chem. Soc., Faraday Trans. 1* **71,** 2051 (1975).
230. Burbridge, B. W., Keen, I. M., and Eyles, M. K., *Adv. Chem. Ser.* No. 102, p. 400 (1971).
231. Lunsford, J. H., *J. Phys. Chem.* **72,** 4163 (1968).
232. Millikan, T. H., Oblad, A. G., and Mills, G. A., "Clays and Clay Technology," p. 314. Calif. Div. Mines, San Francisco, 1955.
233. Brown, G., ed., "The X-Ray Identification and Crystal Structures of Clay Minerals," pp. 51, 143. Mineral. Soc., London, 1972.
234. Theng, B. K. G., "The Chemistry of Clay-Organic Reactions." Wiley, New York, 1974.
235. Solomon, D. H., Swift, J. D., and Murphy, A. J., *J. Macromol. Sci., Chem.* **A5**(3), 587 (1971).
236. Solomon, D. H., and Murray, H. H., *Clays and Clay Miner.* **20,** 135 (1972).
237. Frenkel, M., *Clays and Clay Miner.* **22,** 435 (1974).

238. Mortland, M. M., Fripiat, J. J., Chaussidon, J., and Uytterhoeven, J., *J. Phys. Chem.* **67,** 248 (1963).
239. Russell, J. D., *Trans. Faraday Soc.* **61,** 2284 (1965).
240. Harter, R. D., and Ahlrichs, J. L., *Soil Sci. Soc. Am., Proc.* **31,** 30 (1967).
241. McLaren, A. D., and Seaman, G. V. F., *Soil Sci. Soc. Am., Proc.* **32,** 127 (1968).
242. Mortland, M. M., *Int. Congr. Soil Sci., 9th* **1,** 691 (1968).
243. Mortland, M. M., and Raman, K. V., *Clays and Clay Miner.* **16,** 393 (1968).
244. Pickett, A. G., and Lemcoe, M. M., *J. Geophys. Res.* **64,** 1579 (1959).
245. Ducros, P., and Dupont, M., *C. R. Acad. Sci.* **254,** 1409 (1962).
246. Davidtz, J. C., *J. Catal.* **43,** 260 (1976).
247. Wright, A. C., Granquist, W. T., and Kennedy, J. V., *J. Catal.* **25,** 65 (1972).
248. Hattori, H., Milliron, D. L., and Hightower, J. W., *Am. Chem. Soc., Div. Pet. Chem., Prepr.* **28,** 33 (1973).
249. Swift, H. E., and Black, E. R., *Ind. Eng. Chem., Prod. Res. Dev.* **13,** 106 (1974).

Selective Oxidation of Propylene

GEORGE W. KEULKS, L. DAVID KRENZKE,
AND THOMAS M. NOTERMANN*

Department of Chemistry
Laboratory for Surface Studies
The University of Wisconsin-Milwaukee
Milwaukee, Wisconsin

I. Introduction	183
II. Activation of Propylene	185
A. Allylic Intermediate	185
B. Hydroperoxide Intermediate	187
C. Complete Combustion	190
III. Activation of Oxygen	191
A. General Comments	191
B. Role of Lattice Oxygen	191
C. Chemical Nature of the Lattice Oxygen	195
D. Role of Adsorbed Oxygen	196
IV. Active Catalyst Systems	198
A. General Comments	198
B. The Cuprous Oxide System	199
C. The Bismuth Molybdate System	199
D. The Uranium Antimonate System	204
E. Scheelite-Structured Systems	205
F. The Bismuth Iron Molybdate System	207
G. The Multicomponent System	209
V. Nature of the Active Sites	210
A. The Dual-Site Concept	210
B. Active Sites for Propylene Adsorption	210
C. Other Experimental Evidence for Active Sites	213
D. Active Site Model	219
VI. Conclusions	221
References	222

I. Introduction

The conversion of light hydrocarbons into products containing oxygen or other hetero atoms produces important intermediates for the

* Present address: Union Carbide Corporation, Charleston, West Virginia.

petrochemical industry. A number of such intermediates are derived from propylene, e.g., acrylonitrile, acrolein, acrylic acid, and propylene oxide.

One of the more successful conversions has been the selective oxidation of propylene to acrolein. In 1948, Hearne and Adams (*1*) reported that cuprous oxide produced acrolein from propylene with a yield of about 50% at propylene/oxygen ratios of about one. Even though the yield of acrolein was low, the search for improved catalytic systems provided few catalysts until the development of catalysts based on bismuth and molybdenum.

In 1959, Idol (*2*), and in 1962, Callahan et al. (*3*) reported that bismuth/molybdenum catalysts produced acrolein from propylene in higher yields than that obtained in the cuprous oxide system. The authors also found that the bismuth/molybdenum catalysts produced butadiene from butene and, probably more importantly, observed that a mixture of propylene, ammonia, and air yielded acrylonitrile. The bismuth/molybdenum catalysts now more commonly known as *bismuth molybdate catalysts* were brought to commercial realization by the Standard Oil of Ohio Company (SOHIO), and the vapor-phase oxidation and ammoxidation processes which they developed are now utilized worldwide.

Shortly after the introduction of the bismuth molybdate catalysts, SOHIO developed and commercialized an even more selective catalyst, the uranium antimonate system (*4*). At about the same time, Distillers Company, Ltd. developed an oxidation catalyst which was a combination of tin and antimony oxides (*5*). These earlier catalyst systems have essentially been replaced on a commercial scale by multicomponent catalysts which were introduced in 1970 by SOHIO. As their name implies, these catalysts contain a number of elements, the most commonly reported being nickel, cobalt, iron, bismuth, molybdenum, potassium, manganese, and silica (*6*–*8*).

The investigation of the mechanism of olefin oxidation over oxide catalysts has paralleled catalyst development work, but with somewhat less success. Despite extensive efforts in this area which have been recently reviewed by several authors (*9*–*13*), there continues to be a good deal of uncertainty concerning the structure of the reactive intermediates, the nature of the active sites, and the relationship of catalyst structure with catalytic activity and selectivity. Some of this uncertainty is due to the fact that comparisons between various studies are frequently difficult to make because of the use of ill-defined catalysts or different catalytic systems, different reaction conditions, or different reactor designs. Thus, rather than reviewing the broader area of selective oxidation of hydrocarbons, this review will attempt to focus on a single aspect of selective hydrocarbon oxidation, the selective oxidation of propylene to acrolein, with the following questions in mind:

1. How is propylene activated?
2. What is the nature of the reactive oxygen species?
3. How important is catalyst structure and catalyst preparation to catalytic activity and selectivity?
4. Is there a generalized model which can be used for describing the active sites?

II. Activation of Propylene

A. Allylic Intermediate

The use of isotopic tracers has demonstrated that the selective oxidation of propylene proceeds via the formation of a symmetrical allyl species. Probably the most convincing evidence is presented by the isotopic tracer studies utilizing ^{14}C-labeled propylene and deuterated propylene. Adams and Jennings (*14, 15*) studied the oxidation of propylene at 450°C over bismuth molybdate and cuprous oxide catalysts. The reactant propylene was labeled with deuterium in various positions. They analyzed their results in terms of a *kinetic isotope effect*, which is defined by the probability of a deuterium atom being abstracted relative to that of a hydrogen atom. Letting $z = k_D/k_H$ represent this relative discrimination probability, the reaction paths shown in Fig. 1 were found to be applicable to the oxidation of $1-C_3H_6-3d$ and $1-C_3H_6-1d$.

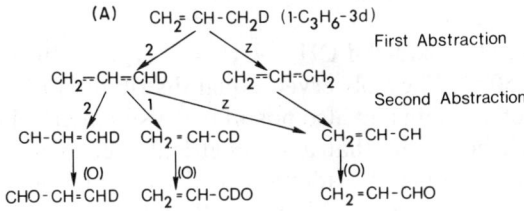

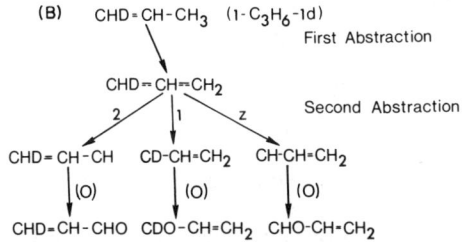

FIG. 1. Reaction paths for $1-C_3H_6-3d$ and $1-C_3H_6-1d$ (*14, 15*).

The value of z was calculated from the distribution of deuterium and acrolein and found to be very close to the theoretical value. It was also noted that the distribution of deuterium in acrolein was the same regardless of which deuterated propylene was used as the starting material and that the deuterium was found only on the terminal carbon atoms of the product acrolein.

These results were interpreted in the following way:

1. The first step in the oxidation of propylene is the abstraction of a hydrogen atom from the methyl group of propylene, and because an isotope effect was observed for this process, this abstraction is the rate-determining step.

2. The hydrocarbon intermediate formed after abstraction of the methyl hydrogen is symmetrical and not cyclic because no deuterium is found on the middle carbon atom of acrolein. The structure of this intermediate is probably similar to a π-allylic species, $CH_2 \text{---} CH \text{---} CH_2$, in which the terminal carbon atoms are rapidly equilibrated.

3. Propylene undergoes two successive hydrogen abstractions before the addition of oxygen.

4. The mechanism of propylene oxidation over bismuth molybdate and cuprous oxide is the same.

Voge et al. (16) demonstrated that a symmetrical intermediate is formed during the oxidation of $^{13}CH_3-CH=CH_2$ over cuprous oxide at 300°C. An analysis of the distribution of ^{13}C in acrolein showed that the ^{13}C content of the carbonyl group was 50 to 55% of that of the original methyl group of propylene.

McCain et al. (17) oxidized $CH_3-CH=^{13}CH_2$ over bismuth phosphomolybdate at 450°C. They observed equal distribution of ^{13}C in the terminal carbons of acrolein and also postulated a symmetrical intermediate. They suggested, however, that a symmetrical species could be formed from propylene by either the addition or the abstraction of a hydrogen atom by the catalyst. This would lead to the following scheme.

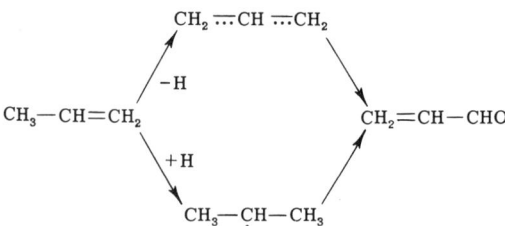

To test this hypothesis, propylene was oxidized on a deuterated catalyst surface. The π-allyl route would yield acrolein containing no deuterium, whereas the isopropyl route would give approximately 50% monodeuterated acrolein. The product that they observed contained virtually no deuterium, which confirmed the allylic route in the formation of acrolein from propylene.

Studies have also been made on the oxidation of propylene labeled with ^{14}C. Sachtler (18) and Sachtler and de Boer (19) oxidized all three ^{14}C-labeled propylenes over bismuth molybdate and concluded from the distribution of ^{14}C in the acrolein that a symmetrical allylic species was involved.

In comparison to the bismuth molybdate and cuprous oxide catalyst systems, data on other catalyst systems are much more sparse. However, by the use of similar labeling techniques, the allylic species has been identified as an intermediate in the selective oxidation of propylene over uranium antimonate catalysts (20), tin oxide–antimony oxide catalysts (21), and supported rhodium, ruthenium (22), and gold (23) catalysts. A direct observation of the allylic species has been made on zinc oxide by means of infrared spectroscopy (24–26). In this system, however, only adsorbed acrolein is detected because the temperature cannot be raised sufficiently to cause desorption of acrolein without initiating reactions which yield primarily oxides of carbon and water.

B. Hydroperoxide Intermediate

Even though there is rather strong evidence that the selective oxidation of propylene proceeds via the formation of a symmetrical allyl intermediate, there is some evidence that allyl hydroperoxide (or peroxide) intermediates may be involved under certain select conditions. Even before the work of Adams and Jennings (14), Margolis (27) proposed such intermediates for the oxidation of propylene over cuprous oxide. Margolis suggested that only at temperatures higher than normal reaction temperatures was propylene able to react with the oxygen of the cuprous oxide lattice. Therefore, at reaction temperatures, the oxidation of propylene was proposed to involve adsorbed molecular surface oxygen according to the following scheme (27).

$$O_2 \rightleftharpoons 2O$$
$$O + e \rightarrow (O)^-$$
$$O_2 + e \rightarrow (O_2)^-$$
$$C_3H_6 + (O_2)^- \rightarrow (C_3H_5OOH)$$
charged complex 1 (hydroperoxide)
$$(C_3H_5OOH)^+ \rightarrow (C_3H_4O) + (H_2O)^+$$
$$(C_3H_4O)^+ + e \rightarrow (C_3H_4O) \text{ gas}$$

$(C_3H_4O) + O_2 \rightarrow (C_3H_4O-O_2)$
charged complex 2
$(C_3H_4O-O_2) \rightarrow CO_2 + H_2O + (RH)^+$
$(RH)^+ + O_2 \rightarrow CO_2 + H_2O + (R'H)^+$
$C_3H_6 - e \rightarrow (C_3H_6)^+$
$(C_3H_6)^+ + 2(O)^- \rightarrow (C_3H_6 \cdot OO)$
charged complex 3
$(C_3H_6 \cdot OO) \rightarrow C_2H_4O + HCHO$
$(C_3H_6 \cdot OO) \rightarrow ROO + (RH)^+$
$(RH)^+ + O_2 \rightarrow R'OO + (R''H)^+$

Cant and Hall (22) studied the oxidation of propylene labeled with deuterium over rhodium and ruthenium catalysts supported on a low surface area α-alumina. The most selective reaction was the oxidation of propylene over rhodium at 200°C where substantial amounts of acrolein and acetone were produced. The oxidation reaction was first order in propylene and inverse first order in oxygen, suggesting that the reaction occurred between two adsorbed species which were competing for surface sites. Whereas Adams and Jennings noted two isotope effects (formation of allyl from propylene and formation of C_3H_4 from the allyl species), Cant and Hall observed only one isotope effect (formation of allyl from propylene). From their results, they proposed the mechanism shown in Fig. 2 for propylene oxidation over supported noble metals. Pathway A requires that the methyl carbon of propylene not lose its identity. It was suggested, therefore, that incorporation of adsorbed molecular oxygen occurs simultaneously with abstraction of an allylic hydrogen forming allyl hydroperoxide which can decompose on the surface to acrolein and water. Approximately 15% of the acrolein is

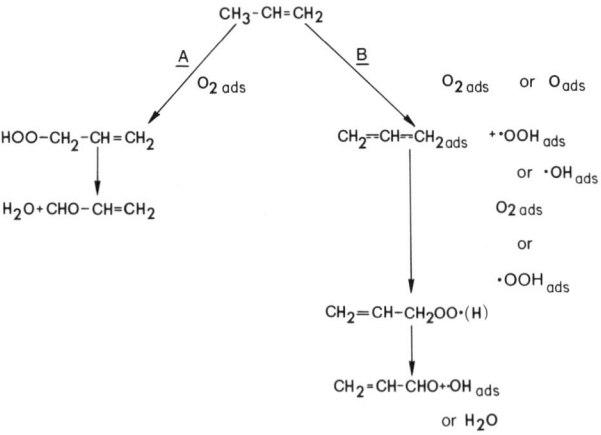

FIG. 2. Reaction mechanism for propylene oxidation over noble metals (22).

formed using pathway A. Pathway B also results in the formation of allyl hydroperoxide, but only after the formation of a symmetrical π-allylic surface species.

The possible formation of allyl hydroperoxide as an intermediate in the oxidation of propylene over metal oxides was also proposed by Daniel and Keulks (28). They studied the oxidation of propylene over bismuth molybdate and several other similar metal oxide catalysts using several different types of reactors. The reactors differed only in the free volume of the post-catalytic zone which was maintained at an elevated temperature. Their results indicated that in addition to the heterogeneous oxidation occurring over the catalysts, a surface-initiated homogeneous reaction was occurring in the post-catalytic zone. Moreover, the homogeneous reaction did not take place when the reactor with the post-catalytic zone contained no catalyst, nor did the homogeneous reaction occur when the same catalyst, under identical conditions, was placed into a reactor with no post-catalytic volume. The reaction occurring in the post-catalytic volume resulted in the increase in conversion of propylene and the formation of propylene oxide and formaldehyde as well as greatly increased amounts of acetaldehyde. This was in agreement with McCain et al. (17), who observed similar reactions under similar conditions.

To explain their results, Daniel and Keulks proposed that allyl hydroperoxide (or peroxide) was the initiator of the gas-phase reaction since hydroperoxides are known to be good radical reaction initiators. In addition, it was felt that the decomposition of allyl hydroperoxide in the gas phase could result in the formation of propylene oxide and even acrolein. They proposed a modified reaction mechanism, which is presented in Fig. 3.

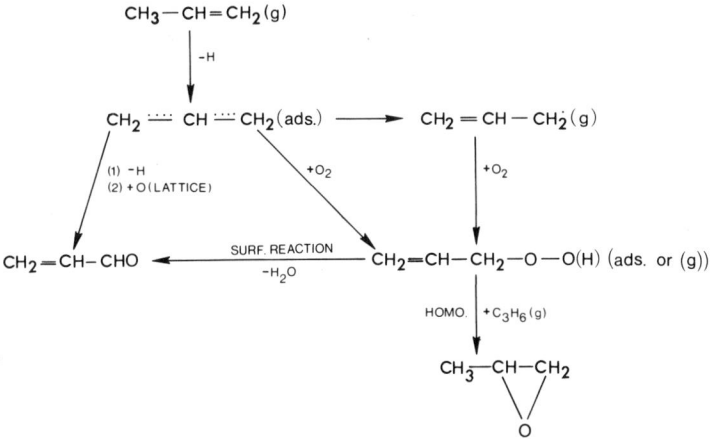

FIG. 3. Modified reaction mechanism for propylene oxidation (28).

C. Complete Combustion

The selective oxidation of propylene is accompanied by small yields of saturated aldehydes, acids, and oxides of carbon. Consequently, it seems appropriate to examine several of the proposals made regarding the origin of these compounds.

Russian workers using ^{14}C-labeled propylene, acrolein, and acetaldehyde (29–31) have determined that carbon oxides are formed chiefly through the further oxidation of acrolein and that acetaldehyde and formaldehyde are produced either from acrolein or directly from the symmetrical π-allylic intermediate. These two saturated aldehydes can then undergo further oxidation (about 20 times more rapidly than acrolein) to CO and CO_2. The overall scheme proposed is given in Fig. 4.

Keulks et al. (32) have also concluded from the oxidation of ^{14}C-labeled and unlabeled acrolein that carbon dioxide is formed almost exclusively from the further oxidation of acrolein. Thus, it can be seen that the initial step in the formation of carbon dioxide and the other side products of propylene oxidation is the formation of a symmetrical π-allyl intermediate. This π-allylic intermediate is responsible for both the selective and nonselective oxidation of propylene, the course of the overall reaction depending on the subsequent reaction pathway of the allylic species.

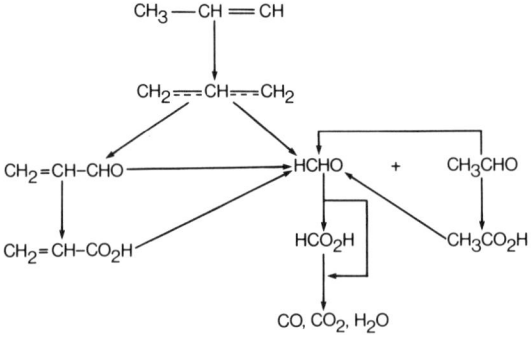

FIG. 4. Reaction scheme for complete combustion (29–31).

III. Activation of Oxygen

A. General Comments

Even though the formation of an allylic intermediate has been relatively firmly established, the nature of the active oxygen species which interacts with the allylic intermediate is less clear. Nevertheless, much attention has been directed toward understanding the various oxygen species found on the surface of oxide catalysts.

The adsorption of oxygen on oxide catalysts can be viewed as occurring according to a stepwise process,

$$O_2(g) \rightarrow O_{2(ads)} \rightarrow O_2^- \rightarrow 2O^- \rightarrow 2O^{2-}$$

The existence of the molecular radical ion O_2^-, of atomic O^-, and of the regular ions in the lattice O^{2-} has been firmly established. A review by Lunsford (33) presents a summary of the experimental evidence which led to the discovery of O_2^- and O^-. The participation of these various forms of oxygen in hydrocarbon oxidation is discussed in a review by Sachtler (11). It seems clear that both adsorbed and lattice oxygen species play an important role in the selective oxidation of hydrocarbons.

B. Role of Lattice Oxygen

A redox mechanism for oxidation catalysis was proposed by Mars and van Krevelen (34) for the oxidation of aromatics over V_2O_5. This mechanism introduced the concept that lattice oxygen of a reducible metal oxide could serve as a useful oxidizing agent for hydrocarbons. Moreover, it formed the basis for the early work at SOHIO which led to the development of the bismuth molybdate catalyst. Since that time there have been many reports which support the redox concept.

Aykan (35) reported that ammoxidation of propylene occurred over a silica-supported bismuth molybdate catalyst in the absence of gas-phase oxygen, although the catalytic activity decreased rapidly with increasing catalyst reduction. The reduction process was followed by X-ray and it was found that phase changes which occurred in the catalyst and the decrease in catalytic activity corresponded quantitatively to the depletion of lattice oxygen.

Batist et al. (36) measured the rate of oxidative dehydrogenation of 1-butene in the presence of oxygen and the rate of catalyst (Bi_2MoO_6) reduction with 1-butene and found the rates and product distribution to be almost the same. From this they concluded that the rate of reaction is controlled by the rate of catalyst reduction and therefore involves the redox mechanism.

Peacock et al. (37) showed that a bismuth molybdate catalyst can be reduced with propylene and that the oxygen appearing in the gaseous products (acrolein, carbon dioxide, and water) can be quantitatively replaced in the lattice. The amount of oxygen removed during reduction corresponds to the participation of many sublayers of oxide ions.

Reports involving other oxide systems, Fe–Sb (38), Bi–W, Sn–Sb, Sb–Mo, Sn–P (39), U–Sb (20), and Ni–Sn–P–K (40), have also shown that these catalysts have the capacity to act as a source of active oxygen and can function for some time in the absence of gas-phase oxygen. Although these data strongly support the redox mechanism, perhaps the most compelling evidence for the participation of lattice oxygen in these reactions comes from studies of propylene oxidation in the presence of isotopic oxygen (41–47).

Keulks (41) examined the reactivity of lattice oxygen in bismuth molybdate toward the oxygen-exchange reaction by circulating oxygen-18 over the catalyst for long periods of time (2 hr) at temperatures between 250° and 500°C. No change in the isotopic composition of gas-phase oxygen was detected and, in addition, circulating a mixture of oxygen-16 and oxygen-18 did not produce the scrambled oxygen species. This implies that the extent of the reversible, dissociative chemisorption of oxygen on bismuth molybdate must be very small, if it occurs at all. During the oxidation of propylene in the presence of gas-phase oxygen-18, only 2 to 2.5% of the oxygen atoms in the acrolein and carbon dioxide produced were isotopically labeled. This lack of extensive incorporation of oxygen-18 into the reaction products implies the participation of lattice oxide ions in the selective and nonselective oxidation reactions.

Wragg et al. (42) studied the roles of gas-phase oxygen and oxide ions in the selective oxidation of propylene to acrolein over bismuth molybdate by enriching the gaseous oxygen or oxide ions with oxygen-18. In preliminary experiments, they attempted to measure the rate of exchange between gaseous oxygen and lattice oxide ions in the temperature range of 475° to 500°C. They found no evidence of this exchange either in the presence or absence of propylene and the products of its oxidation. The possibility of exchange between the oxygen in acrolein and oxide ions was also checked at 500°C and found not to occur. The experiments involved oxidizing propylene in the presence of (1) $^{18}O_2$ with an oxygen-18-enriched catalyst, (2) $^{16}O_2$ with an oxygen-18-enriched catalyst, (3) $^{18}O_2$ with a catalyst containing normal oxygen, and (4) $^{16}O_2$ with a catalyst containing normal oxygen. Experiments (1) and (4) were primarily to reveal the existence of any background impurities at the important mass numbers. The crucial experiments were numbers 2 and 3. These indicated that the oxygen incorporated into acrolein was derived from the lattice and not the gas phase.

Sancier et al. (43) used oxygen-18 to examine the relative role of adsorbed versus lattice oxygen in propylene oxidation over a silica-supported bismuth molybdate catalyst as a function of temperature. At 400°C they observed the formation of predominantly acrolein[^{16}O] rather than acrolein[^{18}O], indicating significant participation of lattice oxygen. However, as the reaction temperature was decreased, the authors concluded that the role of adsorbed oxygen became more important.

A study by Otsubo et al. (44) using isotopically labeled bismuth molybdate catalysts as well as $^{18}O_2$, lends support to the concept of extensive lattice oxygen participation. The author prepared catalysts with the oxygen-18 concentrated in the molybdenum layers and with the oxygen-18 concentrated in the bismuth layers. Their results indicate that propylene is oxidized by a Bi_2O_2 layer and bulk oxygen migration occurs from the MoO_2 layers to the Bi_2O_2 layers. The migration of oxide ions in the bulk produces surface anion vacancies in the MoO_2 layers which serve as reoxidation sites.

Christie et al. (45) and Pendleton and Taylor (46) have recently reported the results of propylene oxidation over bismuth molybdate and mixed oxides of tin and antimony and of uranium and antimony in the presence of gas-phase oxygen-18. Their work indicated that for each catalyst, the lattice was the only direct source of the oxygen in acrolein and that lattice and/or gas-phase oxygen is used in carbon dioxide formation. The oxygen anion mobility appeared to be greater in the bismuth molybdate catalyst than in the other two.

The studies cited above were conducted in recirculation reactors (41, 42, 44), pulse reactors (43), or static systems (45, 46). While such reactors are convenient for isotopic tracer studies, there is the possibility that the data obtained do not represent the behavior of the catalyst in its steady state. Keulks and Krenzke (47) circumvented this difficulty by carrying out the oxidation of propylene in the presence of $^{18}O_2$ over α, and γ-bismuth molybdate in a single-pass integral flow reactor. The reactor was designed to allow the rapid changeover from $^{16}O_2$ to $^{18}O_2$ in the feed gas without changing the partial pressures of the reactants or disrupting the gas flow. Therefore, it was possible to maintain the steady-state conditions. Under these conditions, the rate of oxygen-18 incorporation into products of oxidation and the total oxygen consumption could be related quantitatively to the amount of oxygen which participates in the reaction. The results showed that at 430°C, lattice oxygen participates in the formation of both acrolein and carbon dioxide and that there appeared to be no distinction between the oxide ion which was incorporated into acrolein and the oxide ion which was incorporated into carbon dioxide. The amount of lattice oxygen which was involved in the reaction was calculated to be 100% for the γ phase and 16% for the α phase. This corresponds to the participation of 270 and 85 layers of catalyst oxygen, respectively.

The validity of the conclusions from some of the early oxygen-18 studies (41, 42) was questioned by Novakova and Jirû (48). They observed the ability of the oxygen atom in $H_2^{18}O$ to exchange with the lattice oxygen of several selective oxidation catalysts and suggested that the results of Keulks (41) and Wragg et al. (42) on the selective oxidation of propylene over bismuth molybdate could be explained by the rapid exchange of bulk oxygen with water, rather than by high mobility of lattice oxygen. In addition they suggested that a more detailed investigation might reveal that oxidation occurred via water.

Wragg et al. (49) responded by pointing out (1) that the catalysts used by their group and by Keulks function with the same rate and selectivity for acrolein formation in the absence or presence of gaseous oxygen, which indicates numerous layers of highly mobile oxide ions are involved and (2) that the isotopic composition of the oxidation products in Keulks' work can only be explained by the participation of oxide ions from approximately 500 layers of the catalyst crystals. It was also mentioned that although the oxygen in water may indeed exchange with the oxide ions of the catalyst, the exchange of any fraction of the oxygen in the solid greater than that which corresponds to the movement within the outermost layer implies that an oxygen species is mobile within the lattice. Therefore, under conditions where lattice oxygen diffusion is rapid, individual surface processes involving propylene and water can lead to the appearance of a significant fraction of the oxygen in the solid in the gas-phase products without necessarily implying the water takes part as a reactant in the oxidation process.

The role of water in the oxidation of propylene has been investigated by several workers. Moro-oka et al. (50) oxidized propylene to acrolein and acetone over a SnO_2–MoO_3 catalyst at 365°C in the presence of O_2 and $H_2^{18}O$. They found that the oxygen-18 tracer was incorporated into the acetone produced but did not appear in the acrolein which was produced simultaneously. Novakova et al. (51) studied the interaction of propylene and $H_2^{18}O$ in the absence of gaseous oxygen over a Mo–W–Sn–Te–O catalyst prepared for the production of acrylic acid. It was reported that the formation of an acrolein–lattice oxygen complex was an intermediate step in acrylic acid formation and that this complex reacts further with the oxygen from water to give acrylic acid.

C. CHEMICAL NATURE OF THE LATTICE OXYGEN

The relationship between the strength of the metal–oxygen bond and the catalytic activity and selectivity for oxidation catalysts has been studied by a number of workers using various techniques (*19, 52, 53*). The applicability of the Mars–van Krevelen mechanism to hydrocarbon oxidation led Sachtler and de Boer (*19*) to postulate that the tendency of an oxide to donate its oxygen should be of major importance in determining whether it is a selective oxidation catalyst. If reduction of the oxide is easy, i.e., if the free energy of oxygen dissociation is small, then oxygen could be donated to a hydrocarbon molecule, and the catalyst should be active, but not necessarily selective. On the other hand, if it is difficult to dissociate oxygen because the metal-oxygen bond is strong, then the oxide should have low catalytic activity. In the intermediate range, oxides might be moderately active and selective.

Roiter *et al.* (*52*) suggested that there was a level of reduction beyond which selectivity decreased due to the thermodynamic limitations of reduction since consecutive reduction steps would become more difficult. Support of this suggestion is provided by the work of Boreskov *et al.* (*53*). They calculated the energy of the lattice metal–oxygen bond from the equilibrium gas phase oxygen pressure as a function of temperature for several metal oxides. They found that the lattice oxygen bond energy increased with the amount of lattice oxygen removed. Thus, removal of lattice oxygen became thermodynamically more unfavorable as the oxides became more reduced.

Unfortunately, the absolute value of the metal–oxygen bond energy does not unambiguously define the level of selectivity in the partial oxidation process. In addition, the specific interactions between the substance being oxidized and the catalyst surface and the character of the metal-oxygen bond are important parameters as well.

Investigations by Trifiro *et al.* (*54–57*) using infrared spectroscopy have led to the identification of a covalent metal–oxygen double bond which is present in a great majority of the selective oxidation catalysts. On the other hand, this bond is systematically absent from total oxidation catalysts such as the oxides of iron, nickel, and cobalt (*55, 56*). Trifiro *et al.* (*57*) and Akimoto and Echigoya (*58*) have also reported that there is a direct interaction between adsorbed propylene or acrolein and this double-bonded oxygen species. On the basis of these observations, it has been suggested by Trifiro that this type of surface lattice oxygen is directly involved in the selective oxidation of olefins.

As indicated previously, it is frequently assumed that weakly bound oxygen leads to the complete combustion of hydrocarbons. However, another possible pathway to complete combustion, may involve the activation of an oxide ion (59). Kazanskii (60) studied in detail the elementary steps of the reduction, photoreduction, and reoxidation of the surfaces of oxide catalysts. He reported that upon exposure to light some semiconductor oxides undergo a charge transfer from the O^{2-} ion to the cation, as shown:

$$M^{n+} + O^{2-} \underset{}{\overset{h\nu}{\rightleftharpoons}} M^{(n-1)+} + O^-$$

This process generates the extremely reactive O^- ion, which is capable of reacting with H_2 or CO at liquid nitrogen temperatures, and could certainly participate in the conversion of propylene to carbon dioxide. Margolis (12) has reported a relationship between the rate of total oxidation and the charge-transfer energy for isostructural molybdates. The correlation shows an increase in the rate of combustion with decreasing charge-transfer energy. This is consistent with the idea that the activation of lattice oxygen via charge transfer may play an important role in the complete combustion process.

D. Role of Adsorbed Oxygen

It was mentioned earlier that Sancier et al. (43) concluded from the results of oxygen-18 experiments that the oxidation of propylene over a silica-supported bismuth molybdate catalyst involved adsorbed oxygen at low temperatures (320°C) and bulk oxygen at high temperatures (400°C). Akimoto et al. (61) have shown, also by oxygen-18 experiments, that an adsorbed oxygen species was incorporated into acrolein during the oxidation of propylene over cuprous oxide at 350°C. Although these results support the concept of adsorbed oxygen participation, they are not conclusive because the data were not collected nor treated quantitatively.

The only direct evidence for the participation of adsorbed oxygen in the selective oxidation of propylene under reaction conditions is the work of Cant and Hall (22, 23). They reported that propylene could be oxidized to acrolein over noble metals (Au, Rh, and Ru) supported on low-surface-area α-alumina or silica. These catalysts do not contain any usable lattice oxygen, therefore, only adsorbed or gas-phase oxygen is available as a reactant. The mechanism of propylene oxidation over these catalysts was determined by isotopic labeling techniques. It was shown to involve the interaction of an allylic intermediate with a molecular oxygen species to form a hydroperoxide species which subsequently decomposed into water and acrolein (see Section II,B). It should be noted, however, that these metal catalysts are not typical of the oxidation catalysts usually associated with acrolein formation, in that they operate at a much lower temperature (200°C vs. 350°–500°C and give a much poorer selectivity to acrolein (3–30% vs. 80–95%) than do the typical oxide catalysts.

The types of adsorbed oxygen species which can exist on metal oxides have been extensively studied by a number of workers (25, 26, 33, 60, 62–64). Kazanskii (60) investigated the reoxidation of oxides by ESR and suggested that it occurs by a sequence of steps:

$$O_2 + e^- \rightarrow O_2^-$$
$$O_2^- + e^- \rightarrow 2O^-$$
$$O^- + e^- \rightarrow O^{2-}$$

rather than by one elementary step. For example, if oxygen is adsorbed at $-196°C$ on a previously reduced vanadium oxide on silica gel catalyst, only O_2^- radicals are observed in the ESR spectrum. However, at 160°C the O_2^- is slowly converted to O^-. The rate of this process increases with increasing temperatures, and finally at 400°C all the adsorbed oxygen is completely converted into regular diamagnetic oxide ions.

Krylov (62) studied the adsorption of oxygen and propylene on vanadium oxide/MgO and molybdenum oxide/MgO catalysts by ESR and IR at 25°C. He observed the formation of O_2^- radicals and π-allyl complexes during the simultaneous adsorption of O_2 and C_3H_6. The data indicated that an electron transfer took place from the olefin to the oxygen through the transition metal ion forming the following complex:

$$\begin{array}{ccc} O_2^- & H^+ \cdots C_3H_5 \\ | & \vdots & | \\ Mg\!\!-\!\!\!&\!\!\!O\!\!\!&\!\!\!-Mo^{5+} \end{array}$$

This step is followed by the stabilization of O_2^- and the subsequent interaction between O_2^- and C_3H_5 yielding acrolein and a surface hydroxyl group. The mechanism for complete oxidation was proposed to involve a high concentration of nonstabilized, highly mobile surface oxygen species.

Kugler et al. (25, 26) directly observed the formation and reaction of a π-allyl complex and O_2^- at room temperature over zinc oxide by ESR and IR. The reaction produced primarily adsorbed acrolein. However, as indicated earlier, the products could not be desorbed without causing reactions which yielded carbon dioxide and water.

IV. Active Catalyst Systems

A. General Comments

From the data presented earlier, several general conclusions may be reached with regard to catalysts for the selective oxidation of propylene. For a catalyst to be active and selective for the partial oxidation of propylene, it must be capable of performing several functions: (1) provide appropriate adsorption sites for propylene to be transformed into an intermediate surface allylic species; (2) act as a source and sink of oxygen, i.e., it must provide active oxygen as a reactant and must adsorb or absorb gaseous oxygen; and (3) act as a source and sink of electrons, i.e., it must supply electrons to reduce gaseous oxygen and must absorb electrons at some stage before the hydrocarbon–oxygen complex can desorb as a neutral molecule. These properties can be influenced by catalyst structure which, in turn, is controlled to a large degree by the preparative method.

It is difficult to predict *a priori* which preparative method will produce the most active and selective catalyst or which preparative method will affect which, if any, of the previously mentioned properties. A great number of recipes have appeared in the patent literature, but any detailed description of the methods which yield the most active and selective catalyst, at least from a commercial viewpoint, remains proprietary. Of course, this makes it very difficult to make comparisons between "experimental" catalysts and "commercial" catalysts. Nevertheless, a number of general chemical variables have been identified as important in attempting to produce a specific catalyst. For example, for molybdate catalysts prepared by precipitation, these variables include the temperature of the precipitation, the concentration of the reagents, the aging of the precipitate, and the temperature of the calcination (65). For supported catalysts, the nature of the support also becomes an important variable in determining the final catalytic activity and selectivity.

While it is recognized that preparative methods play a vital role in determining catalytic activity and selectivity, few in-depth investigations, relating catalyst structure to catalytic activity and selectivity and comparing one system with another, have appeared in the literature. Consequently, the approach used in this section will be to examine the catalyst systems which have been associated with the historical development of the selective oxidation of propylene.

B. The Cuprous Oxide System

This system is more accurately defined as a multiphase ($Cu-Cu_2O-CuO$) system. Consequently, it has proven to be very difficult to characterize. The catalyst undergoes changes in its bulk chemical composition, activity, and selectivity during the oxidation of propylene. The final stationary-state composition is a function of the reaction mixture, the temperature, and the duration of the experiment.

Isaev et al. (66) observed via in situ X-ray diffraction that cupric oxide is reduced to a mixture of cuprous oxide and copper, while metallic copper is partially oxidized to cuprous oxide. However, for sufficiently prolonged operation with an unsupported powder catalyst at temperatures of 350°C or greater, the X-ray analysis revealed that only metallic copper was present regardless of the initial composition. Electron diffraction studies of thin-copper-film catalysts by the same authors suggest that a thin layer of cuprous oxide is actually the active phase. This result is supported by the later work of Wood et al. (67), who made simultaneous in situ measurements of the electrical conductivity and optical properties of a single crystal of cuprous oxide and the kinetics of acrolein formation at 350°C in a differential flow reactor. It was found that changes in the feed-gas composition caused great changes in the conductivity of the crystal suggesting that only a relatively thin surface layer participates in the reaction. Results of reflectance spectroscopy and kinetic measurements indicate that the active solid phase for acrolein formation is stoichiometric or copper-rich cuprous oxide. Oxygen-rich cuprous oxide or cupric oxide favors the complete combustion of propylene. This is in agreement with the work of Gorokhovatskii et al. (68) who reported that the selectivity of a copper oxide catalyst increased with the decrease in the degree of oxidation of the copper.

C. The Bismuth Molybdate System

As a catalyst for propylene oxidation, Bi_2O_3 itself has fairly low activity and yields primarily the products of complete oxidation. Pure molybdenum trioxide has an even lower activity, but is fairly selective. In combination, however, remarkable activity and selectivity for propylene oxidation is obtained. Although industrial catalysts contain silica and phosphate as well as Bi_2O_3 and MoO_3, many fundamental studies have employed catalysts containing only bismuth and molybdenum oxides in an attempt to determine the structure of the catalytically active phase. As a result of such studies, it is now known that bismuth molybdate catalysts display their superior properties only if the catalyst composition lies within the composition range of $Bi/Mo = \frac{2}{3}$ to $\frac{2}{1}$ (atomic ratio).

1. Structural Information

a. *The α Phase.* The crystal structure of the α phase of bismuth molybdate, $Bi_2Mo_3O_{12}$, has been investigated using X-ray diffraction techniques by several investigators. The early reports by Zambonini (69) indicated a tetragonal, defect scheelite structure, while the single-crystal study of Mekhtiev et al. (70) indicated a monoclinic structure (a = 7.89 Å, b = 11.70 Å, c = 12.24 Å, β = 116°20′) which was related to the scheelite-type structure. Erman et al. (71, 72) also reported a monoclinic structure (a = 7.85 Å, b = 11.70 Å, c = 12.25 Å, β = 116°20′) with space group $P2_1/c$. An X-ray diffraction study with a high-temperature camera showed that the crystal structure remained unchanged up to its melting point at 650°C.

The X-ray data of Aykan (35) was generally in good agreement with the data of Erman et al. (71) and Bleijenberg et al. (73) although the pattern contained numerous additional reflections. The Guinier pattern could not be indexed satisfactorily with the lattice parameters of Mekhtiev, however the pattern was completely indexed on the basis of computer-refined monoclinic lattice parameters of a = 7.719 Å, b = 11.516 Å, c = 11.985 Å, and β = 115°25′. On the basis of 4 $Bi_2O_3 \cdot 3MoO_3$ per unit cell, the theoretical density was calculated to be 6.197 g/cm³ in comparison to the observed density of 6.14 g/cm³ at 25°C.

More recent single-crystal studies (74–76) have confirmed the monoclinic structure and parameters determined by Aykan. The monoclinic symmetry of the $Bi_{2/3}\phi_{1/3}MoO_4$ superstructure, derived from the $CaWO_4$ (scheelite) structure, was determined to result from an ordered arrangement of vacancies (ϕ) in the Ca position of the scheelite structure. Additionally, each molybdenum ion was determined to be tetrahedrally surrounded by 2 oxygen neighbors at an average distance of 1.72 Å and 2 oxygens at 1.87 Å. Because of the significant contacts among neighboring tetrahedra, another oxygen ion occurs in the vicinity of the oxomolybdenum tetrahedron at a distance of 2.2 Å. If the coordination of molybdenum by oxygen were considered to be fivefold, the coordination polyhedron would be an irregular trigonal bipyramid with one of the apical oxygens at a distance of 2.2 Å. In the $Bi_{2/3}MoO_4$ structure, all MoO_4 tetrahedra occur in pairs to form Mo_2O_8 groups in which each of the two MoO_4 tetrahedra share a common edge. Not all these groups have exactly the same shape, however. One group possesses a center of symmetry, while the other does not. Although bismuth has 8 oxygen neighbors, the bismuth–oxygen contacts are not all equivalent: only four have normally accepted values for the bismuth–oxygen bonds, ranging from 2.09 to 2.35 Å; the other four range from 2.60 to 2.93 Å.

b. *The β Phase.* The β phase of bismuth molybdate, $Bi_2Mo_2O_9$, was first identified by Erman et al. (*71, 72, 77, 78*) and the single-crystal studies indicated an orthorhombic crystal structure with lattice parameters $a = 10.79$ Å, $b = 11.89$ Å, $c = 11.86$ Å, and space group $Pnmm$ or $Pnm2_1$. The results, using a high-temperature powder camera, indicated that the metastable modifications, $β'$ and $β''$, transformed in the temperature range of 300° to 500°C to the β phase, which remained unaltered until its melting point at 662°C. Chen (*74*) was also able to index the X-ray diffraction pattern of $Bi_2Mo_2O_9$ on the basis of an orthorhombic structure with lattice parameters $a = 11.95$ Å, $b = 10.81$ Å, and $c = 11.89$ Å. In agreement with 8 $Bi_2Mo_2O_9$ per unit cell, the density was observed to be 6.52 g/cm³.

In contrast, Van den Elzen and Rieck (*79*) reported a monoclinic structure for $Bi_2Mo_2O_9$ with space group $P2_1/c$ and lattice parameters $a = 11.946$ Å, $b = 10.795$ Å, $c = 11.876$ Å, and $β = 90.15°$. The inability of several researchers to prepare single crystals and the existence of metastable modifications of the β phase of bismuth molybdate are apparently the main determents to additional clarity concerning its structure.

c. *The γ Phase.* An earlier report by Zemann (*80*) showed that the crystal structure of natural koechlinite, Bi_2MoO_6, was orthorhombic, $a = 5.50$ Å, $b = 16.24$ Å, $c = 5.49$ Å, and space group $Cmca$ (D_{2h}^{18}). Aykan (*35*) indexed the X-ray data from the γ phase of bismuth molybdate with parameters which were in excellent agreement with those reported by Zemann. The observed density, 8.26 g/cm³, concurred with the density calculated on the basis of 4 Bi_2MoO_6 per unit cell and the melting point of this phase was determined to be 933°C.

Recently, Van den Elzen and Rieck (*81*) investigated the crystal structure of the γ phase of bismuth molybdate. The parameters, $a = 5.487$ Å, $b = 16.226$ Å, and $c = 5.506$ Å, determined from single-crystal data, were in agreement with those reported by Zemann; however, the space group was determined to be $Pca2_1$. The structure was described as alternate layers of $(BiO^+)_{2n}$ and $(MoO_4^{2-})_n$ perpendicular to the y direction, an arrangement which is strongly related to the $LiBi_3O_4Cl_2$ structure and its isotypes (*82*). The $(MoO_4^{2-})_n$ layer consists of deformed oxomolybdenum octahedra, sharing four corners with neighboring oxomolybdenum octahedra in the same layer perpendicular to the y direction. The molybdenum–oxygen bond distances in such a layer can be divided in two classes, the first with bond lengths of approximately 1.75 Å and the second with bond lengths of approximately 2.24 Å. The other two apex ions of the oxomolybdenum octahedra point toward a $(BiO^+)_{2n}$ layer, one below and one above. These molybdenum–oxygen bonds have intermediate lengths, 1.85 Å and 1.93 Å respectively. The bismuth is bonded to six oxygens, the four bismuth–oxygen distances in the BiO layer range from 2.15 to 2.50 Å. Each bismuth shares two more oxygens at approximate distances of 2.33 Å and 2.66 Å with two neighboring oxomolybdenum octahedra.

A structural modification was observed when the γ phase of bismuth molybdate was subjected to temperatures in the range of 750°C (83). The transition from the low- to the high-temperature modification resulted in the stabilization of a tetragonal structure, $a = 3.95$ Å and $c = 17.21$ Å, which contained oxomolybdenum tetrahedra having no anions in common in the MoO_4^{2-} layer. The structural transformations of the γ phase of bismuth molybdate during heat treatment were also probed by Erman et al. (71, 72, 77, 84). The work, involving a high-temperature vacuum camera, showed that the b parameter remained unchanged on heating whereas the a and c parameters increased considerably. The structural modifications which accompany heat treatment were schematically represented as follows:

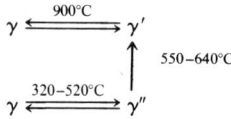

The γ phase of bismuth molybdate underwent a reversible transformation to the metastable tetragonal γ″ modification. This metastable modification was observed in the temperature range of 520° to 550°C and underwent an irreversible transformation to the γ′ modification which readily formed at 700°C. The results indicated that the γ′ modification corresponds to that reported by Blasse (83). However, refinement of the crystal data utilizing a single crystal revealed that this γ′ modification was orthorhombic with lattice parameters $a = 15.99$ Å, $b = 15.92$ Å, and $c = 17.43$ Å. An additional observation was the reversible transformation of the γ′ modification to γ at 900°C.

2. Catalyst Reduction

The establishment of the structures and thermal transformations of the catalytically active phases of bismuth molybdate resulted in research directed toward investigating the stability of the structure under reducing conditions. Fattore et al. (38) investigated an unsupported bismuth molybdate catalyst with composition $Bi_2O_3 \cdot 2.66MoO_3$ during propylene oxidation in the absence of gaseous oxygen. Examination of the bismuth molybdate catalyst by X-ray diffraction showed that it was initially composed of $Bi_2Mo_3O_{12}$ together with a small amount of Bi_2MoO_6 and that phase modifications occurred as a function of time of reduction. The $Bi_2Mo_3O_{12}$ phase concentration decreased with time while the concentration of a new phase and MoO_2 increased with time of reduction. The chemical composition of this new phase was approximately Bi_4MoO_9 and appeared to have a pseudotetragonal monoclinic structure with parameters $a = 5.58$ Å, $b = 5.58$, Å, $c = 11.62$ Å, and $\gamma = 89.00$. This compound was an intermediate formed during the reduction of bismuth molybdate. Metallic bismuth was observed only after extended reduction which resulted in total structure collapse, forming molybdenum dioxide and metallic bismuth.

Silica supported bismuth molybdate catalysts were examined by Dalin et al. (85) during extended steady-state ammoxidation of propylene. X-Ray diffraction measurements indicated that a significant alteration in the phase composition of supported catalysts occurred during operation. The α and β phases of bismuth molybdate, which were originally present, transformed into the γ phase of bismuth molybdate as described by the equation

$$Bi_2O_3 \cdot m MoO_3 \rightarrow Bi_2O_3 \cdot MoO_3 + (m - 1)MoO_3$$

The molybdenum trioxide initially resulting from this interaction was in an amorphous form. However, an increase in the operating temperature favored molybdenum trioxide crystallization. In a similar manner, Aykan (35) studied the reduction of a silica supported bismuth phosphomolybdate catalyst, consisting of the α phase of bismuth molybdate, bismuth phosphate, and silica, during the ammoxidation of propylene in the absence of gaseous oxygen. X-Ray diffraction measurements revealed that the overall reduction of the catalytically active α phase of bismuth molybdate to metallic bismuth and molybdenum dioxide proceeded according to the following two consecutive reactions, where the γ phase of bismuth molybdate is an intermediate.

$$Bi_2Mo_3O_{12} \xrightarrow{C_3H_6, NH_3} Bi_2MoO_6 + 2MoO_2 + O_2$$

$$Bi_2MoO_6 \xrightarrow{C_3H_6, NH_3} 2Bi + MoO_2 + 2O_2$$

An alumina supported bismuth molybdate catalyst with a bismuth to molybdenum atomic ratio equal to one was examined by high-temperature X-ray diffraction techniques during propylene oxidation (86). According to the results, abrupt compositional changes occurred between 500° and 600°C and between 600° and 660°C. An unidentifiable structure was detected, which had been suggested by Fattore et al. (38) to correspond to the new phase observed during their experiments.

The combination of these experimental findings indicate that active bismuth molybdate catalysts undergo phase transformations when exposed to reducing conditions similar to the conditions of catalysis. The phase transformations are highly dependent upon both temperature and the severity of the reducing atmosphere. However, the occurrence of solid state reactions in the catalysts suggests that the bulk structure of the catalysts plays an important role in catalytic reaction.

D. THE URANIUM ANTIMONATE SYSTEM

In 1966, a catalyst based on a complex uranium antimonate system was developed and brought into commercial use (4, 87). Several physical methods of analysis were used in an attempt to clarify relationships between the structure and properties of the uranium antimonate system and its catalytic properties (20, 88, 89). X-Ray diffraction and infrared analysis demonstrated that the optimum selectivity for acrylonitrile formation coincided with the maximum concentration of the USb_3O_{10} compound. The crystal structure of USb_3O_{10} was shown to consist of layers of heavy atoms and oxygen ions alternated by layers of oxygen ions. Measurements by ESCA indicated that the surface layers contained U^{5+} and Sb^{5+} with intensities corresponding to the USb_3O_{10} formula.

Based on experimental evidence, Grasselli (20) proposed a possible mechanism for propylene oxidation. It was envisioned that propylene chemisorbs on an active site, postulated as an oxygen-ion vacancy on Sb^{5+}, and loses an α-hydrogen (proton or radical) to an uranyl oxygen. The addition of a proton to an uranyl oxygen results in a U^{6+} atom and the coordination of the π-allyl to antimony results in a Sb^{4+} atom. The two atoms then disproportionate to produce two metal atoms with a $+5$ oxidation state. The π-allylic moiety, planar and parallel to the surface of the catalyst, finds equivalent lattice oxygens at opposite corners of the distorted octahedron about the antimony atom to which it is attached. The second hydrogen abstraction from either of the equivalent terminal methylenes of the allyl complex occurs via a bridging, loosely held oxygen and a C_3H_4 moiety results. Loss of water from the new complex creates an anion vacancy that can be readily filled with an oxygen from the underlying layers by rapid lattice oxygen transfer. The oxygen inserts into the C_3H_4 complex and after electronic rearrangement forms chemisorbed acrolein which desorbs and the site is regenerated. The oxygens involved in the reaction were suggested to reach the active sites primarily by means of lattice oxygen transfer while gaseous oxygen incorporates into the crystal lattice through other regenerated sites and transfers to the active site.

The pentavalent antimony ion in an octahedral environment was postulated as necessary for the partial oxidation of propylene, and the uranium ion was indicated to be important in stabilizing the Sb^{5+} structurally and in aiding its regeneration with gaseous oxygen during the course of catalytic operation.

E. Scheelite-Structured Systems

Improved catalytic performance, selectivity and resistance to fusion, over bismuth molybdate catalysts was reported by McClellan (*90*) for catalysts obtained by chemically combining bismuth, molybdenum, phosphorus, and silica. After calcination at 450°C, the bismuth phosphomolybdate-on-silica catalyst showed an X-ray pattern of mainly crystalline $Bi_2(MoO_4)_3$ which subsequently was converted to a new, "substantially amorphous," phase after calcination at 800°C. *Substantially morphous* meant that the X-ray diffraction lines were broad diffuse bands of low intensity. The pattern of lines for this novel phase indicated a scheelite structure. A special interaction of silica with bismuth molybdate was also suggested by Callahan *et al.* (*91*).

Cesari *et al.* (*75, 92*) obtained new bismuth molybdovanadate compounds corresponding to the formula $Bi_{1-x/3}Mo_xV_{1-x}O_4$, which form a solid solution with composition limits of bismuth vanadate, ($x = 0$), and the α phase of bismuth molybdate, ($x = 1$). When an increasing proportion of vanadium atoms were replaced by molybdenum in $BiVO_4$, a tetragonal scheelite-type structure was stabilized and persisted over almost the entire range of x from about 0.1 to values close to 1. The lattice parameters changed regularly with x, and it was concluded that the replacement of vanadium by the molybdenum atoms resulted in random vacancies of the bismuth cations in order to balance the difference in negative charges between the tetrahedral anion groups $(VO_4)^{3-}$ and $(MoO_4)^{2-}$.

Point defects in the form of cation vacancies (ϕ) were introduced by Aykan *et al.* (*93–95*) into molybdates, tungstates, and vanadates with scheelite-type crystal structures. The authors studied the catalytic properties of more than 30 scheelite-structure phases represented by the formula $A_{1-x}\phi_xMO_4$ (M = molybdenum, tungsten, and/or vanadium and A may include Li, Na, K, Ag, Ca, Sr, Ba, Cd, Pb, Bi, and/or a rare earth element in quantities appropriate to achieve charge balance for the normal oxidation states). It was found that the defects can be introduced according to the formulas $A^+_{0.5-3x}A^{3+}_{0.5+x}\phi_{2x}MoO_4$ and $A^{2+}_{1-3x}A^{3+}_{2x}\phi_x MoO_4$, where ϕ represents a vacant lattice site which would normally be occupied by an A cation.

In the system $A^+_{0.5-3x}A^{3+}_{0.5+x}\phi_{2x}MO_4$, the authors (96) reported that the dimensions of the unit cells of these scheelite phases vary significantly with changes in either type or number of cations in the A sites and that the number of A cation sites which can be vacant is less than approximately 15% before additional phases appear. The results for the oxidation of propylene over the composition $A^+_{0.5-3x}A^{3+}_{0.5+x}\phi_{2x}MoO_4$ (A^+ = Li, Na, or Ag) showed that when $x = 0$ the activity is very low. The activity dramatically increased when 4% of the A cation sites were made vacant and continued to increase with the defect level until a two-phase mixture resulted in decreased activity.

In scheelite-type systems containing divalent ions, the tolerance for vacancies was more limited than in systems containing substantial amounts of monovalent ions. A vacancy limit of about 7.5% appeared to prevail at calcination temperatures between 550° and 800°C (97). The results for the oxidation of propylene over $A^{2+}_{1-3x}Bi^{3+}_{2x}\phi_x MoO_4$, ($A^{2+}$ = Pb, Cd, or Ca), compositions showed that when $x = 0$, the activity was very low but increased rapidly with increasing defect concentration. When bismuth was absent, the activity and selectivity were very poor on comparison with bismuth containing defect scheelites.

Additionally, Aykan et al. (98) reported the results for scheelite-type systems in which A sites are occupied by divalent elements and bismuth, and M sites contain vanadium and molybdenum. The tolerance for vacancies in this system was reported to be 15% of the A cation sites. Good yields of acrolein were obained when bismuth and defects were present in the scheelite-structured catalysts.

Although it was not suggested that defects are required for selective oxidation over other catalysts, the results indicated that defects and bismuth must be present for high activity and selectivity over scheelite-type catalysts. The authors concluded that the defects which were introduced into the bulk of these phases must manifest themselves in some manner at the surface. The question of how the introduction of defects into these phases affected their catalytic properties was not resolved. However, the active site for catalysis was suggested as a cation vacancy which could abstract a proton from an olefin to form the well-established allyl intermediate and should offer considerable stabilization to a surface hydroxyl group.

Alkhazov et al. (99–101) compared the physicochemical properties and catalytic activities of Group II element molybdates. The calcium, strontium, cadmium, and barium molybdates have a scheelite structure while the zinc and magnesium molybdates have a monoclinic structure. The vibrational frequency of the ν_3 band of the molybdate ion in the infrared spectra was found to decrease linearly with an increase in the ionization potential of the Group II element, indicating a weakening of the MoO bond. The activity for complete oxidation of propylene decreased linearly with an increased Mo—O bond strength for the scheelite-type molybdates. The monoclinic molybdates, however, deviated from the correlation line determined for the scheelite-structured molybdates. The authors concluded that the catalytic activity for complete combustion was dependent both on the stability of the oxygen bond in the catalyst and on the crystallographic structure of the catalyst.

The charge transfer bands in the UV spectra decreased in energy linearly with increased ionization potential of the Group II metal of the molybdate (12, 100). Conductivity measurements showed a regular increase in conductance with lower charge-transfer energy for the scheelite-structured molybdates. The authors suggested that the scheelite structure was responsible for the formation of orbitals and quasi-bands that allow the transport of electrons and holes through larger distances than normally expected. Alkhazov et al. (101) also showed that the catalytic activity for propylene oxidation increases with increased electrical conductivity of the transition metal molybdate.

F. THE BISMUTH IRON MOLYBDATE SYSTEM

Voge et al. (102) patented a series of catalysts consisting of bismuth, iron, oxygen, and an element selected from the group of phosphorus, molybdenum, or tungsten. Although the catalysts were not characterized, the compositions $Bi_4Fe_2Mo_4O_{21}$, $Bi_4Fe_2Mo_6O_{27}$, and $Bi_4Fe_2Mo_9O_{36}$ exhibited very pronounced activity and selectivity at low reaction temperatures for the oxidative dehydrogenation of olefins. One comparison, the catalytic behavior of ferric molybdate and bismuth ferrite, was greatly inferior. Enhanced performance was also observed by Morita et al. (103) upon addition of iron to bismuth molybdate catalysts.

Daniel and Keulks (104) investigated Bi–Fe–Mo oxide catalysts prepared by reacting the α-bismuth molybdate with ferric hydroxide. Comparison of these catalysts with bismuth molybdate and ferric oxide indicated that mechanistically the Bi–Fe–Mo oxide catalysts resembled bismuth molybdate in their ability to form an allyl species. Under the same reaction conditions, the composition with Bi–Fe–Mo atomic ratio equal to 6:9:10 exhibited higher conversion than and the same selectivity as the bismuth molybdate catalysts. In contrast to bismuth molybdate, the Bi–Fe–Mo oxide catalysts were found to maintain their activity and selectivity for acrolein under highly reducing conditions. Furthermore, the authors suggested that the combination of Bi–Fe–Mo–O resulted in the formation of a distinct chemical compound which possibly was responsible for the observed catalytic behavior.

Annenkova et al. (105) studied both the physicochemical and catalytic properties of the Bi–Fe–Mo oxide system. The X-ray diffraction, infrared spectroscopic, and thermographic measurements indicated that the catalysts were heterogeneous mixtures consisting principally of ferric molybdate, α-bismuth molybdate, and minor amounts of bismuth ferrite and molybdenum trioxide. The Bi–Fe–Mo oxide catalysts were more active in the oxidation of butene to butadiene and carbon dioxide than the bismuth molybdate catalysts. The addition of ferric oxide to bismuth molybdate was also found to increase the electrical conductivity of the catalyst.

Batist et al. (106) examined the ability of $BiPO_4$, Fe_2O_3, and Cr_2O_3 to promote the α phase of bismuth molybdate in the oxidation of butene to butadiene. These compounds when present in quantities equivalent to 10% (mole ratio) showed a pronounced promoting effect and in the case of $BiPO_4$ and Fe_2O_3 this effect was suggested to be caused by the formation of the γ-bismuth molybdate within the system. The properties of the Bi–Fe–Mo oxide system in particular were studied over the compositional range from α-bismuth molybdate to ferric molybdate. The X-ray patterns within this range were complex but were resolved by assuming the presence of $Bi_2(MoO_4)_3$, $Fe_2(MoO_4)_3$, Bi_2MoO_6, and a new compound, which they described as "compound X," existing over the entire range of Bi/Fe atomic ratios 8:2 to 1:9. The γ-bismuth molybdate was found only in the sample with Bi/Fe atomic ratio equal to 9:1, while the α-bismuth molybdate decreased and the ferric molybdate increased with iron concentration. Since the concentrations of both α-bismuth molybdate and ferric molybdate were less than expected in a simple mixture, the new compound was thought to contain both bismuth and iron. This "compound X" had never been observed in their earlier studies with bismuth molybdate samples, but exhibited an X-ray pattern similar to that reported by McClellan (90) in the patent literature for a bismuth phosphomolybdate-on-silica catalyst. Activity measurements as a function of the Bi/Fe atomic ratio showed that all samples were more active than expected for a mixture of the α-bismuth molybdate and ferric molybdate. Although the activity of the new compound fluctuated to a certain extent on going from the Bi-rich to the Fe-rich side, the activity of the sample Bi/Fe = 5:5, with a minimum concentration of the new compound, exhibited low activity. Moreover, the promoted catalysts did not lose activity during the catalytic reaction as did the α-bismuth molybdate. The adsorption studies showed that the promoted catalysts possessed a substantially increased number of adsorption sites and an increased tendency, surpassing the γ-bismuth molybdate, to dissociate O_2. The authors suggested that the promoting action of $BiPO_4$, Fe_2O_3, and Cr_2O_3 was caused by their tendency to increase the γ-bismuth molybdate concentration in α-bismuth molybdate. They considered, however, that other factors were involved since the new compound exceeded the α-bismuth molybdate in activity to a considerable degree while maintaining high selectivity.

Sleight and Jeitschko (*107*) studied the Bi–Fe–Mo oxide system by X-ray diffraction. They reported the formation of a ternary compound, $Bi_3(FeO_4)(MoO_4)_2$. The X-ray pattern of this compound was similar to that reported for "compound X" by Batist (*106*).

LoJacono *et al.* (*108*) also utilized X-ray diffraction methods to study the structural and phase transformations which occurred in the Bi–Fe–Mo oxide system. They detected two ternary compounds containing bismuth, molybdenum, and iron. One of the compounds formed when the atomic ratio Bi/Fe/Mo = 1:1:1; the other formed when the atomic ratio Bi/Fe/Mo = 3:1:2. The X-ray data indicated a close structural relationship of the bismuth iron molybdate compounds with the scheelite structure of α-phase bismuth molybdate. Moreover, their structures were similar to "compound X." The structure of the Bi/Fe/Mo = 3:1:2 compound was identical to the compound reported by Sleight and Jeitschko (*107*). The authors proposed that the structures of both of the compounds could be viewed as resulting from the substitution of Fe^{3+} in the α-phase lattice. In the Bi/Fe/Mo = 1:1:1 compound, 1 Mo^{6+} ion is replaced by 2 Fe^{3+} ions: one Fe^{3+} ion occupies a Mo^{6+} site; the other Fe^{3+} ion occupies one of the vacant bismuth sites. In the Bi/Fe/Mo = 3:1:2 compound, the Fe^{3+} ion replaces one Mo^{6+} ion while the additional Bi^{3+} ion occupies one of the vacant bismuth sites.

G. The Multicomponent System

As mentioned earlier, the multicomponent oxide catalysts currently commercialized contain bismuth, iron, and molybdenum, in addition to several other cations. Although few reports concerning multicomponent catalysts have appeared in the literature, there is agreement that iron affects several aspects of the catalyst system. Measurements on multicomponent catalysts by Wolfs *et al.* (*109–111*) showed that Fe^{3+} was partially reduced to Fe^{2+} after the catalytic reaction, indicating that Fe^{3+} ions are involved in the reaction mechanism. The observed Fe^{3+}/Fe^{2+} redox couple was associated with the increased activity of the catalyst.

Although Wolfs indicated that the catalyst particles are covered by a skin of bismuth molybdate, Batist (*112*) recently found bismuth, molybdenum, and iron in the surface layers of multicomponent catalysts. Additional data are needed to determine if multicomponent catalysts gain their activity as a result of the formation of compounds such as bismuth iron molybdate, or by surface enhancement of an active component such as γ-phase bismuth molybdate, or by creation of low-energy electronic transitions. Of course, due to their complexity, all of these factors may be important.

V. Nature of the Active Sites

A. The Dual-Site Concept

In the previous sections of this review, it has been shown that most effective catalysts for the selective oxidation of propylene contain at least two types of metal oxides—an amphoteric or low-valence oxide, such as bismuth, tin, iron, or cobalt, and an oxide of a high valence metal, such as molybdenum or antimony. Moreover, it has been suggested several times that each of these metal oxide components may give rise to an active site; for example, propylene may adsorb on an active site associated with one of the metal oxide components, and oxygen may adsorb on an active site associated with another metal oxide component. This problem has been studied using spectroscopic, adsorption, and kinetic techniques. It now seems appropriate to consider some of these studies in detail, attempting to relate the solid structure of the catalyst to the active sites wherever possible.

B. Active Sites for Propylene Adsorption

Unfortunately, general agreement concerning the adsorption site for propylene has not been reached. One school of thought has developed which postulates the low-valence cation as a site of propylene activation. Another school of thought proposes the high-valence cation as the site of propylene activation.

Haber (*59, 113*) has reported data supporting the concept that the low-valence cation is responsible for propylene activation. He studied the kinetics of catalyst reduction for α-, β-, and γ-bismuth molybdates using both hydrogen and propylene as reductants. Because the coordination of molybdenum varies among the three phases (tetrahedral in α phase, tetrahedral and octahedral in β phase, and octahedral in γ phase), the energy of activation for reduction should also vary if the molybdenum and associated oxygens are directly involved. The results of hydrogen reduction follow this pattern, and the value of E_A for reduction decreases in the expected order,

$$Bi_2(MoO_4)_3 > Bi_2Mo_2O_9 > Bi_2MoO_6$$

However, the reduction with propylene shows no correlation with molybdenum coordination. This suggests that the bismuth centers are responsible for the adsorption of propylene.

Further evidence supporting the bismuth center as a site of propylene activation comes from the analysis of the rates of formation and product distribution of propylene oxidation over bismuth oxide, bismuth molybdate, and molybdenum oxide. Bismuth molybdate is highly active and selective for the conversion of propylene to acrolein. However, the interaction of propylene with its component oxides yields very different results. Haber and Grzybowska (*113*), Swift *et al.* (*114*), and Solymosi and Bozso (*115*) showed that in the absence of oxygen, propylene is converted to 1,5-hexadiene over bismuth oxide with good selectivity and at a high rate, whereas molybdenum oxide is known to be a fairly selective but a nonactive catalyst for acrolein formation. The formation of 1,5-hexadiene over bismuth oxide can be explained if the adsorption of propylene on a bismuth site yields a π-allylic species. Two of these allylic intermediates can then combine to give 1,5-hexadiene.

These observations suggest a reaction scheme for bismuth molybdate catalysts where the allylic species is formed initially at a bismuth center and then reacts further at a molybdenum site to produce acrolein. Thus, once the allylic complex is formed, the MoO polyhedra are highly active and selective for acrolein formation. This hypothesis was tested by investigating the oxidation of bromoallyl (C_3H_5Br) over molybdenum oxide (*116*). Since the C—Br bond in bromoallyl is much weaker than the C—H bond in propylene, the ease of formation of the allylic species should be significantly enhanced with bromoallyl compared with propylene. If the initial propylene activation occurs on bismuth, then the reaction of bromoallyl over molybdenum oxide should approach the activity and selectivity of propylene over bismuth molybdate. This was the observed result, and the authors concluded that the bismuth site was responsible for the formation of the allylic intermediate.

A somewhat different picture emerges from the adsorption studies of Matsuura and Schuit (*117*). They have attempted to elucidate the active adsorption sites on γ-bismuth molybdate by measuring the adsorption equilibria of butene, propylene, oxygen, water, butadiene, acrolein, and ammonia. They studied both fully oxidized and partially reduced bismuth molybdate catalysts. Their work led to the assignment of two different types of adsorption sites:

1. *A sites* that strongly adsorbed butadiene in an activated process according to a single-site Langmuir isotherm. The site was suggested as a surface oxygen anion (O_A) which, if removed by previous reduction, adsorbed water on the residual vacancy. Molecular oxygen adsorbed even at room temperature on pairs of A vacancies in a fast irreversible process. The presence of the A site appeared to be connected with the presence of bismuth, however its properties varied with the amount of molybdenum present.

2. *B sites* that weakly adsorbed butene and also butadiene according to a dual site Langmuir isotherm. The B sites were suggested to be a cluster of two sites and to consist also of surface oxygen anions (O_B). The concentration of the B site, which was suggested to be connected with molybdenum, decreased on comparison from Bi_2MoO_6 to $Bi_2Mo_2O_9$ to $Bi_2Mo_3O_{12}$ to MoO_3, but its properties showed little change. Catalytic oxidation was suggested to occur on a cluster of O_A and O_B, named the *R site*, containing five oxygen anions of which at least one is an O_A and the others O_B.

Matsuura and Schuit (*118*) also studied adsorption as a function of sintering time and temperature. The data indicated that rearrangement of the bulk structure is accompanied by a change in the surface structure. Samples that were reduced at 350°C lost only O_A sites while B sites were also eliminated at 400°C. The phenomena during bulk reduction and reoxidation were explained in the following manner. Reduction at 350°C removes the oxygens from the active rows in the Bi_2O_2 layers and reoxidation starts at the A vacancies to fill these rows again by diffusion into the interior. At temperatures above 400°C, however, the rows are filled from the layers above and below the Bi_2O_2 layers. Thus, the reoxidation above 400°C follows another diffusion path and even more importantly a different port of entrance. From the data, the authors proposed that the rate-determining reaction leads to the formation of an allylic species bonded to a vacancy on the Bi_2O_2 edge (V_{Bi}). The vacancy also has two O_B, which belong to two different B sites, as nearest neighbors. An allylic intermediate would then be in a position with the terminal carbon atoms symmetrically oriented with respect to V_{Bi}. The reaction of butene to butadiene was visualized as follows: Adsorption of the olefin on B-site-activated complex-adsorbed state in which a proton or a hydrogen atom is on an oxygen belonging to this B site but the allyl is on V_{Bi} dissociation of the second H atom to an oxygen belonging to a second B-site desorption of product.

Recently, Batist et al. (119) measured the adsorption of acrolein and propylene on the γ phase of bismuth molybdate in order to ascertain whether the sites proposed for butadiene formation from butene were also pertinent for the formation of acrolein from propylene. The results showed that acrolein is adsorbed on the γ phase of bismuth molybdate according to two different types of processes apparently related to those of butadiene. There was a fast, weak adsorption similar to the weak butadiene adsorption and therefore connected with the B sites. The second type was a slow, activated process comparable to the strong adsorption of butadiene and therefore associated with the A sites. The weak adsorption also occurred on MoO_3, while an adsorption with similar characteristics to the strong adsorption was observed on Bi_2O_3. These results supported the assumption that B sites occur on edges of the MoO_2 layers and the A sites on edges of the Bi_2O_2 layers.

Another communication has presented a comparison of the adsorption properties of the α, β, and γ phases of bismuth molybdate (120). The β phase exhibited two types of adsorption: a slow, activated type for butadiene and a fast, weak type for butadiene and butene. The two types of adsorption were similar, but not completely identical, to the A and B sites on the γ phase of bismuth molybdate. The slow and strong adsorption on the β phase was stronger than on the γ phase and the weak adsorptions were single site instead of dual site.

Weak- and single-type adsorptions of butene and butadiene were reported for the α phase of bismuth molybdate. The total amount of adsorption was low and the strong adsorption of butadiene was not observed. A previous comparison of the adsorption properties of the α phase and the γ phase also revealed that for the α phase, the number of B sites undergoes a drastic decrease without changing their specific properties (106). On the other hand, the concentration of A sites on the α phase was observed to remain essentially the same but their properties were changed substantially from the γ phase. In particular, the A sites on the α phase lost their property to adsorb butadiene.

C. Other Experimental Evidence for Active Sites

1. Infrared Spectroscopy

As indicated previously, the nature of the oxomolybdenum species of the bismuth molybdate catalysts has been determined mainly by the efforts of Trifiro et al. (54–57, 121, 122).

Initial results concerning the characterization of the bismuth molybdates by this technique showed an important contrast between the three active phases. The α phase of bismuth molybdate was shown to consist of edge-sharing oxomolybdenum tetrahedral groups while the γ phase was shown to consist of corner-sharing oxomolybdenum octahedra (*121, 123*). The β phase was characterized by infinite chains of oxomolybdenum octahedra sharing corners within the chain and by oxomolybdenum tetrahedra bridging adjacent oxomolybdenum groups.

Mitchell and Trifiro (*122*), utilizing IR and UV reflectance spectroscopy, showed the presence of both octahedral and tetrahedral oxomolybdenum species on the surface of supported and unsupported bismuth molybdate catalysts. A MoO_6 species with three terminal oxygens and three bridged oxygens was shown to have maximum concentration in the most active catalyst. This species, formed by corner-sharing oxomolybdenum octahedra, was suggested as the catalytically active species. The authors related the spectroscopic results to the adsorption data of Schuit. The B sites, on which the olefin adsorbed, were identified with the terminal oxygen since multiply bonded, terminal oxygens are more electrophilic and more strongly bound to molybdenum than bridging oxygen. The A sites were identified with bridging oxygen involving two molybdenum or a molybdenum and a bismuth. The authors concluded that the bifunctional nature of the bismuth molybdate catalysts is due to the presence of two types of molybdenum oxygen bonds in the particular structure rather than to the presence of two metal oxide components. Also consistent with the preceding conclusions are the results of Trifiro *et al.* (*57*) and Akimoto and Echigoya (*58*) which showed a definite interaction between adsorbed propylene or acrolein and oxygen which is multiply bonded to molybdenum.

2. *Electron Spin Resonance*

Because of its ability to detect paramagnetic species, several workers have used the ESR technique to determine the oxidation state of molybdenum in molybdate catalysts.

Sancier *et al.* (*124*) investigated the oxidation of propylene over silica-supported bismuth molybdate and silica-supported molybdenum trioxide by measurements of the ESR spectra of the catalysts during propylene oxidation at 325° and 390°C. The intensity of the Mo^{5+} signal was found to be inversely proportional to the total propylene conversion, viz., the higher the Mo^{5+} signal, the lower the propylene conversion. As the O_2/C_3H_6 ratio of the reactant gases was increased, the Mo^{5+} signal intensity decreased and the conversion of propylene increased. Moreover, the Mo^{6+}/Mo^{5+} ratio varied linearly with the O_2/C_3H_6 ratio of the reactant gases. Therefore, the composition of reactant gases governed the distribution of Mo^{5+} and Mo^{6+} ions on the surface. In agreement with an earlier report by Peacock et al. (125), the Mo^{5+} signal intensity was considerably greater for the inactive molybdenum trioxide catalyst at 325°C than for the active bismuth molybdate catalyst. Furthermore, the Mo^{5+} signal intensity increased with increasing temperature for the molybdenum trioxide catalyst whereas for the bismuth molybdate catalyst the Mo^{5+} signal intensity decreased with increasing temperature.

These results show that Mo^{6+} sites, and/or the oxygen species associated with the Mo^{6+} site, are essential for propylene oxidation and that the conversion rate is high when the relative Mo^{6+} surface density is high. Besides providing positive evidence for the reduction of Mo^{6+} to Mo^{5+} during the oxidation of propylene, these results also reveal that the Mo^{5+} species can be oxidized back to Mo^{6+} ions via two routes. One route involves molecular oxygen and the other involves bismuth. Of these two possibilities, the latter is more efficient and provides greater catalyst stability. Burlamacchi et al. (126) demonstrated that the oxidation of Mo^{5+} to Mo^{6+} via a route involving a partner cation with redox properties such as bismuth or iron is a necessary property for good catalytic activity.

Likewise, Novakova et al. (127) suggested that the Fe^{3+} in ferric molybdate catalysts has the role of maintaining molybdenum in the higher valence state (Mo^{6+}). During an experiment on the oxidation of methanol on molybdenum trioxide in the absence of gaseous oxygen, the reaction was selective to formaldehyde in the initial stages, but as the reaction progressed, complete oxidation occurred. The explanation was that Mo^{6+} is selective to partially oxidized products while Mo^{4+} is responsible for complete oxidation. By analogy, the selectivity to acrolein was observed to decrease as the bismuth molybdate catalyst became highly reduced (125). The decrease in selectivity was explained as resulting from strong interactions of the allyl species with molybdenum ions in lower oxidation states (Mo^{4+}) which would result in a weakening of carbon–carbon bonds.

The ESR technique also yielded information concerning the role of the bulk structure during propylene oxidation (128, 129). It was possible to reduce silica-supported bismuth molybdate and silica-supported molybdenum trioxide by simple evacuation and by exposure to propylene. The removal of oxygen from the surface resulted in the formation of pentacoordinated Mo^{5+} ions with a tetragonal pyramidal arrangement (129). In addition to the surface reduction, the authors reported bulk reduction, which was governed by the state of the surface and dependent upon bulk to surface diffusion. The presence of oxygen vacancies at the surface was suggested to induce a rapid diffusion of O^{2-} ions from the subsurface layers. The results further led to the suggestion that after a certain concentration of lattice defects form, corner sharing oxomolybdenum octahedra undergo a partial switch towards edge-sharing octahedra. The authors proposed that like the external reduced sites, even the internal reduced sites have a role in the catalytic activity, possibly in the mechanism of oxygen transfer from the bulk solid to the surface.

Matsuura (130) measured the ESR spectra of bismuth molybdate catalysts both before and after reduction of butene. A board signal of high intensity was observed for the γ phase of bismuth molybdate. The authors proposed that in the layered Bi_2MoO_6 structure, a (MoO_4^{2-}) layer shifts with respect to the nearest layers and causes the formation of Mo–Bi–Mo sites. The results supported the earlier proposed reaction site model, based on adsorption measurements, which consisted of an A site (Bi) and two B sites (Mo).

3. Electrical Conductivity

The electrical conductivities of silica-supported bismuth molybdate and silica-supported molybdenum trioxide were examined by Peacock et al. (131) during reduction and reoxidation. On exposure to propylene, 1-butene, or hydrogen, pressed disks of the catalysts showed rapid and reproducible increases of electrical conductivity at temperatures between 400° and 550°C. The conductivities decreased to their original values on reoxidation by gaseous oxygen. In the presence of an excess of a 1:1 mixture of propylene and oxygen, bismuth molybdate showed a nearly constant conductivity which varied only slightly as the proportions of propylene and oxygen were changed from 9:1 to 1:9. A notable feature of the results was the reproducibility of the conductance measurements. This was shown by concordant values after many cycles of reduction and oxidation and by the stability of the conductances of partially reduced catalysts. This stability indicated a virtual absence of effects from transient or chemisorbed species, and suggested that major changes in the lattice ions themselves cause the changes in conductance.

In a similar manner, Sancier et al. (132) examined the charge-transfer processes in silica-supported bismuth molybdate catalysts by means of electrical conductivity measurements. Like Peacock, the authors observed an increase in conductivity during reduction and a decrease in conductivity upon reoxidation of the catalyst by gaseous oxygen. The temperature dependence of the charge-transfer rates during initial oxidation and reduction yielded apparent activation energies of 18 kcal/mole for oxidation and 23 kcal/mole for reduction. The rate-limiting step associated with the 18 kcal/mole activation energy was the dissociation of the molecular oxygen. The 23 kcal/mole activation energy was associated with the chemisorption of propylene.

Morrison (133) measured the relative Fermi energy of titanium dioxide-supported bismuth molybdate, cuprous oxide, vanadium pentoxide, and ferric molybdate catalysts used for the partial oxidation of olefins and alcohols. The results showed that the bulk Fermi energy of these catalysts has a common value, and indicated that the bulk Fermi energy of catalysts is limited to a certain range of values if high activity is to be exhibited. The bulk Fermi energy of bismuth molybdate, intermediate between the Fermi energies of Bi_2O_3 and MoO_3, was stable against treatments by propylene. It was therefore suggested that oxygen vacancies form an impurity band which "pins" the Fermi energy of the bulk. The stability of the conductances of partially reduced bismuth molybdate catalysts was also observed by Peacock.

In addition to the microscopic redox processes of bismuth ions and molybdenum ions, the combination of these conductivity measurements leads to the conclusion that the macroscopic, bulk conductivity properties of the bismuth molybdate catalyst affect the catalytic reaction.

4. Mössbauer Spectroscopy

In order to directly observe changes in the partner cation of molybdenum, Mössbauer spectroscopy has been used (134–136). Although this tool is not sensitive to bismuth, it can detect other ions with redox properties such as tin or iron.

Firsova et al. (134) investigated the formation of hydrocarbon complexes on the surface of a supported tin molybdate catalyst during propylene and acrolein adsorption. The measurements were made after the sample was subjected to hydrocarbon adsorption at 200° and 400°C. Propylene adsorption resulted in the formation of a surface complex at 200°C only after preliminary treatment of the catalyst with oxygen. At 400°C, however, the same complex was formed without preliminary oxygen adsorption. Acrolein adsorption also resulted in the formation of a surface complex which was more strongly bonded to the surface. The changes observed in the resonance spectra during propylene and acrolein adsorption indicated that Sn^{4+} was partially reduced to Sn^{2+}. Analysis of the spectrum also indicated that a surface complex between the hydrocarbon and the tin ion formed through oxygen (Sn—O—C—) when propylene or acrolein were chemisorbed. Since the surface complex was not observed on supported tin dioxide, it was concluded that the presence of the molybdenum ion in the tin molybdate catalyst plays a role in the formation of the tin–hydrocarbon complex.

Maksimov *et al.* (*135*) studied a cobalt molybdate catalyst containing a small amount of Fe^{3+}, directly under catalytic reaction conditions. Mössbauer measurements were made while a mixture of propylene and oxygen were flowed over the catalyst at 310°C. During the catalytic reaction, the spectrum of the catalyst showed the presence of Fe^{2+} which, when the reaction was stopped and the products evacuated, returned to Fe^{3+}. The introduction of additional iron into the cobalt molybdate catalyst increased not only the catalytic activity and selectivity but also the electrical conductivity. Other workers also showed that a decrease of charge-transfer energy in the lattice of the catalyst increased the activity and selectivity for propylene oxidation (*137*). Maksimov considered that the formation of Fe^{2+} resulted from a charge transfer from Mo^{5+} to Fe^{3+} which would yield Mo^{6+} and Fe^{2+}. Since Fe^{2+} only existed during the reaction, the authors concluded that the Fe^{2+} readily transmits this charge to acceptor levels of molecular oxygen.

Firsova *et al.* (*136*) also investigated a cobalt molybdate catalyst containing a small amount of Fe^{3+}, after exposure to a reaction mixture of propylene and oxygen. The authors observed the valence change of Fe^{3+} to Fe^{2+} and the formation of a surface complex between the hydrocarbon and the iron (Fe—O—C—). In contrast to pure iron molybdate which also forms a surface Fe—O—C— complexes, the electronic transitions in the cobalt iron molybdate were reversible. The observed valence change showed that iron ions increase the electronic interaction between ions in the catalyst and the components of the reaction mixture.

In addition to confirming that the partner cation of molybdenum undergoes a valence change during the catalytic reaction, the combination of these experimental results indicates that the partner cation is involved in the formation of a surface hydrocarbon complex. However, there is not general agreement concerning the stage in the reaction mechanism that the hydrocarbon becomes associated with the partner cation of molybdenum.

5. *Electron Spectroscopy for Chemical Analysis*

Matsuura and Wolfs (*110*) examined the surface composition of the α, β, and γ phases of bismuth molybdate utilizing ESCA. The authors observed the presence of both Mo^{6+} and Bi^{3+} in the surface layers of the three bismuth molybdate phases. The relative intensity of the bismuth signal increased while the molybdenum signal remained constant with an increasing concentration of bulk-phase bismuth in the catalysts.

Grzybowska *et al.* (*138*) performed similar ESCA measurements on the three bismuth molybdates, bismuth oxide, and molybdenum trioxide. The results indicated that the surface layer composition of the three bismuth molybdates corresponded with the bulk composition, viz., the relative intensity of bismuth decreased while the relative intensity of molybdenum increased for the series Bi_2MoO_6, $Bi_2Mo_2O_9$, and $Bi_2Mo_3O_{12}$. Reduction of the bismuth molybdate samples at 400°C resulted in a similar surface layer composition of bismuth and molybdenum for all three of the bismuth molybdate phases. This composition corresponded to the original surface layer composition found for $Bi_2Mo_3O_{12}$.

One possible conclusion is that under reducing conditions, metal cation movement occurs. Another possible conclusion is that despite the similar surface layer composition of bismuth and molybdenum for the three phases of bismuth molybdate, the three bismuth molybdate phases possess different catalytic activities, catalytic selectivities, adsorption properties, surface oxomolybdenum species, and reducibilities because the surface properties of the active bismuth molybdates are dependent upon the foundation upon which they exist, i.e., upon the bulk structure and its chemical and electronic properties.

D. Active Site Model

Even though the studies discussed have brought about considerable advancement in the level of our understanding of propylene oxidation and selective oxidation catalysts, not all of the details concerning active sites have been reconciled. However, it appears that one reasonable schematic representation of the active sites for propylene oxidation on molybdate catalysts may be given by

$$A\overset{O}{\diagdown}Mo=O$$

This model correlates with a major fraction of the experimental results. As discussed, the nature of the oxomolybdenum species may be correlated with the A and B sites of Schuit. The B sites (O_B), involved in olefin adsorption, were identified with terminal molybdenum oxygens ($Mo-O_t$) since multiply bonded terminal oxygens are more electrophilic and more strongly bonded to molybdenum than bridging oxygen. The A sites (O_A) were identified with bridging oxygen. The results which showed a definite interaction between adsorbed propylene and oxygen which is multiply bonded to molybdenum provided substantial support for this model.

Considerations concerning the atomic arrangements within the scheelite structure reveal that the removal of cations from the A positions would result in fewer bridged oxygens (Mo—O—Bi) and more terminal molybdenum oxygens (Mo—O_t). Therefore, the increased activity caused by the removal of A-site cations may be a result of the increased concentration of terminal molybdenum oxygens.

The molybdenum–oxygen bond strength has been shown to weaken in direct proportion to an increased ionization potential of the A cations in several scheelite structured molybdates. Thus, it would be reasonable to assume that the nature of the A cation will have a profound effect on the molybdenum–oxygen bond strength which, in turn, will influence the activity and the selectivity of the catalysts.

Dewig (139) has established that there also is a direct correlation between the ease of charge transfer between molybdenum and oxygen, i.e.,

$$\underset{Mo^{6+}}{\overset{O}{\|}} \longrightarrow \underset{Mo^{5+}}{\overset{O^{2-}}{\|}} \text{ or } \underset{Mo^{5+}}{\overset{O^{-}}{|}}$$

and the activation energy of the allylic hydrogen abstraction during the rate-determining step in the oxidation of propylene. This implies that the rate-determining step for acrolein formation involves a charge transfer process during which the allylic hydrogen of propylene is abstracted by lattice oxygen.

It also appears that the reactive oxygen acts as a radical oxygen. Akimoto and Echigoya (58) have shown that the double-bond-type oxygen of molybdenum is similar to a free radical oxygen in nature, in contrast to the bridged oxygens (Mo—O—M). This free radical character was observed to increase as a result of electron transfer from an adsorbed olefin species to a nonbonding atomic orbital of Mo^{6+}. Indeed, when the electronegativity of the molybdenum ion in Mo=O decreases by electron transfer, the electrons are expected to be more localized around the double-bond oxygen, with an approach of its oxidation number to -1, i.e.,

$$\underset{Mo^{6+}}{\overset{O}{\|}} \longrightarrow \underset{Mo^{5+}}{\overset{O^{-}}{|}}$$

Additionally, the reactivity of the Mo=O was observed to increase inversely with the electronegativity of the molybdenum ion. Thus, one may reasonably conclude that Mo^{6+} accepts an electron during the oxidation of olefins forming Mo^{5+}=O or Mo^{5+}—O^-, the double-bond-type oxygen behaving like a radical oxygen and abstracting the α-hydrogen of propylene.

There exists a good deal of evidence that the initially adsorbed olefin becomes coordinated to a metal cation (59, 140). The above mode of action of the Mo^{6+}—O_t species, does not necessitate that the olefin adsorb on the Mo^{6+} ion to form an Mo^{5+} ion. There may be a direct transfer of electrons from the adsorbed propylene to the Mo^{6+} ion. However, it is generally accepted that at high temperatures, many free electrons in the bulk are formed by the reduction of the catalyst. The Mo^{5+} formation may result from electron transfer through the bulk (58, 141, 142). Thus, it is also likely that Mo^{6+} accepts an electron through the lattice during the oxidation of propylene, forming the Mo^{5+}—O_t^- species while the allylic species becomes coordinated to another active site.

This model also suggests that the bridged oxygen will be active. It is reasonable, however, that the bridged oxygen bond strength, and hence its activity, will be a function of the nature of the A cation as well as the catalyst structure. According to the results of Schuit and Otsubo, the bridged oxygen appears in the product acrolein. Incorporation of the gas-phase oxygen into the catalyst most likely occurs at both A and B sites. A high degree of lattice oxygen mobility, as shown by isotopic oxygen experiments, allows for oxygen migration between the two active sites. Of course, the oxygen mobility is a function of catalyst structure and temperature. This, in turn, can affect the ease of reoxidation (level of reduction) of the catalyst, which, in turn, affects the activity and selectivity.

VI. Conclusions

The selective oxidation of propylene is an important model reaction for studying oxidation reactions over oxide catalysts. Much information has been gathered over the past three decades that helps to answer the four questions raised at the outset of this review.

Considerable evidence exists that indicates the selective oxidation of propylene proceeds via the formation of a symmetrical allyl species. Subsequent steps may vary as a function of the catalyst. Some catalyst systems may abstract a second hydrogen atom before the insertion of oxygen. Others may add molecular oxygen, forming a hydroperoxide intermediate, which may then subsequently decompose into acrolein and water.

While a number of oxygen species can exist on the surface of oxides, the reactive oxygen for the selective oxidation of propylene is lattice oxygen. In addition, rapid diffusion of oxygen from the surface into the bulk and from the bulk to the surface often is observed with many selective oxidation catalysts. The nature of the reactive oxygen on the surface, however, appears to be the multiply bonded oxygen and the bridged-type oxygen.

Under reaction conditions solid-state reactions occur that are a function of the temperature and the composition of the reacting gas mixture. In addition, it appears that movement of vacancies, anions, and, possibly, cations occur. Consequently, the catalyst bulk structure is an important parameter in determining activity and selectivity. It follows then that precautions need to be taken in the preparation of the catalyst so that the desired structure can be achieved.

While complete agreement has not yet been reached with regard to the exact nature of the active sites, it does seem relatively well established that the selective oxidation reaction occurs via a dual-site mechanism. The hydrocarbon interacts with a reduction site; the oxygen interacts with a reoxidation site. Moreover, these sites may be somewhat independent of one another, allowing a simple redox cycle to be used to describe the overall reaction.

REFERENCES

1. Hearne, G. W., and Adams, M. L., U.S. Patent 2,451,485 (1948).
2. Idol, J. D., Jr., U.S. Patent 2,904,580 (1959).
3. Callahan, J. L., Foreman, R. W., and Veatch, F., U.S. Patent 3,044,966 (1962).
4. Callahan, J. L., and Gertisser, B., U.S. Patent 3,198,750 (1965).
5. Bethall, T., and Hadley, D. T., U.S. Patent 3,094,565 (1963).
6. Nippon Kayaku Co., Ltd., Dutch Patent 7,006,454 (1970).
7. Grasselli, R. K., Miller, A. F., and Hardman, A. F., Ger. Patent 2,203,710 (1972).
8. Shiraishi, T., Kishiwada, S., Shimizu, S., Honmaru, S., Ichihashi, H., and Nagaska, Y., Ger. Patent 2,133,110 (1972).
9. Sampson, R. J., and Shooter, D., *Oxid. Combust. Rev.* 1, 225 (1965).
10. Voge, H. H., and Adams, C. R., *Adv. Catal.* 17, 151 (1967).
11. Sachtler, W. M. H., *Catal. Rev.* 4, 27 (1970).
12. Margolis, L. Ya., *Catal. Rev.* 8, 241 (1973).
13. Hucknall, D. J., "Selective Oxidation of Hydrocarbons." Academic Press, New York, 1974.
14. Adams, C. R., and Jennings, T. J., *J. Catal.* 2, 63 (1963).
15. Adams, C. R., and Jennings, T. J., *J. Catal.* 3, 549 (1964).
16. Voge, H. H., Wagner, C. D., and Stevensen, D. P., *J. Catal.* 2, 58 (1963).
17. McCain, C. C., Gough, G., and Godin, G. W., *Nature (London)* 198, 989 (1963).
18. Sachtler, W. M. H., *Rec. Trav. Chim. Pays-Bas* 82, 243 (1963).
19. Sachtler, W. M. H., and De Boer, N. K., *Proc. Int. Congr. Catal., 3rd* 1, 252 (1965).
20. Grasselli, R. K., and Suresh, D. D., *J. Catal.* 25, 273 (1972).
21. Godin, G. W., McCain, C. C., and Porter, E. A., *Proc. Int. Congr. Catal., 4th* 1, 271 (1971).
22. Cant, N. W., and Hall, W. K., *J. Catal.* 22, 310 (1971).
23. Cant, N. W., and Hall, W. K., *J. Phys. Chem.* 75, 2914 (1971).

24. Kokes, K. J., *Catal. Rev.* **6**, 1 (1972).
25. Kugler, B. L., and Kokes, R. J., *J. Catal.* **32**, 170 (1974).
26. Kugler, B. L., and Gryder, J. W., *J. Catal.* **44**, 126 (1976).
27. Margolis, L. Ya., *Adv. Catal.* **14**, 429 (1963).
28. Daniel, C., and Keulks, G. W., *J. Catal.* **24**, 529 (1971).
29. Gorshkov, A. P., Kolchin, I. K., Gribov, I. M., and Margolis, L. Ya., *Kinet. Katal.* **9**, 1086 (1968).
30. Gorshkov, A. P., Kolchin, I. K., Isgulyants, G. V., Derbentsev, Yu. I., and Margolis, L. Ya., *Dokl. Akad. Nauk SSSR* **186**, 827 (1959).
31. Gogarin, S. G., Kolchin, I. K., and Margolis, L. Ya., *Neftekhimiya* **10**, 59 (1970).
32. Keulks, G. W., Rosynek, M. P., and Daniel, C., *Ind. Eng. Chem., Prod. Res. Dev.* **10**, 138 (1971).
33. Lunsford, J. H., *Catal. Rev.* **8**, 135 (1973).
34. Mars, P., and van Krevelen, D. W., *Chem. Eng. Sci. Suppl.* **3**, 41 (1954).
35. Aykan, K., *J. Catal.* **12**, 281 (1968).
36. Batist, Ph. A., Prette, H. J., and Schuit, G. C. A., *J. Catal.* **15**, 267 (1969).
37. Peacock, J. M., Parker, A. J., Ashmore, P. G., and Hockey, J. A., *J. Catal.* **15**, 398 (1969).
38. Fattore, V., Fuhrman, Z. A., Manara, G., and Notari, B., *J. Catal.* **37**, 223 (1975).
39. Niwa, M., and Murakami, Y., *J. Catal.* **26**, 359 (1972).
40. Marak, E. J., Moffat, A. J., and Waldrop, M. A., *Proc. Int. Congr. Catal., 6th* **1**, 376 (1977).
41. Keulks, G. W., *J. Catal.* **19**, 232 (1970).
42. Wragg, R. D., Ashmore, P. G., and Hockey, J. A., *J. Catal.* **22**, 49 (1971).
43. Sancier, K. M., Wentreck, P. R., and Wise, H., *J. Catal.* **39**, 141 (1975).
44. Otsubo, T., Miura, H., Morikawa, Y., and Shirasaki, T., *J. Catal.* **36**, 240 (1975).
45. Christie, J. R., Taylor, D., and McCain, C. C., *J. Chem. Soc., Faraday Trans. 1* **72**, 334 (1976).
46. Pendleton, P., and Taylor, D., *J. Chem. Soc., Faraday Trans. 1* **72**, 1114 (1976).
47. Keulks, G. W., and Krenzke, L. D., *Proc. Int. Congr. Catal., 6th* **2**, 806 (1977).
48. Novakova, J., and Jirû, P., *J. Catal.* **27**, 155 (1972).
49. Wragg, R. D., Ashmore, P. G., and Hockey, J. A., *J. Catal.* **28**, 337 (1973).
50. Moro-oka, Y., Takifa, Y., and Ozaki, A., *J. Catal.* **27**, 177 (1972).
51. Novakova, J., Dolejsek, Z., and Habersbernia, K., *React. Kinet. Catal. Lett.* **4**, 389 (1976).
52. Roiter, V. A., Golodets, G. I., and Pyatnitzkii, Yu. I., *Proc. Int. Congr. Catal., 4th* **1**, 466 (1971).
53. Boreskov, G. K., Popovskii, V. V., and Sazonov, V. A., *Proc. Int. Congr. Catal., 4th* **1**, 439 (1971).
54. Trifiro, F., Centola, P., and Pasquon, I., *J. Catal.* **10**, 86 (1968).
55. Trifiro, F., Centola, P., Pasquon, I., and Jirû, P., *Proc. Int. Congr. Catal., 4th* **1**, 252 (1971).
56. Trifiro, F., and Pasquon, I., *J. Catal.* **12**, 412 (1968).
57. Trifiro, F., Kubelkova, L., and Pasquo, I., *J. Catal.* **19**, 121 (1970).
58. Akimoto, M., and Echigoya, E., *J. Catal.* **35**, 278 (1974).
59. Haber, J., *Z. Chem.* **13**, 241 (1973).
60. Kazanskii, V. B., *Kinet. Katal.* **14**, 72 (1973).
61. Akimoto, M., Akiyama, M., and Echigoya, E., *Bull. Chem. Soc. Jpn.* **49**, 3367 (1976).
62. Krylov, O. V., *Kinet. Katal.* **14**, 24 (1973).
63. Krylov, O. V., Pariiskii, G. B., and Spiridonov, K. N., *J. Catal.* **23**, 301 (1971).

64. Krylov, O. V., and Kaduskin, A. A., *J. Catal.* in press (1978).
65. Trifiro, F., Forzatti, P., and Villa, P. L., in "Preparation of Catalysts" (B. Delmon, P. A. Jacobs, and G. Poncelet, eds.), p. 147. Elsevier, Amsterdam, 1976.
66. Isaev, O. K., Kushnarev, M. Ya., and Margolis, L. Ya., *Dokl. Akad. Nauk SSSR* **119**, 104 (1958).
67. Wood, B. J., Wise, H., and Yolles, R. S., *J. Catal.* **15**, 355 (1969).
68. Gorokhovatskii, Ya. G., Vovyanko, I. I., and Rubanik, M. Ya., *Kinet. Katal.* **7**, 65 (1966).
69. Zambonini, F., *Z. Kristallogr. Mineral. Petrogr.* **58**, 226 (1923).
70. Mekhtiev, K. M., Gamidov, R. S., Mamedov, Kh. S., and Belov, N. V., *Dokl. Akad. Nauk SSSR* **162**, 397 (1965).
71. Erman, L. Ya., Gal'perin, E. L., Kolchin, I. K., Dobrzhanskii, G. F., and Chernyshev, K. S., *Russ. J. Inorg. Chem.* **9**, 1174 (1964).
72. Erman, L. Ya., and Gal'perin, E. L., *Russ. J. Inorg. Chem.* **15**, 441 (1970).
73. Bleijenberg, A. C. A. M., Lippens, B. C., and Schuit, G. C. A., *J. Catal.* **4**, 581 (1965).
74. Chen, T., *J. Cryst. Growth* **20**, 29 (1973).
75. Cesari, M., Perego, G., Zazzetta, A., Manara, G., and Notari, B., *J. Inorg. Nucl. Chem.* **33**, 3595 (1971).
76. Van den Elzen, A. F., and Rieck, G. D., *Acta Crystallogr., Sect. B* **29**, 2433 (1973).
77. Erman, L. Ya., and Gal'perin, E. L., *Russ. J. Inorg. Chem.* **11**, 122 (1966).
78. Erman, L. Ya., and Gal'perin, E. L., *Russ. J. Inorg. Chem.* **13**, 487 (1968).
79. Van den Elzen, A. F., and Rieck, G. D., *Mat. Res. Bull.* **10**, 1163 (1975).
80. Zemann, J., *Struct. Rep.* **20**, 449 (1956).
81. Van den Elzen, A. F., and Rieck, G. D., *Acta Crystallogr., Sect. B* **29**, 2436 (1973).
82. Wilson, A. J. C., *Struct. Rep.* **11**, 305 (1948).
83. Blasse, G., *J. Inorg. Nucl. Chem.* **28**, 1124 (1966).
84. Erman, L. Ya., Gal'perin, E. L., and Sobolev, B. P., *Russ. J. Inorg. Chem.* **16**, 258 (1971).
85. Dalin, M. A., Mangasaryan, N. A., Serebryakov, B. R., Mekhtieva, V. L., Portyanskii, A. E., and Mekhtiev, K. M., *Dokl. Akad. Nauk SSSR* **200**, 624 (1971).
86. Rashkin, J. A., and Pierron, E. D., *J. Catal.* **6**, 332 (1966).
87. Callahan, J. L., and Gertisser, B., U.S. Patent 3,308,151 (1967).
88. Grasselli, R. K., and Callahan, J. L., *J. Catal.* **14**, 93 (1969).
89. Grasselli, R. K., Suresh, D. D., and Knox, K., *J. Catal.* **18**, 356 (1970).
90. McClellan, W. R., U.S. Patent 3,415,886 (1968).
91. Callahan, J. L., Szabo, J. J., and Gertisser, B., U.S. Patent 3,362,998 (1968).
92. Cesari, M., Perego, G., Zazzetta, A., Manara, G., and Notari, B., Ital. Patent 769,558 (1965).
93. Aykan, K., Sleight, A. W., and Rogers, D. B., *J. Catal.* **29**, 185 (1973).
94. Aykan, K., Halvorson, D., Sleight, A. W., and Rogers, D. B., *J. Catal.* **35**, 401 (1974).
95. Sleight, A. W., Aykan, K., and Rogers, D. B., *J. Solid State Chem.* **13**, 231 (1975).
96. Aykan, K., Rogers, D. B., and Sleight, A. W., U.S. Patent 3,806,470 (1974).
97. Aykan, K., Rogers, D. B., and Sleight, A. W., U.S. Patent 3,843,553 (1974).
98. Aykan, K., Rogers, D. B., and Sleight, A. W., U.S. Patent 3,843,554 (1974).
99. Alkhazov, T. G., and Adzhamov, K. Yu., *Kinet. Katal.* **15**, 201 (1974).
100. Alkhazov, T. G., Adzhamov, K. Yu., Dolgov, V. Ya., Krylov, O. V., and Margolis, L. Ya., *Dokl. Akad. Nauk SSSR* **204**, 624 (1972).
101. Alkhazov, T. G., and Adzhamov, K. Y., *React. Kinet. Catal. Lett.* **2**, 45 (1975).
102. Voge, H. H., Armstrong, W. E., and Ryland, L. B., U.S. Patent 3,110,746 (1963).
103. Morita, Y., Nishikawa, E., and Kiguchi, S., *Mem. Sch. Sci. Eng., Waseda Univ.* No. 32, p. 55 (1968); (*Chem. Abstr.* **71**, 105715f).

104. Daniel, C., and Keulks, G. W., *J. Catal.* **29,** 475 (1973).
105. Annenkova, B., Alkhazov, T. G., and Belen'kii, M. S., *Kinet. Katal.* **10,** 1076 (1969).
106. Batist, Ph. A., van de Moesdijk, C. G. M., Matsuura, I., and Schuit, G. C. A., *J. Catal.* **20,** 40 (1971).
107. Sleight, A. W., and Jeitschko, W., *Mat. Res. Bull.* **9,** 951 (1974).
108. LoJacono, M., Notermann, T., and Keulks, G. W., *J. Catal.* **40,** 19 (1975).
109. Wolfs, M. W. J., and Batist, Ph. A., *J. Catal.* **32,** 25 (1974).
110. Matsuura, I., and Wolfs, M. W. J., *J. Catal.* **37,** 174 (1975).
111. Wolfs, M. W. J., and van Hooff, J. H. C., in "Preparation of Catalysts" (B. Delmon, P. A. Jacobs, and G. Poncelet, eds.), p. 161. Elsevier, Amsterdam, 1976.
112. Batist, Ph. A., personal communication. 1975.
113. Haber, J., and Grzybowska, B., *J. Catal.* **28,** 489 (1973).
114. Swift, H. E., Bozik, J. E., and Ondrey, J. A., *J. Catal.* **22,** 427 (1971).
115. Solymosi, F., and Bozso, F., *Proc. Int. Congr. Catal., 6th* **1,** 365 (1977).
116. Gamid-Zade, E. G., Kuliyev, A. R., Mamedov, E. A., Rizayev, R. G., and Sokolovski, V. D., *React. Kinet. Catal. Lett.* **3,** 191 (1975).
117. Matsuura, I., and Schuit, G. C. A., *J. Catal.* **20,** (1971).
118. Matsuura, I., and Schuit, G. C. A., *J. Catal.* **25,** 314 (1972).
119. Batist, Ph. A., Bouwens, J. F. H., and Matsuura, I., *J. Catal.* **32,** 362 (1974).
120. Matsuura, I., *J. Catal.* **33,** 420 (1974).
121. Trifiro, F., Hoser, H., and Scarle, R. D., *J. Catal.* **25,** 12 (1972).
122. Mitchell, P. C. H., and Trifiro, F., *J. Chem. Soc. A* p. 3183 (1970).
123. Grzybowska, B., Haber, J., and Komorek, J., *J. Catal.* **25,** 25 (1972).
124. Sancier, K. M., Dozono, T., and Wise, H., *J. Catal.* **23,** 270 (1971).
125. Peacock, J. M., Sharp, M. J., Parker, A. J., Ashmore, P. G., and Hockey, J. A., *J. Catal.* **15,** 379 (1969).
126. Burlamacchi, L., Martini, G., and Trifiro, F., *J. Catal.* **33,** 1 (1974).
127. Novakova, J., Jirû, P., and Zavadil, V., *J. Catal.* **17,** 93 (1970).
128. Burlamacchi, L., Martini, G., and Ferroni, E., *J. Chem. Soc., Faraday Trans. 1* **68,** 1586 (1972).
129. Burlamacchi, L., Martini, G., and Ferroni, E., *Chem. Phys. Lett.* **9,** 420 (1971).
130. Matsuura, I., personal communication. 1976.
131. Peacock, J. M., Parker, A. J., Ashmore, P. G., and Hockey, J. A., *J. Catal.* **15,** 387 (1969).
132. Sancier, K. M., Aoshima, A., and Wise, H., *J. Catal.* **34,** 257 (1974).
133. Morrison, S. R., *J. Catal.* **34,** 462 (1974).
134. Firsova, A. A., Khovanskaya, N. N., Tsyganov, A. D., Suzdalev, I. P., and Margolis, L. Ya., *Kinet. Katal.* **12,** 792 (1971).
135. Maksimov, Yu. V., Suzdalev, I. P., Gol'danskii, V. I., Krylov, O. V., Margolis, L. Ya., and Nechitailo, A. E., *Dokl. Akad. Nauk SSSR* **221,** 880 (1975).
136. Firsova, A. A., Imshennik, V. K., Maksimov, Yu. V., Margolis, L. Ya., and Suzdalev, I. P., *Dokl. Akad. Nauk SSSR* **214,** 621 (1974).
137. Adzhamov, K. Yu., Alkhazov, T. G., Dolgov, V. Ya., Krylov, O. V., and Margolis, L. Ya., *Dokl. Akad. Nauk SSSR* **209,** 314 (1973).
138. Grzybowska, B., Haber, J., Marczewski, W., and Ungier, L., *J. Catal.* **42,** 327 (1976).
139. Dewig, J., *Chem. Soc. Symp. Hydrocarbon Oxid., London, 1970.*
140. Haber, J., Sochacka, M., Grzybowska, B., and Golebiewski, A., *J. Mol. Catal.* **1,** 35 (1975).
141. Krylov, O. V., and Margolis, L. Ya., *Kinet. Katal.* **11,** 432 (1970).
142. Naccache, C., Bandiera, J., and Dufaux, M., *J. Catal.* **25,** 335 (1972).

σ−π Rearrangements and Their Role In Catalysis

BARRY GOREWIT* AND MINORU TSUTSUI

Chemistry Department
Texas A & M University
College Station, Texas

- I. Introduction 227
- II. The σ−π Rearrangement 228
  - A. Driving Force for the σ−π Rearrangement 228
- III. Classification of σ−π Rearrangements 230
  - A. Reaction on the Metal 230
  - B. Reaction of the Ligand 234
- IV. Role of σ−π Rearrangements in Catalysis 235
  - A. Coupling of Olefins 235
  - B. Oxidation of Olefins 238
  - C. Hydroformylation 244
  - D. Activation of C—H Bonds 245
  - E. Olefin Metathesis 246
  - F. Carbene Complexes 249
  - G. Stereospecific Reactions 249
  - H. Catalysis by Vitamin B_{12} 256
- V. Conclusion 261
  - References 261

I. Introduction

In the relatively short time since the establishment of the π-bonded "sandwich" structure of Hein's polyaromatic chromium compounds (*1*) the discovery of a large number of similar compounds has been reported (*2–11*). An equally significant outgrowth of this pioneering work, however, has been the establishment of a new and important class of organometallic reactions, the σ−π rearrangement. With the recent resurgence of interest in catalysis by metals and metal-organic compounds, one sees the σ−π rearrangement constantly invoked in the litera-

---

* Present address: Department of Chemistry, University of Southern California, Los Angeles, California 90007.

ture to explain the action of these compounds, whether the process under study is industrial or biological. It is the intent of this review to trace the progress made in the understanding of the σ–π interconversion and its role in catalysis.

II. The σ–π Rearrangement

The term σ–π *rearrangement* is defined as an intramolecular process in which an organic group that is σ bonded (*monohapto*, h') to a transition metal becomes π bonded (*n-hapto*, h^n) to the metal. The π–σ rearrangement is the opposite change (h^n–h').

A. Driving Force for the σ–π Rearrangement

Since the σ and π forms of a transition metal complex would be expected to differ only slightly in such physical properties as ligation and solvation and lattice energies, the predominance of one form or the other must be primarily due to the nature of the metal–ligand bonding. There are three alternate explanations of the proclivity to σ–π or π–σ rearrangement in terms of bonding. These are summarized as follows.

A comparison of the molecular orbital distributions of the σ and π compounds is shown in Fig. 1 (*12*). The difference becomes apparent considering that σ bonding involves an overlap of σ orbitals on the ligand with $d\sigma$ orbitals on the metal, while π bonding involves σ overlap between π orbitals of the ligand with $d\sigma$-acceptor orbitals on the metal as well as

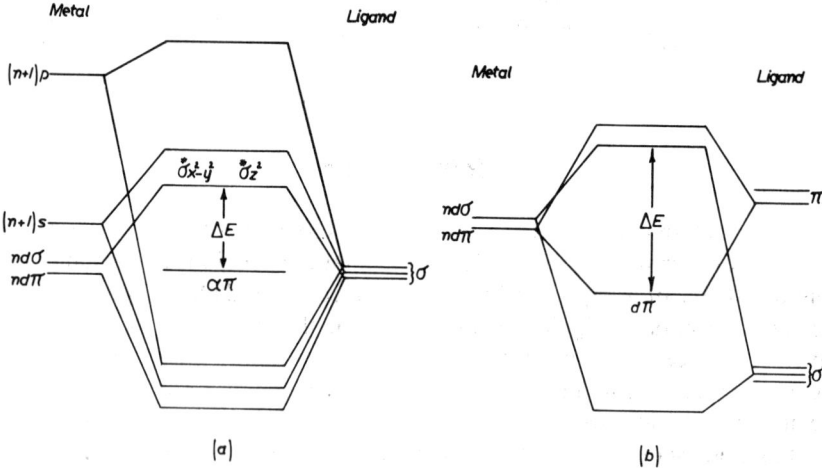

FIG. 1. Schematic representation of energy levels for organotransition metal compounds (a) without π bonding and (b) with π bonding (*12*).

"back bonding" by a flow of electron density from filled metal dxy or other $d\pi-p\pi$-hybrid orbitals to antibonding orbitals on the ligand. For σ-bonded compounds there is a correlation between stability and the probability of an electron transition from the relatively high-energy-filled d orbitals to the lowest antibonding orbitals. Therefore, the stability of a σ-bonded organometallic compound is enhanced as the energy of this transition, ΔE, is increased. An increase in ΔE is the result of lowering the energy of the d-nonbonding orbitals by combination with low-energy π-type orbitals in suitable ligands. For π complexes, ΔE is increased by back bonding from nonbonding metal d to antibonding ligand π orbitals. Therefore, on losing a ligand the σ structure will be destabilized relative to the π state and a $\sigma-\pi$ rearrangement can occur. The reverse will occur on addition of a ligand to a π complex. In the latter case, ligands with large donor capacities increase the electron density on the metal, causing a rise in the energy of the d orbitals which would favor the σ form.

The relative ease of rearrangement can also be correlated with "hardness" or "softness" of the ligand as compared to the metal (12). The σ-bonded organic groups are "harder" than the corresponding π-bonded groups because of the small polarizability of the localized σ orbitals as compared with the delocalized π-electron cloud. Thus, "hardening" the metal will favor the σ compound, and "softening" it will favor the π complex. Hardening or softening the ligand can also initiate rearrangement. Therefore, the addition of strongly π-bonding ligands, such as CO, CN^-, PR_3, or dimethylsulfoxide, will result in a $\pi-\sigma$ rearrangement, and the removal of such ligands will cause the reverse reaction.

If the σ-bonded ligand is considered as occupying one coordination site and the π-bonded ligand is considered as occupying two (or more) coordination sites of the metal, then the $\sigma-\pi$ rearrangement might be considered as an intramolecular addition to a coordinatively unsaturated metal atom (13), while the $\pi-\sigma$ rearrangement can be envisioned as the reverse or an intramolecular elimination from a coordinatively saturated metal. Any change in the metal which renders it coordinatively unsaturated, such as the expulsion of a labile ligand, would promote a $\sigma-\pi$ rearrangement of a remaining ligand to fill the coordination requirements of the metal, and conversely, the aquisition of an additional ligand would promote a $\pi-\sigma$ process.

The driving force for a $\sigma-\pi$ rearrangement can be understood therefore in terms of ligand field stabilization energy ΔE, the concept of "hard" and "soft" acids and bases, and preferred coordination number of the metal atom. Keeping these concepts in mind, specific types of $\sigma-\pi$ rearrangements may be discussed.

III. Classification of σ–π Rearrangements

It is possible to separate σ–π rearrangements into two general classes: those induced by a reaction on the metal and those induced by a reaction of the ligand.

A. Reaction on the Metal (14)

As an example of rearrangements induced by a reaction on the metal, the rearrangement of tri-σ-phenylchromium (1) to bis(biphenyl)chromium(0) (3) and benzenebiphenylchromium(0) (2) (Hein's complexes) can be discussed (1, 12). The σ complex (1) is prepared by the reaction of three equivalents of phenylmagnesium bromide with one equivalent of tris-tetrahydrofuranchromium(III) chloride.

(1)

Compound (1), a bright-red crystalline solid, was converted to a black air-sensitive pyrophoric solid by the loss of tetrahydrofuran by either warming or washing with ether. Treatment of the black solid with oxygen-free water produced (2) and (3). A large body of evidence such as the observation of the decomposition of the solid with electron paramagnetic resonance spectroscopy, showing homolytic cleavage of the chromium–carbon bonds, as well as the observation of monodeuterated benzene as the only deuterated product when hydrolysis is carried out in D_2O, supports the mechanism involving π-arene radicals.

The existence of a π-radical transition state in which the organic groups are quite close to the metal is supported by the observation that the yield of π-arenechromium complexes decreases significantly in going from phenyl to benzyl to phenethyl moieties, and comparative studies of reactions of the phenyl and benzyl Grignard reagents and tris-tetrahydrofuranchromium(III) chloride followed by hydrolysis and deuterolysis indicates that π-radical formation, hydrogen abstraction, and hydrogen transfer occur within the π-radical transition state. In the case of the π-arene–chromium complexes, the σ-arene precursors are stabilized by the coordinating ability of the tetrahydrofuran ligands and the synergic stabilizing effect of the π orbitals of the phenyl groups.

Fischer and Hafner reported an alternate synthesis of bis-π-benzene chromium using a reductive Friedel–Crafts method which probably also proceeds via a σ-bonded intermediate which was not isolated (3, 4).

$$3CrCl_3 + 2Al + AlCl_3 + 6C_6H_6 \rightarrow 3[\pi-C_6H_6)_2Cr^+][AlCl_4]^-$$

$$(\pi-C_6H_6)_2Cr^+ \xrightarrow[\text{or OH}^-_{aq}]{S_2O_4^{2-}} (\pi-C_6H_6)_2CR$$

If the phenyl groups are replaced by cyclopentadienyl groups, the corresponding σ compounds, di-σ-cyclopentadienyliron chloride monotetrahydrofuranate (4) and tris-σ-cyclopentadienyliron (6) are quite unstable and rearrange to ferrocenium chloride (5) and ferrocene (7), respectively, on loss of tetrahydrofuran above $-50°$, on treatment with an organic solvent at $-60°$, or under vacuum at $-60°$ (15, 16).

An irreversible π–σ rearrangement has been observed to occur with the treatment of titanocene dichloride (**8**) with dimethyl sulfoxide (*17*). The σ complex (**9**) has been isolated, however, treatment of the reaction mixture with benzene produces the titanium complex of dimethyl sulfoxide (**10**), and treatment of the reaction mixture with maleic anhydride produces the cyclopentadiene addition product (**11**). Monitoring the reaction mixture by proton NMR spectroscopy demonstrates the π–σ change in the bonding of the cyclopentadienyl ligands.

Titanocene dichloride (**8**) is readily reduced to the $[(\pi\text{-}C_5H_5)_2Ti(III)]^+$ species which in turn can be converted into alkyl and hydrido derivatives (*18–20*). These titanocene derivatives have been found to be active catalysts in the low-pressure polymerization of ethylene (*21, 22*) and the

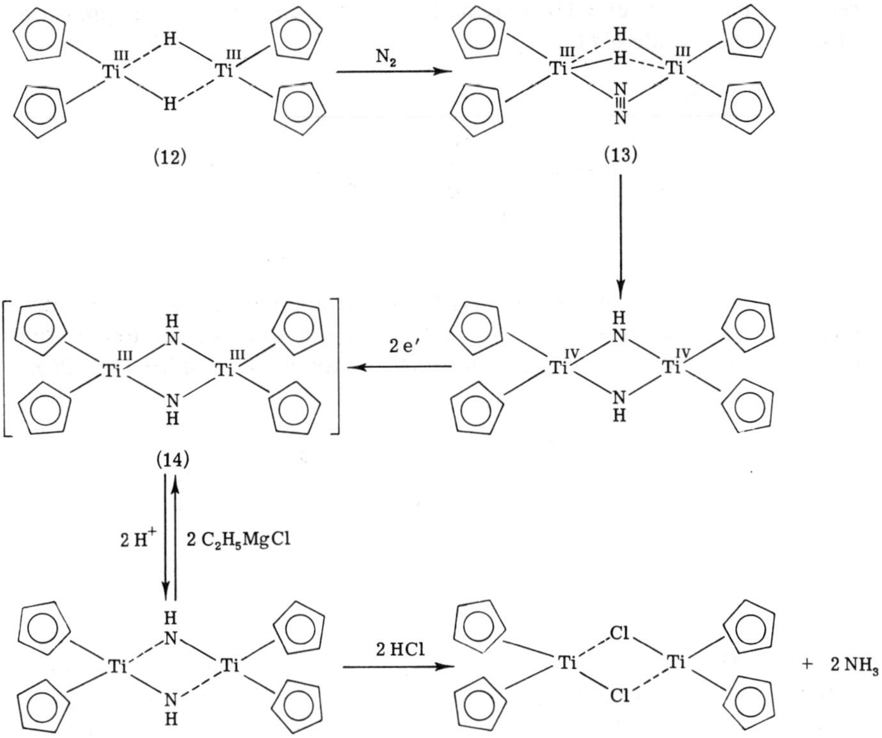

FIG. 2. Proposed mechanism for conversion of molecular nitrogen to ammonia by di-μ-hydrido-bis(dicyclopentadienyl) titantium(III) (20).

fixation of molecular nitrogen (20, 23, 24). Brintzinger has proposed a nitrogen fixation mechanism, as shown in Fig. 2 (20), in which a key step is the formation of a π complex with the molecular nitrogen bridging the two Ti atoms in the dimeric hydrido complex (12) to give structure (13). A π–σ rearrangement then occurs with oxidation of the metal atoms and reduction of the nitrogen atoms by a concomitant migration of the bridging hydrides to form the σ-bonded nitrido complex (14), and eventually ammonia.

A number of readily reversible σ–π rearrangements have been observed wherein a labile ligand such as carbon monoxide is lost by pyrolysis or photolysis, producing a coordinatively unsaturated metal center, which then regains coordinative saturation by means of a σ–π rearrangement. For example, irradiation of σ-alkyl-π-cyclopentadienyl-molybdenum tricarbonyl (15) produces the π-allene complex (16) (25). These reactions may be reversed by placing the π compound under a high pressure of carbon monoxide (14).

$$\underset{(15)}{\text{OC}-\underset{\underset{\text{CO}}{|}}{\overset{\overset{\text{Mo}}{|}}{\text{Mo}}}-\text{CH}_2\text{CH}=\text{CH}_2} \quad \underset{\text{high pressure}}{\overset{h\nu}{\rightleftharpoons}} \quad \underset{(16)}{\text{OC}-\overset{\overset{\text{Mo}}{|}}{\underset{\underset{\text{CO}}{}}{}}-\overset{\text{H}_2\text{C}}{\underset{\text{H}_2\text{C}}{}}\text{CH}} \quad + \quad \text{CO}$$

A number of rearrangements involving loss and gain of cyanide groups has also been observed (14), for example, the interconversion of the pentacyanocobalt complex (17) and tetracyanocobalt complex (18) (26).

$$\underset{(17)}{\left[\begin{array}{c}\text{H}_2\text{C}\overset{\text{H}}{\underset{\text{C}}{\diagup}}\text{CHR}\\\text{NC}-\underset{|}{\overset{|}{\text{Co}}}-\text{CN}\\\text{NC}\end{array}\right]^{3-}} \quad \underset{+\text{CN}^-}{\overset{-\text{CN}^-}{\rightleftharpoons}} \quad \underset{(18)}{\left[\begin{array}{c}\overset{\text{H}}{|}\\\text{H}\diagdown\text{C}\diagup\text{C}\diagdown\text{H}\\\text{H}\diagup\quad\diagdown\text{R}\\\text{Co}\\\text{NC}\quad\text{NC}\quad\text{CN}\quad\text{CN}\end{array}\right]^{2-}}$$

An important $\sigma-\pi$ interconversion which is thought to be the basis for homogeneous (and perhaps heterogeneous) catalytic hydrogenation processes involves the intramolecular transfer of a nucleophile such as hydride from the metal to the ligand.

$$\underset{\underset{\text{M}}{|}}{\text{H}_2\text{C}=\text{CHR}} \quad \overset{\text{H}^-}{\rightleftharpoons} \quad \text{H}_2\text{C}=\underset{\text{M}-\text{H}}{\overset{|}{\text{C}}}\overset{\text{H}}{\diagdown\text{H}} \quad \rightleftharpoons \quad \text{H}_2\text{C}-\underset{\underset{\text{M}}{|}}{\text{C}}\overset{\text{H}}{\diagdown\text{R}}$$

Examples of these interconversions will be discussed in detail.

B. Reaction of the Ligand

A change from σ to π bonding of a ligand may also be brought about by a reaction of that ligand (14). For example, protonation of a σ-bonded ligand is the driving force in several important $\sigma-\pi$ rearrangements such as that of the σ complex (19) which has been converted to the salt (20), which in turn produces acetone on hydrolysis as evidence for the formation of the allene moiety in (20) (27).

Elimination of a hydride ion may also cause a $\sigma-\pi$ rearrangement (14), for example, the treatment of the σ-ethyl complex (21) with triphenylmethylperchlorate produces the π-ethylene complex (22) which

$$\underset{(19)}{\text{Cp(OC)(Fe)(CO)CH}_2\text{CH}\equiv\text{CH}} \xrightarrow{\text{HCl} \atop \text{SbCl}_5} \underset{(20)}{\left[\text{Cp(OC)(Fe)(CO)}\!=\!\!\begin{array}{c}\text{CH}_2\\ \|\\ \text{C}\\ \|\\ \text{CH}_2\end{array}\right]^+ \text{SbCl}_6^-}$$

may be converted back to the σ complex (21) by treatment with sodium borohydride (28).

$$\underset{(21)}{\text{Cp(OC)(Fe)(CO)CH}_2\text{CH}_3} \underset{\text{NaBH}_4}{\overset{(\text{C}_6\text{H}_5)_3\overset{+}{\text{C}}(\text{ClO}_4)^-}{\rightleftarrows}} \underset{(22)}{\left[\text{Cp(OC)(Fe)(CO)}\!\!=\!\!\begin{array}{c}\text{CH}_2\\ \|\\ \text{CH}_2\end{array}\right]^+} \text{ClO}_4^- + (\text{C}_6\text{H}_5)_3\text{CH}$$

The addition of a nucleophile, B, either coordinated to the metal as previously mentioned or from outside its coordination sphere, to a coordinated olefin may promote a π–σ rearrangement.

$$\underset{M}{H_2C\!=\!\!\begin{array}{c}B\\|\\C\\ \diagup\!\diagdown\\ H\ \ R\end{array}} \rightleftarrows \underset{M}{H_2C\!-\!\!\begin{array}{c}B\\|\\C\\ \diagup\!\diagdown\\ H\ \ R\end{array}}$$

These reactions will also be discussed in detail.

IV. Role of σ–π Rearrangement in Catalysis

In a large number of transition metal-catalyzed processes, whether homogeneous or heterogeneous, there is mounting evidence pointing to the involvement of one or more σ–π rearrangement steps. Some of the more widely known of these processes, both of industrial and biological significance, will be discussed briefly in light of the more recent developments in the study of σ–π rearrangements.

A. Coupling of Olefins

The polymerization of ethylene via the Zeigler–Natta process is extremely important commericially and serves as an excellent example of the role of the σ–π rearrangement in catalysis (14). The catalyst in this process consists of a mixture of a transition metal halide, such as TiCl$_4$, and a nontransition metal alkyl, such as Al(C$_2$H$_5$)$_3$. As shown in Fig. 3, the process is thought to begin with an alkylation of the transition metal to give (**23**). Addition of ethylene to the coordinatively unsaturated complex (**23**) gives the coordinatively saturated π complex (**24**). Migration of the alkyl group to the π-ethylene ligand, rendered slightly electron difficient by withdrawal of π-electron density by the metal, with π–σ rearrangement produces a new coordinatively unsaturated complex (**25**), which adds another ethylene ligand and rearranges. The alkyl group R, less a hydride atom, may then be reacquired by the nontransition metal atom, thus regenerating the catalyst, so that in these cases no net addition of R is observed. Through a series of similar steps, a long-chain polyethylene polymer is produced. This mechanism is supported by molecular orbital considerations including correlations of the catalytic activity of the transition metal with the ionization potential of its d electrons and the electronegatives of its negative ions (*29*). In those cases where heterogeneous catalysts are utilized, the polymerization process has been shown

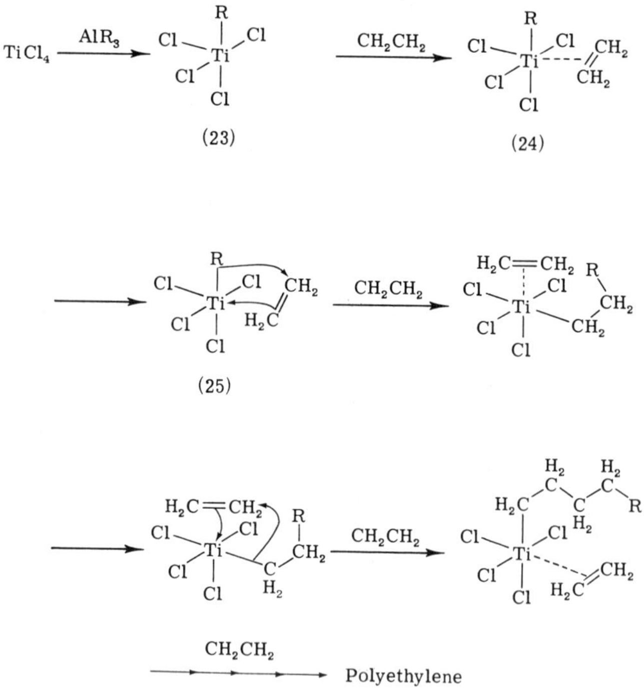

FIG. 3. Ziegler–Natta process for polyethylene (*14*).

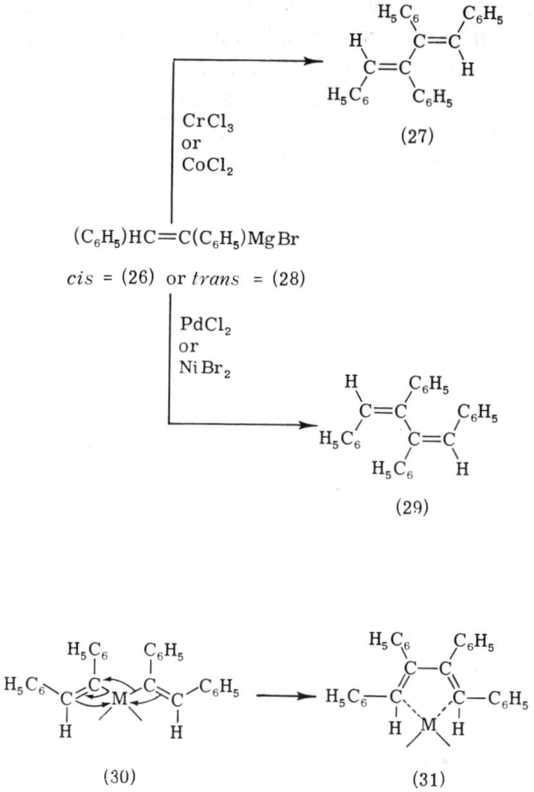

FIG. 4. Stereospecific coupling of vinyl groups (*32, 33*).

to occur by the same mechanism on the surface of the transition metal halide crystal, specifically at those sites where there is a deficiency of halide, and stereospecifically depending upon the surface geometry of the crystal (*30, 31*).

Mechanisms of a nature similar to that of the Ziegler–Natta polymerization may also be envoked to explain the coupling and polymerization reactions of a number of other alkenes which are catalyzed by a number of transition metal compounds. For example, as shown in Fig. 4, stereospecific coupling of 1,2-disubstituted vinyl groups has been observed with the vinyl Grignard, catalyzed by $CrCl_3$, $PdCl_2$, and $NiBr_2$ (*32, 33*).

Catalysis by chlorides of Cr(III) and Co(II), metals which normally exhibit the d^2sp^2 octahedral geometry directs the coupling of either *cis*- or *trans*-diphenylvinylmagnesium bromide, (**26**) or (**28**) to form the *cis-cis-*

1,2,3,4-tetraphenylbutadiene (**27**), while catalysis by compounds such as PdCl$_2$ in NiBr$_2$, in which the metals normally exhibit the dsp^2 square planar or sp^3 tetrahedral geometries direct couplings of (**26**) or (**28**) to form the *trans-trans*-1,2,3,4-tetraphenylbutadiene (**29**). In this case, the first intermediate is probably the σ-vinyl metal complex (**30**) which can rearrange to a π-butadiene complex (**31**).

(30) (31)

This type of coupling reaction is extended a further step to a "double-coupling" reaction if 4 moles of vinylmagnesium bromide and 1 mole of titanium tetrachloride are combined (*34*). A possible mechanism for this reaction is summarized in Fig. 5.

Ethylene may also be dimerized using nickelocene as a catalyst (*35*). In this case it has been suggested that the cyclopentadiene ligands are displaced by the ethylene directly. The reaction is run at high temperature and pressure so that no σ-bonded complex is involved. This process is the first commercial catalytic preparation for the industrially important 1-butene specifically, rather than the thermodynamically favored 2-butene isomers. The chromium analog of nickelocene, dibenzenechromium, is a very effective catalyst for the polymerization of ethylene at high temperature and pressure (*36*). Again, as in the case of nickelocene, it is thought that the initial step of the reaction is the replacement of the π-bonded ligands with ethylene. This is a reasonable assumption since in both cases the reaction temperatures correspond closely to the decomposition temperatures of the two catalysts. In the case of the dibenzene chromium-catalyzed reaction, however, the difference in coordination geometry of the metal atom (octahedral for chromium vs. tetrahedral for nickel) allows the relatively uncrowded coordination of several ethylenes which can then couple by either an ionic or radical process to produce the polymer of molecular weight 34,000.

B. Oxidation of Olefins

The oxidation of olefins to aldehydes using a palladium chloride–copper(II) chloride catalyst, the Wacker Process, is a well-established industrial reaction. The mechanism of this reaction has not been established in detail, but it most probably involves a σ–π rearrangement

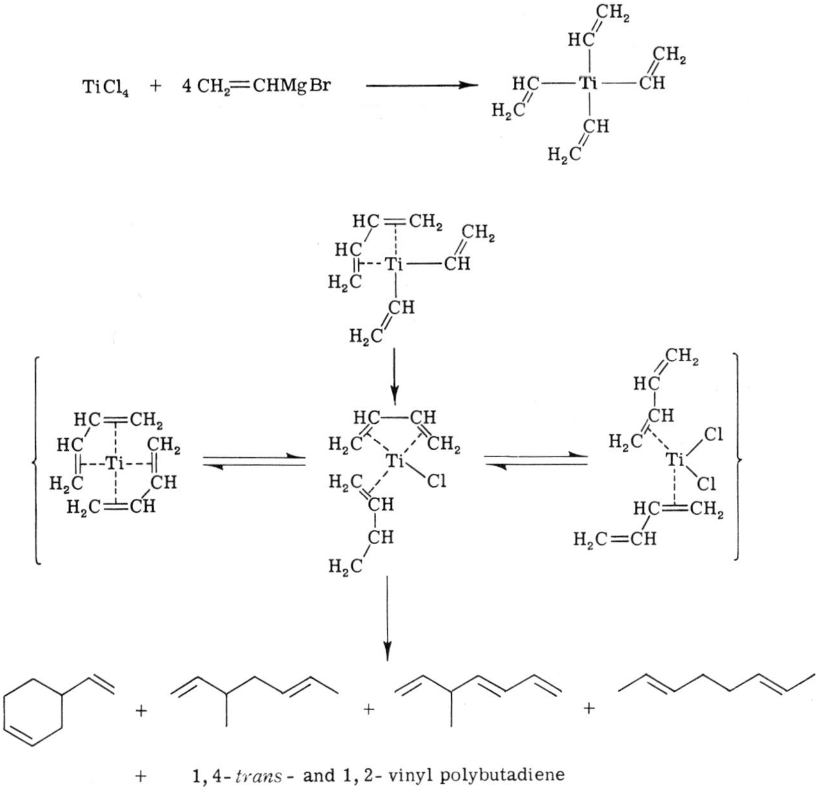

TiCl$_4$ + 4 CH$_2$=CHMgBr → [Ti complex structure]

FIG. 5. Double-coupling reaction of vinyl groups (34).

(14, 37). As shown in Fig. 6, the reaction is most probably initiated by the addition of ethylene to the tetrachloropalladium anion with displacement of chloride to form to π-ethylene complex (32). Addition of water with loss of chloride to form (33), followed by a π → σ rearrangement with migration of hydroxyl, loss of a proton, and a water molecule to form the σ complex (34). A migration of a hydride to the metal with σ–π rearrangement would produce the complex (35), which then rearranges to produce the complex (36). Decomposition of (36) would then produce acetaldehyde and a palladium hydride complex which is in turn reoxidized to the tetrachloropalladium(II) anion by copper(II). The resulting copper(I) chloride is air oxidized to regenerate copper(II) chloride. Stereochemical investigations of hydroxypalladation reactions similar to that proposed to lead to intermediate (35) have shown that when the nucleophile comes from outside of the coordination sphere of the metal,

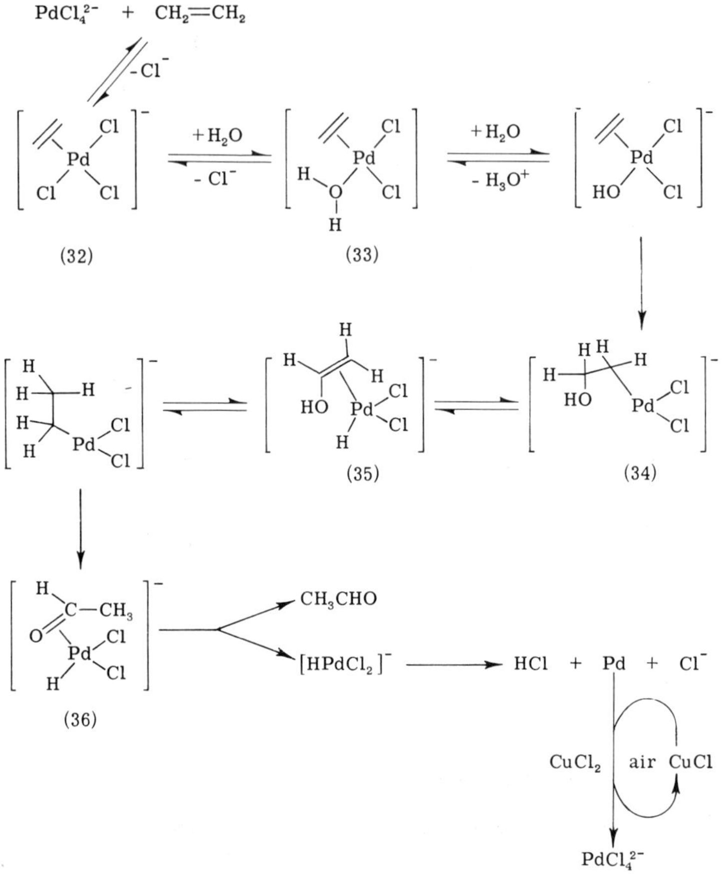

FIG. 6. The Wacker process (37).

attack is trans in a Markovnikov direction; and that if the nucleophile is coordinated prior to attack on the olefin, the addition proceeds cis in an anti-Markovnikov direction (38, 39).

In order to support the proposed mechanism for the Wacker Process, there has recently been growing interest in the preparation of stable π-vinyl alcohol complexes, preferably containing palladium or one of the other metals of the platinum group which show activity as Wacker catalysts. Until recently, the only well-characterized π complexes of vinyl alcohol have contained iron. One of the first reports of the formation of a stable π-vinyl alcohol complex by Ariyaratne and Green described the preparation of π-cyclopentadienyldicarbonyl(β-oxoethyl)iron (39) from the reaction of sodium cyclopentadienyldicarbonyliron (37) with chloroacetaldehyde (40).

Protonation of the σ-acetaldehyde complex (38) yielded the π-vinyl complex (39).

Recently, Cutler, Raghu, and Rosenblum have reported an alternate preparation which produces (38) in very high yield, via the σ-dimethylacetal complex (40) (41).

The σ-dimethylacetal complex (40) is then hydrolyzed on passage through an alumina column, producing the σ-acetaldehyde complex (38) which is converted to the π-vinyl alcohol complex (39) by protonation. Wakatsuki, Nojakura, and Murahashi reported the synthesis of 1,3-bis(π-ethenol)2,4-dichloro-μ-dichloroplatinum(II) (41) (42), however, Thyret, who recently reported the NMR evidence for the formation of tetracarbonyl(π-ethenol)iron (42) at low temperature, could not reproduce the synthesis of (41) (43).

A stable, readily soluble platinum π-vinyl alcohol complex has now been synthesized and well characterized. Initially, this compound, chloro(acetylacetonato)(π-ethenol)platinum(II) (**45**), was synthesized from the π-ethylene compound (**43**) via a vinyl trimethylsilyl ether complex (**44**) using the same general procedure employed in the synthesis of (**41**) and (**42**) (*44*).

Subsequently, a simpler procedure was devised for the preparation of (**45**) and the methyl derivative, chloro(acetylacetonato)-(π-propen-2-ol)platinum(II) (**48**), which involved simply treating (**43**) with acetaldehyde or acetone in the presence of aqueous potassium hydroxide (*45, 46*). The respective σ complexes are obtained as the potassium salts (**46**) and (**47**), most probably via attack by the α carbanion of the acetaldehyde or acetone on the platinum.

The potassium salts (**46**) and (**47**) were then treated with acid to yield the respective π-vinyl alcohol complexes (**45**), R = H, and (**48**), R = CH$_3$. The olefinic nature of the π-vinyl alcohol complex (**45**) has been investigated via X-ray crystalography (*47*), and the double bond, while perpendicular to the molecular plane, is not bisected by it. The principle dimensions are given in Fig. 7.

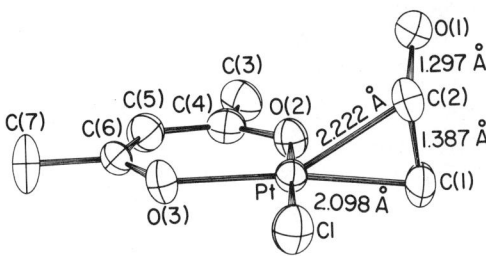

(45) R = H
(48) R = CH$_3$

The fact that the platinum–carbon-bond distances vary considerably suggests that there is some σ-bonded character between the β carbon and the metal (41, 42). The existence of this σ-bond character also supports the explanation for the anomalous NMR spectra of these complexes in polar solvents (44, 46). While the NMR spectrum of the π-vinylsilyl ether complex (44) exhibits the expected ABX pattern for the vinyl protons in

FIG. 7. Bond distances in chloro(acetylacetonato)(π-ethenol)platinum(II) (47).

acetone solution, the π-vinyl alcohol complex (45) exhibits an A$_2$X pattern for its vinyl protons under similar conditions. This apparent anomaly is easily explained by envoking a facile equilibrium between the π-vinyl alcohol compound (45) and its conjugate base, the σ-acetaldehyde complex (46). Indeed, such an equilibrium has been demonstrated and its has been shown that (45) is a monobasic acid with pK_a = 3.5 (46). This sort of equilibrium, to which (45) appears to be structurally predisposed as shown by the X-ray data of Fig. 7, would in effect scramble the β protons, H$_a$ and H$_b$, producing the A$_2$X pattern.

The NMR data for compounds (**47**) and (**48**) are given in Table I. A similar NMR spectrum has been reported by Ariyaratne and Green for the π-vinyl iron complex (**39**) (*40*).

TABLE I
NMR Data[a] for (**47**), (**49**), and a 1:2 Mixture

(*t*)	(**47**)	(**48**)	(**47**):(**48**) (1:2)	% shift
δ-H_c, ppm	7.27	9.16	8.48	64
$J_{H_{ab}-H_c}$, Hz	8	5.5	6.5	60
J_{Pt-H_c}, Hz	71	20	38	65
δ-H_{ab}, ppm	3.89	3.28	3.49	66
$J_{Pt-H_{ab}}$, Hz	76	113	99	64
δ-CH_3, ppm	1.99	1.68	1.88	61
δ-CH_3, ppm	1.89	1.64	1.80	64
δ-H, ppm	5.64	5.32	5.41	72

[a] At 100 MHz in aqueous acetone.

In addition to the σ-π equilibrium, an exchange between the π complex (**45**) and free acetaldehyde has been demonstrated using ^{14}C-labeled acetaldehyde (*46, 48*). After 45 hours at room temperature, 0.46% exchange was observed. While this appears quite small, one must remember that the free vinyl alcohol/acetaldehyde ratio has been estimated to have an upper limit of only 10^{-7} and that in the Wacker Process the equilibrium between π-coordinated and free vinyl alcohol would be shifted considerably in favor of free vinyl alcohol by the overpressure of ethylene. Thus, the behavior of the π-vinyl alcohol complexes (**45**) and (**48**) seem to support the importance of such complexes as intermediates in the Wacker and similar olefin oxidation processes.

C. Hydroformylation

The catalytic addition of hydrogen and carbon monoxide to olefins to give C_6-C_9 alcohols via the aldehyde is the industrially important hydroformylation or oxo reaction. The proposed mechanism of this reaction when carried out under high pressure and high temperature with a cobalt catalyst is summarized in Fig. 8 (*14, 49*). Recent investigations of this process suggests that cobalt hydridocarbonyl (**49**) is produced which then forms a π complex (**50**) with the olefin. Rearrangement of the π complex (**50**) to form the σ-alkyl cobalt tricarbonyl (**51**) occurs, followed by reaction of (**51**) with carbon monoxide to form the alkyltetracarbonyl (**52**) which is in equilibrium with the acylcobalt tricarbonyl (**53**). The

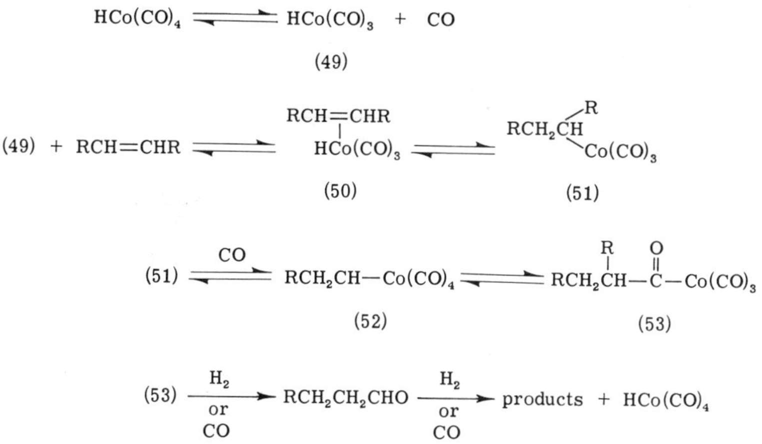

$$HCo(CO)_4 \rightleftharpoons HCo(CO)_3 + CO$$
$$(49)$$

$$(49) + RCH{=}CHR \rightleftharpoons \underset{(50)}{\underset{|}{\overset{RCH{=}CHR}{HCo(CO)_3}}} \rightleftharpoons \underset{(51)}{RCH_2\overset{R}{\overset{|}{CH}}{-}Co(CO)_3}$$

$$(51) \underset{}{\overset{CO}{\rightleftharpoons}} \underset{(52)}{RCH_2CH{-}Co(CO)_4} \rightleftharpoons \underset{(53)}{RCH_2\overset{R}{\overset{|}{CH}}{-}\overset{O}{\overset{\|}{C}}{-}Co(CO)_3}$$

$$(53) \xrightarrow[\text{or}]{H_2} RCH_2CH_2CHO \xrightarrow[\text{or}]{H_2} \text{products} + HCo(CO)_4$$
$$\text{CO} \qquad \qquad \text{CO}$$

FIG. 8. Proposed mechanism of hydroformylation reaction (14, 49).

acylcobalt tricarbonyl (53) is then reduced to the aldehyde by cobalt hydridocarbonyl or molecular hydrogen.

Recently, it has been discovered that catalysis by rhodium compounds is more effective than by the older cobalt catalyst; when tris(triphenylphosphine)rhodium chloride is treated with carbon monoxide, the catalyst bis(triphenylphosphine)rhodium carbonyl chloride is formed. This catalyst is very effective under very mild conditions (49–51). It is believed that the σ–π rearrangement is also important with this catalyst and operates in a manner analogous to that in the cobalt-catalyzed process, since stablization of the σ complex has been shown to lead to olefin isomerization and lower linear selectivity (52).

D. Activation of C—H Bonds

In a recent review, Parshall has discussed the activation of C—H bonds, including the interesting case of the exchange between benzene-d_6 and all of the protons of cyclopentadienerhodium diethylene (53). The C—D bond of benzene is being activated by the rhodium complex and actually broken. The potential application of this sort of process to synthesis is clearly apparent. The proposed mechanisms for the exchange between benzene-d_6 and coordinated ethylene and the subsequent exchange between the ethylene and the cyclopentadiene are given in Figs. 9 and 10, respectively. The importance of the σ–π rearrangement in these processes is quite apparent.

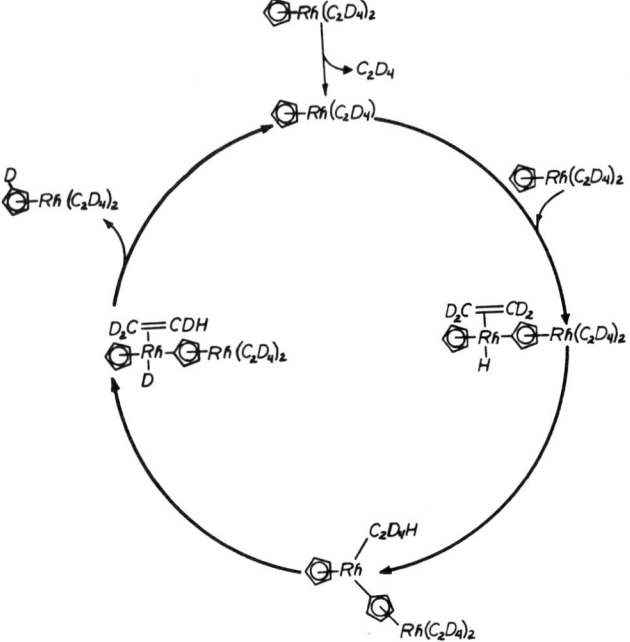

FIG. 9. Proposed mechanism for H-D exchange between coordinated ethylene and cyclopentadienide ligands (53).

E. OLEFIN METATHESIS

The olefin metathesis reaction is a relatively recent development in the chemistry of olefins. This reaction is in essence the catalyzed redistribution of alkylidene groups between olefins (54, 55).

$$\begin{matrix} R^1 \diagdown_C \diagup^H \\ \| \\ R^1 \diagup^C \diagdown_H \end{matrix} + \begin{matrix} R^2 \diagdown_C \diagup^H \\ \| \\ R^2 \diagup^C \diagdown_H \end{matrix} \xrightleftharpoons{\text{catalyst}} R^1HC{=}CHR^2 + R^1HC{=}CHR^2$$

The first catalysis of an olefin metathesis reaction was reported by Banks and Bailey in 1964 (56). They reported that activated molybdenum hexacarbonyl on alumina converted propylene, for example, into ethylene and 2-butene at 150°C and 30 atm. Oxides of rhenium are also powerful heterogeneous catalysts.

The first homogeneous catalysis of an olefin metathesis reaction

$$CH_2{=}CHCH_3 \xrightleftharpoons{Mo(CO)_6} CH_2{=}CH_2 + CH_3{-}CH{=}CH{-}CH_3$$

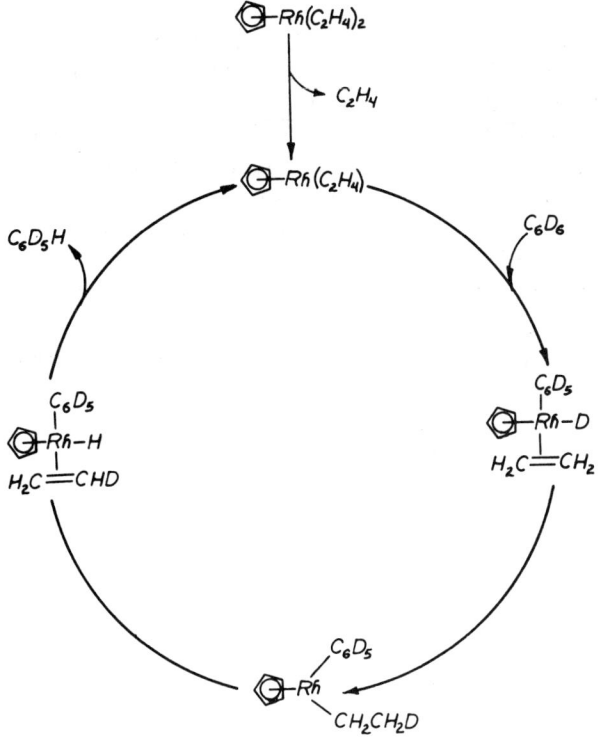

FIG. 10. Mechanism for exchange between benzene-d_6 and coordinated ethylene (53).

was reported Calderon and co-workers in 1967, using [WCl$_6$]—C$_2$H$_5$OH—C$_2$H$_5$AlCl$_2$ as a catalyst (57, 58). The relative rates of reaction of acyclic olefins is decreased as the bulk of substituents on the double bond increases. The rates are also decreased with the substitution of chlorines for the vinylic hydrogens. Cyclic olefins undergo metathesis producing larger (and smaller) rings.

In the presence of small amounts of acyclic olefins, a number of interesting polymers can be obtained. Beginning with cyclic olefins, only ring-opening polymerization can also occur.

The variety of olefins (and acetylenes) which undergo metathesis and the variety of active heterogeneous and homogeneous catalysts which have been used make it difficult to propose one general mechanism for this reaction. There are currently several possibilities which are being considered. These are shown in Fig. 11 (54, 55).

The syncronous making and breaking of a double bond is symmetry-forbidden (59), but the "quasi-butadiene" transition state is made possible by the presence of the metal atom (55). The most serious difficulty with this mechanism is that, at least under heterogeneous conditions,

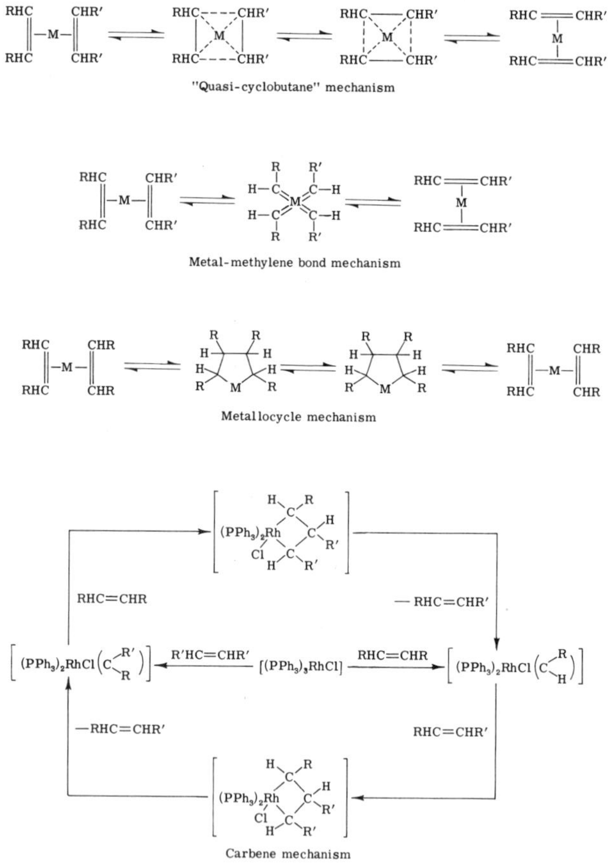

FIG. 11. Possible mechanisms for olefin metathesis reaction (54, 55).

even-chain-length olefins can react to form odd-chain-length products. These products might be explained by double-bond isomerizations followed by subsequent metathesis steps. The presence of cyclobutane which might persist in small quantities has not been reported. The other proposed mechanisms, which involve metal-carbene intermediates are substantiated to some extent by the observations that diazomethane reacts with metathesis catalysts producing dinitrogen and ethylene and that carbene complexes can generate metathesis catalysts. Metallocycles of the type proposed as intermediates in the metallocycle and carbene mechanisms have been isolated. A $\pi-\sigma$-type rearrangement is quite probable directly after coordination leading to the various proposed transition states, or alternatively, carbenes might be formed more directly. While the most recent studies of the chemistry of metal carbene complexes tend to support the carbene mechanism strongly, it is quite possible that the various metathesis process proceed by different mechanisms depending on the nature of the particular catalyst and the reaction conditions.

F. Carbene Complexes

The role of carbenes and metal carbene complexes in transition metal-catalyzed processes is suspected of being quite extensive (61). For example, the role of carbenes in the olefin metathesis reaction as described in the previous section is probably important (55, 60). It is quite possible that the $\sigma-\pi$ rearrangement is important in these reactions also, but this has not been investigated in detail.

G. Stereospecific Reactions

In many synthetic processes, it is desirable to obtain only one of several possible isomers of a new compound, and the most effective catalyst for such a process would be the one exhibiting the maximum stereospecificity. In order to design a stereospecific catalyst for a reaction, one may vary the size and preferred coordination geometry of the metal, or one may vary the electronic properties or geometries of the ligands attached to the metal.

As discussed in connection with olefin-coupling reactions and shown in Fig. 4, the coupling of vinyl Grignard reagents is stereospecific and dependent upon the transition metal catalyst used (32, 33). The dimerization of ethylene, shown in Fig. 6, was also shown to produce primarily the terminal olefin 1-butene (35). The size of the metal has also been shown to influence the course of the catalyzed oligomerization reactions of butadiene. When bis-(π-allyl) metal complexes are used as sources of "bare" metal atoms, the results differ as shown in Fig. 12 (*62*). Stereospecificity can be significantly increased by employing mono-π-allyl nickel chloride rather than the bis-π-allyl complex as a catalyst in the polymerization of butadiene and the dimerization of propylene, for example. The mono-π-allyl catalyst is formed by cracking the dimeric π-allyl nickel chloride with a trialkyl or triaryl phosphine and combining this with a Lewis acid such as aluminum chloride. The catalytic complex has been isolated. Utilization of a less basic phosphine such as triphenylphosphine favors less-branched products (*63*).

$$C_4H_6 \xrightarrow{\text{Ni}(PR_3)(AlCl_4)} 1,4\text{-}cis\text{-polybutadiene}$$

$$CH_3CH=CH_2 \xrightarrow{\text{Ni}(P(C_6H_{11})_3)(AlCl_4)} \underset{30-40\%}{CH_3CH_2CH_2\overset{CH_3}{\underset{|}{C}}=CH_2} + \underset{60-70\%}{CH_2=\overset{H_3C}{\underset{|}{C}}-\overset{CH_3}{\underset{|}{CH}}-CH_3}$$

$$\xrightarrow{\text{Ni}(P(C_6H_5)_3)(AlCl_4)} \underset{20\%}{CH_3CH_2CH_2CH=CHCH_3} + \underset{80\%}{CH_3CH=CH\overset{CH_3}{\underset{|}{CH}}CH_3}$$

Stereospecific coupling reactions of butadiene are catalyzed by the π-allyl titanium and chromium compounds also, in this case giving, respectively, *trans,trans,cis*-1,5,9-cycooctatriene and a 60:40 mixture of *trans,trans,trans*- and *trans,trans,cis*-1,5,9-cyclooctatriene (*62*). While triallylchromium catalyzes the coupling of butadiene to form the 1,2-polybutadiene exclusively, triallylchromium iodide produces only cyclododecatriene. In the presence of triallylcobalt, butadiene reacts to form 5-methylhepta-1,3,6-triene, but if triallylcobalt iodide is substituted, 1,4-*cis*-polybutadiene is formed. A comparison of the activities of π-allyl (*62*) and π-cyclopentadienyl metal catalysts (*64, 65*) reveals that the π-allyl compounds tend to be the more active catalysts. This is thought to be due to lower delocalization energy of the allyl anion (0.828β vs. 2.472β for

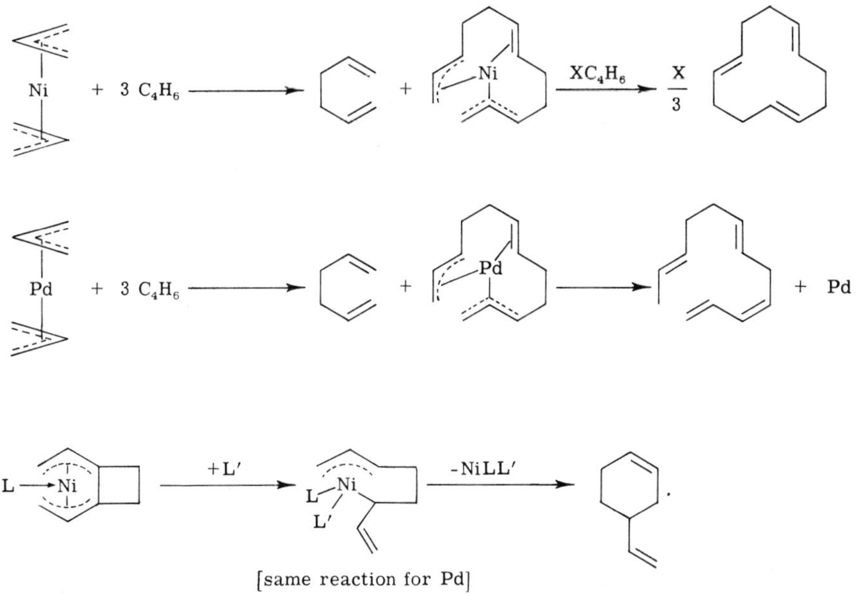

FIG. 12. Metal-catalyzed coupling of butadiene (62).

the cyclopentadienide anion, based on simple Hückel MO theory) which allows more facile $\sigma-\pi$ interconversion, and to fact that the allyl group has an electron affinity closer to that of the metal as compared to the cyclopentadienyl group which facilitates electron flow between the metal and ligand, thus favoring easily reversible redox changes during catalysis (62).

The effect of the ligand on the stereochemical course of a reaction catalyzed by a transition metal compound can be significant. For example, in the nickel-catalyzed cross-coupling reactions of alkyl Grignard reagents and aryl halides, the effect of changing the ligands on the metal is summarized in Fig. 13 (66, 67).

These reactions are believed to proceed via three intermediates whose relative importance (and hence the product distribution) depends upon the ability of the ligands to promote an isomerization through a $\sigma-\pi-\sigma$-rearrangement sequence as shown in Fig. 14 (66). The product distribution appears to depend more on the electron-donating ability of the ligands than their steric requirements. The stereochemical properties of the ligand are, however, critical in cases where optical activity is to be maintained, for example, the product, optically pure (+)-(s)-2-methylbutylbenzene (55) is obtained from the cross coupling of (+)-(s)-

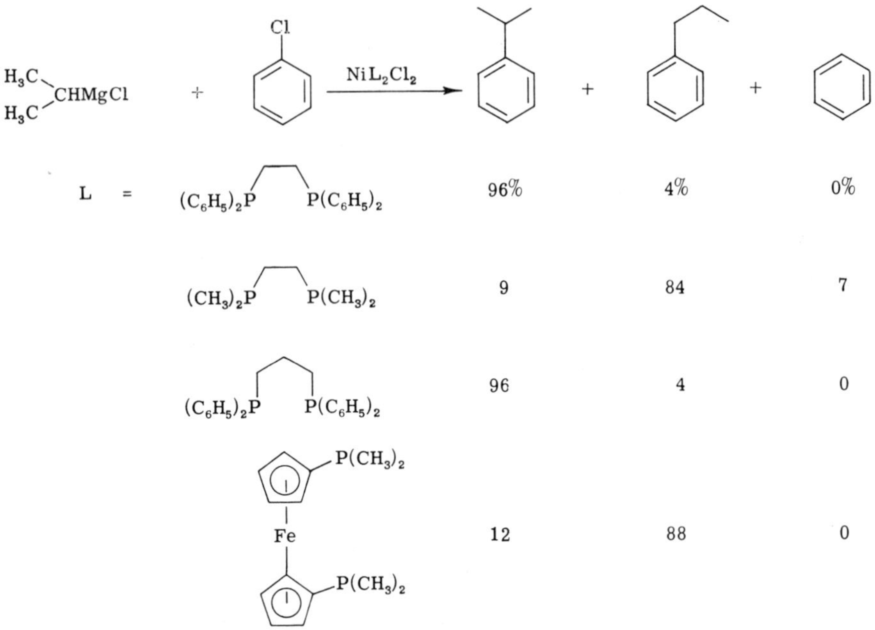

FIG. 13. Dependence of product distribution on ligands in nickel-catalyzed cross-coupling reactions (66).

L	isopropylbenzene	n-propylbenzene	benzene
$(C_6H_5)_2P\frown P(C_6H_5)_2$	96%	4%	0%
$(CH_3)_2P\frown P(CH_3)_2$	9	84	7
$(C_6H_5)_2P\frown P(C_6H_5)_2$	96	4	0
ferrocene-1,1'-diyl-bis-$P(CH_3)_2$	12	88	0

2-methylbutyl chloride (**54**) with chlorobenzene when the ligand dpp (**56**) is used, but when the same reaction is run with the ligand dmpe (**57**) the optical purity of the product is much lower (66).

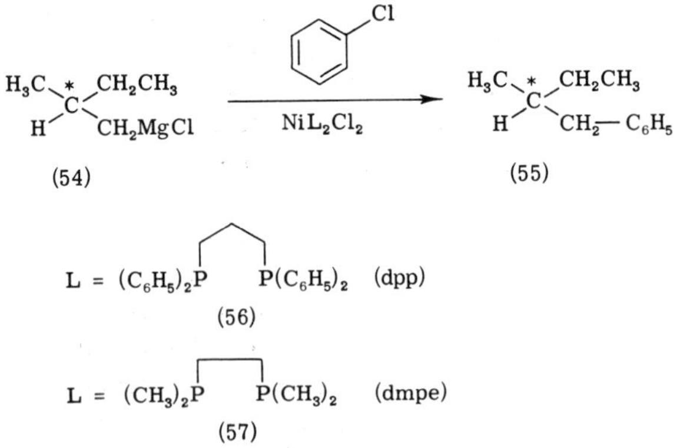

$$H_3C\overset{*}{\underset{H}{C}}\begin{array}{c}CH_2CH_3\\CH_2MgCl\end{array} \;+\; C_6H_5Cl \xrightarrow{NiL_2Cl_2} H_3C\overset{*}{\underset{H}{C}}\begin{array}{c}CH_2CH_3\\CH_2-C_6H_5\end{array}$$

(54) → (55)

L = $(C_6H_5)_2P\frown P(C_6H_5)_2$ (dpp)
(56)

L = $(CH_3)_2P\frown P(CH_3)_2$ (dmpe)
(57)

An optically active cross-coupling product may also be generated by a nickel catalysts having an optically active ligand. For example, (−)-

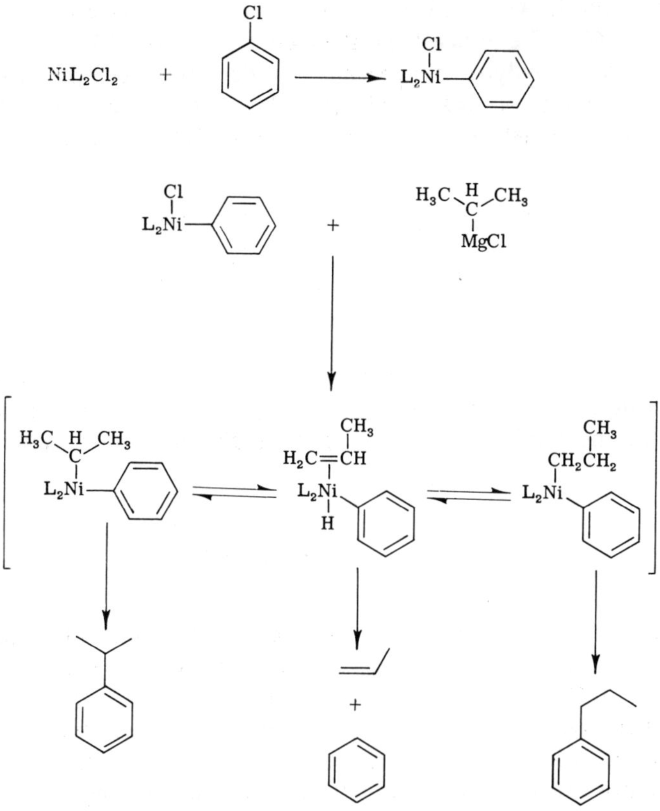

FIG. 14. Proposed mechanism for nickel-catalyzed cross-coupling reaction (66).

2,3-*O*-isopropylidene-2,3-dihydroxy-1,4-bis(diphenylphosphino)butane, (−)-diop (**58**).

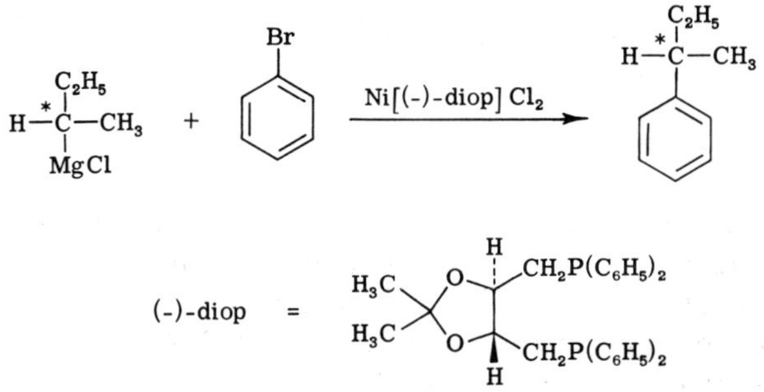

(**58**)

In the reaction shown the (R) product was obtained in 14.8% enantiomeric excess (ee) *(68)*. The optically active ligand, (−)-diop **(58)**, has been used in asymmetric catalytic hydrogenation. For example, the use of the rhodium(I) complex **(61)** converts the alkene **(59)** to (R)-N-acetylphenylalanine **(60)** with an optical yield of 72% and a chemical yield of 95% *(69)*.

$$\underset{(59)}{\underset{H_5C_6}{\overset{H}{\diagdown}}C=C\underset{NHCOCH_3}{\overset{CO_2H}{\diagup}}} \xrightarrow[\underset{(S \,=\, solvent)}{Rh(I)\,[(-)\text{-diop}]\,Cl\,[S]}]{H_2} \underset{(60)}{\overset{H_3CCONH}{\underset{C_6H_5}{H_2C-\overset{|}{C}H-CO_2H}}}$$

(59)   (61)   (60)

Apparently, the large optically active (−)-diop ligand **(58)** acts as a template during either the initial π-coordination step or π–σ-rearrangement step in the mechanism proposed for hydrogenation via a homogeneous catalyst *(69)*.

$$L_nM\!-\!\!\overset{H}{\underset{}{|}}\!\!\diagdown\!\!\!\diagup \longrightarrow L_nM-\overset{|}{\underset{|}{C}}-\overset{|}{\underset{|}{C}}-H$$

This step has been shown to be stereospecific in the case of hydrogenation (deuteration) catalyzed by dicyclopentadienylmolybdenumdihydride (dideuteride) **(62)** with cis addition of the metal hydride to the double bond *(70)*. In these systems there is spectral evidence for the initial formation of σ- and π-electron donor–acceptor complexes between the catalyst and substrate prior to π-complex formation and π–σ rearrangement *(70, 71)*. This catalyst **(62)** has also been used for the selective homogeneous hydrogenation of 1,4- or 1,3-dienes to monoenes, for example, cyclopentadiene **(63)**. The reactions are run at elevated temperature *(72)*.

$$\underset{(63)}{\bigcirc\!\!\!=} \xrightarrow[\underset{(C_5H_5)_2MoH_2}{no\ solvent}]{H_2,\ 140-180°} \underset{}{\bigcirc\!\!\!-}$$

(63)   (62)

Acyclic compounds such as **(64)** react with 1,2 addition to give monoenes *(72)*.

Hydrogenation by (π-arene)chromium tricarbonyl complexes has also been shown to convert dienes to *cis*-monoenes, but these catalysts appear to add to conjugated dienes in a 1,4 manner *(73, 74)*.

The stereospecific synthesis of a trisubstituted olefin has been reported via catalysis by tris(triphenylphosphine)rhodium hydride. This reaction is summarized in Fig. 15 *(75)*.

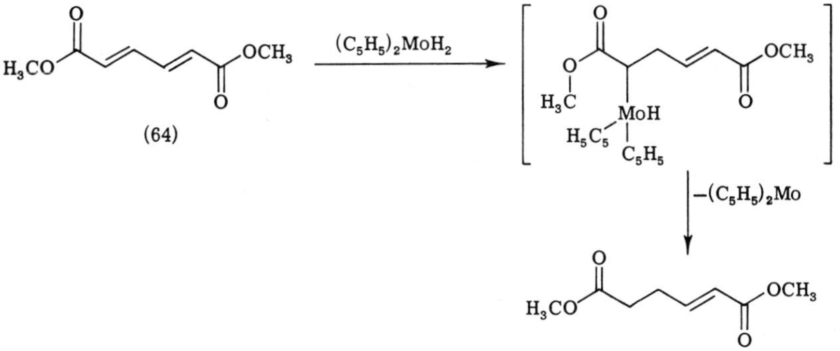

[Scheme: reaction of (64) H$_3$CO-diester diene with (C$_5$H$_5$)$_2$MoH$_2$ giving intermediate then −(C$_5$H$_5$)$_2$Mo to product]

Several π-allyl palladium chloride complexes have been prepared from steroidal olefins. Reaction of cholest-4-ene (65), for example, with bis(benzonitrile)palladium dichloride produced dimeric complexes of structures (66) and (67) (76).

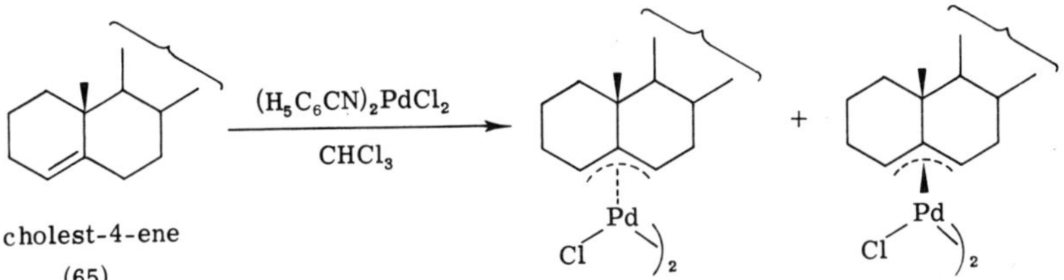

[Scheme: cholest-4-ene (65) + (H$_5$C$_6$CN)$_2$PdCl$_2$ / CHCl$_3$ → (66) + (67)]

The metal tends to orient itself such that it is on the less sterically hindered side of the steroid. The bulk of the metal atom then directs subsequent reduction or oxidation, thus making these reactions quite stereoselective (76, 77).

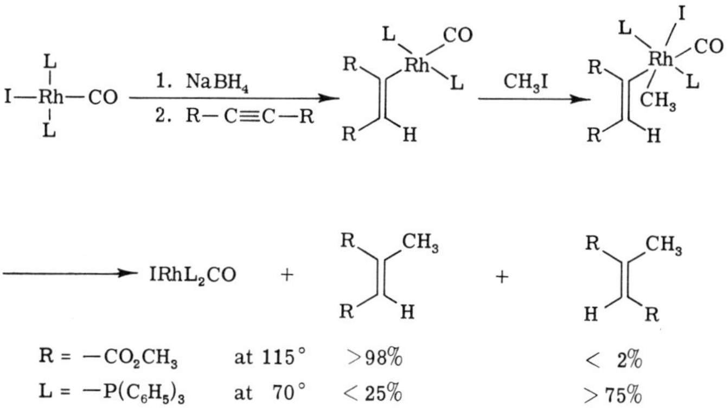

[Scheme for Fig. 15]

$R = -CO_2CH_3$ at 115° >98%   < 2%
$L = -P(C_6H_5)_3$ at 70°   < 25%   > 75%

FIG. 15. Stereospecific synthesis of a trisubstituted olefin (75).

H. Catalysis by Vitamin B_{12}

Traditionally when considering catalysis in biological systems primary emphasis has been placed on the enzymes—very high-molecular-weight proteins. However, recently there has been growing interest in the role of the transition metals, particularly as found in coenzymes such as B_{12}, in biochemical reactions.

The structures of the biologically active forms of B_{12} were solved relatively recently (1961) (78) and were shown to contain a cobalt atom surrounded by a corrin ring as shown in Fig. 16 (80). The crystal structure also showed a cobalt–carbon σ bond which was quite surprising since the few compounds with cobalt–carbon σ bonds known at that time were quite unstable (79). The corrin ring is similar to the porphyrin ring, but its greater saturation imports less rigidity than the porphyrin. Corrinoids with the axial 5,6-dimethylbenzimidazole substituent are called *cobalamins*. Vitamin B_{12} with Co(III) and CN^- in the top axial position is

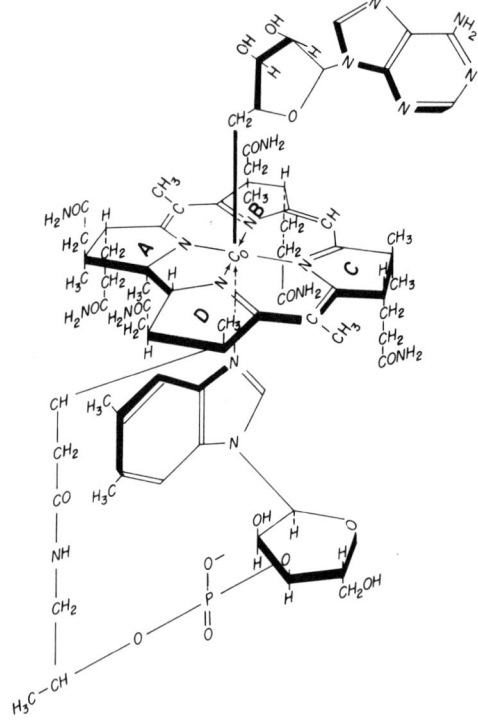

FIG. 16. Structure of vitamin B_{12} (80).

cyanocobalamin; vitamin B_{12a} with Co(III) and H_2O in the top axial position is aquocobalamin. The reduced cobalamins are B_{12r} (Co(III), no top axial substituent) and B_{12s} (Co(I), no top axial substituent). Co-factor B_{12} is called *5′-deoxyadenosylcobalamine*. In methylcobalamin, the top axial substituent is CH_3.

A number of model compounds have been used to study the reactions of B_{12}. The ligands of these compounds are usually oximes and Schiff's bases and pyridine is a common axial base. Cobalt-bisdimethylglyoxime, $Co(dmg)_2$ (68) is the most popular model and is commonly called *cobaloxime (81)*.

$$\underset{(68)}{\begin{array}{c}H_3C-C=N\diagdown\underset{|}{\overset{O\cdots H\cdots O}{|}}\diagup N=C-CH_3\\H_3C-C=N\diagup\underset{O\cdots H\cdots O}{\overset{Co}{|}}\diagdown N=C-CH_3\end{array}}$$

The reactions catalyzed by B_{12} may be grouped into two classes: those catalyzed by methylcobalamin and those catalyzed by cofactor B_{12}. The former reactions include formation of methionine from homocysteine, methanogenesis (formation of methylmercury is an important side reaction), and synthesis of acetate from carbon dioxide (82). The latter reactions include the ribonucleotide reductase reaction and a variety of isomerization reactions (82). Since dehydration and deamination have been studied quite extensively and very possibly proceed via $\sigma-\pi$ rearrangements, these processes will be considered in detail.

The dehydration and deamination reactions appear to operate in a parallel fashion and will be considered together. Schrauzer and Silbert propose a base-catalyzed cleavage of the carbon–cobalt bond in the B_{12}-catalyzed diol dehydration reaction as shown in Fig. 17 (81), based on demonstrated lability of the beta hydrogens in alkyl cobaloximes with electronegative groups in this position.

$$\underset{\underset{Co(dmg)_2}{|}}{CH_2-\overset{\overset{H}{|}}{C}H-R}\xrightarrow{OH^-}[Co(I)(dmg)_2+CH_2=CHR]\longrightarrow CH_3CHO$$

R = OH, CN when R = OH

It is possible that the cobalt becomes coordinated to the double bond during this process, thus implicating a $\sigma-\pi$ rearrangement. That this is

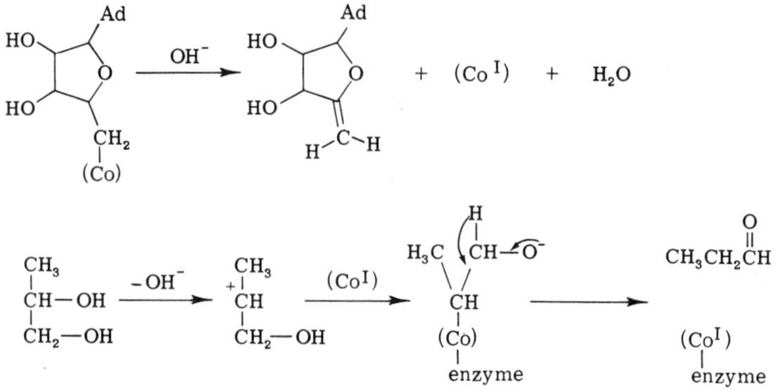

FIG. 17. Proposed mechanism for B_{12}-catalyzed diol dehydration (81).

the mode of action in the enzyme system has been questioned since an attempt to trap the proposed B_{12s} intermediate has been unsuccessful, and further, an authentic sample of formylmethylcobalamin, synthesized by several different routes, does not behave as an intermediate would be expected to in this reaction (83). Recently, reactions of vinyl ethers with cobalamines and cobaloximes have been reported (84). These reactions produced products (70) and (72), as shown in Fig. 18, in which the new ligands are σ bonded to the cobalt. These reactions are thought to proceed through π complexes (69) and (71). It is possible that the σ-bonded compounds are formed directly without π-complex formation by direct electrophilic addition of the vinyl group with concomitant deprotonation by triethylamine, as shown in Fig. 19. Several of the mechanisms proposed for B_{12}-catalyzed reactions which invoke π-complex intermediates are summarized in Figs. 20, 21 and 22 (85–88).

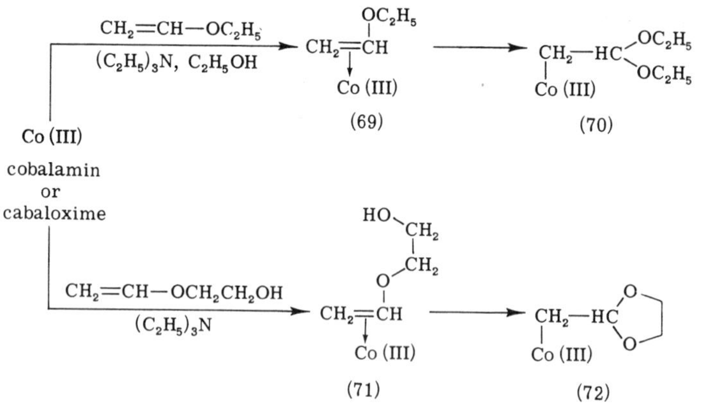

FIG. 18. Reaction of cobalamine and coboloxime with vinyl ethers (84).

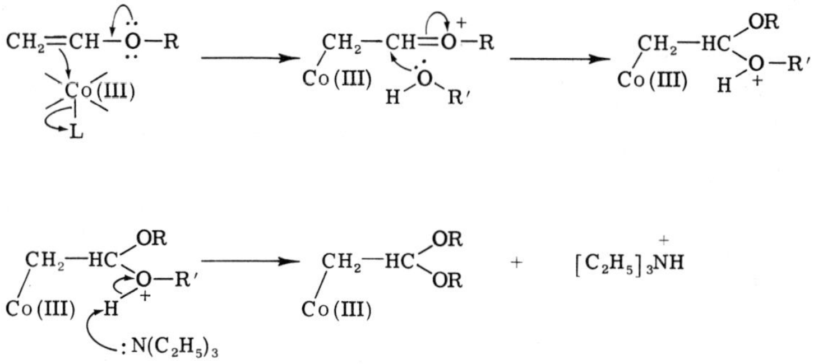

$\sigma-\pi$ REARRANGEMENTS AND THEIR ROLE IN CATALYSIS 259

FIG. 19. Direct-addition mechanism for addition of vinyl ethers to cobalamin and cobaloximes.

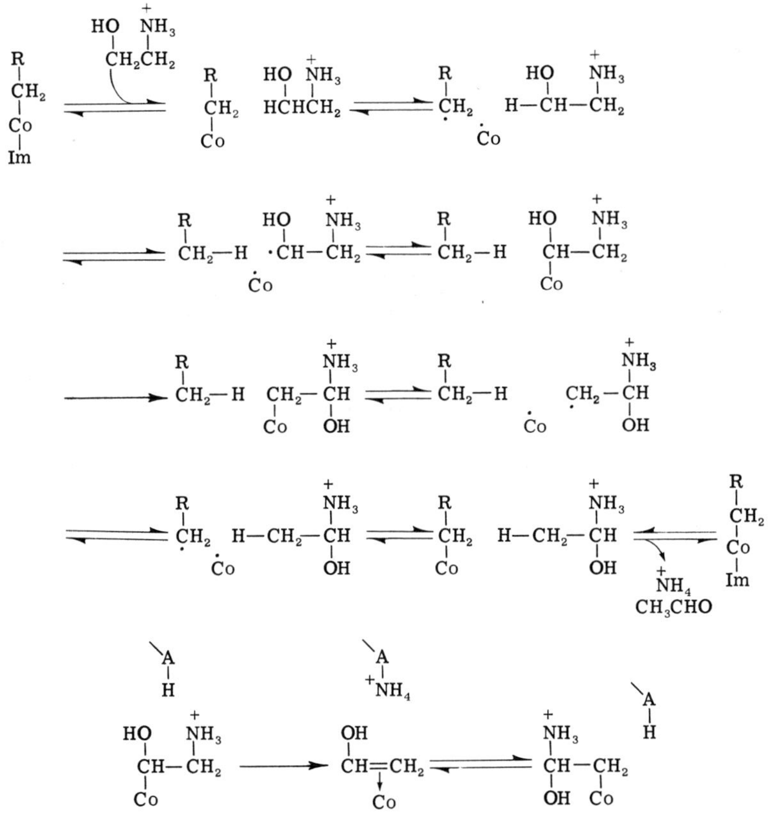

FIG. 20. Proposed mechanism for B_{12}-catalyzed deamination (85).

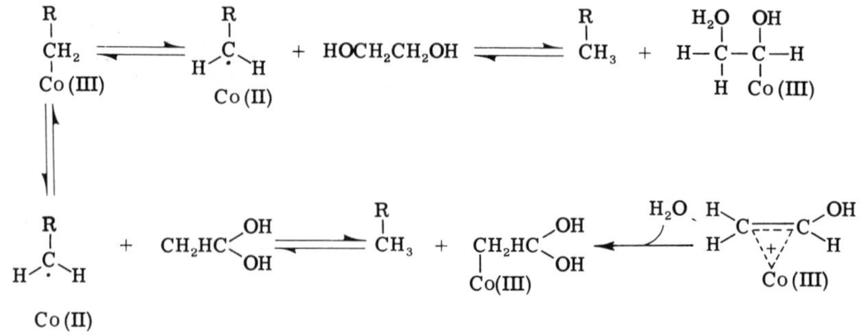

FIG. 21. Scheme of Silverman and Dolphin for vitamin B_{12}-catalyzed diol dehydration (87).

These mechanisms are supported by a good deal of evidence for the symmetric intermediate of type (**73**) which would react via a π–σ rearrangement.

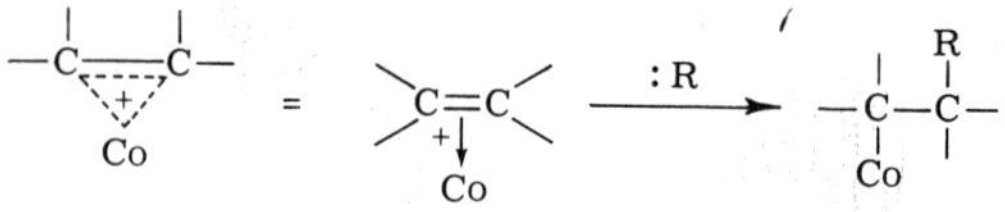

(73)

Thus, while there is some controversy as to the mode of action of B_{12} in its enzyme systems, it is quite possible that the σ–π interconversion which appears to be ubiquitous in metal-catalyzed nonbiological processes is also important in the processes of life.

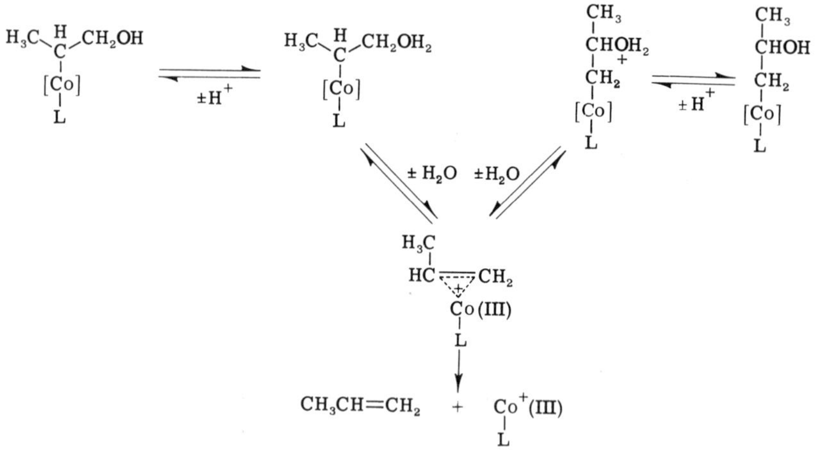

FIG. 22. Proposed mechanism diol dehydrase reaction (88).

V. Conclusion

The preceeding discussion has attempted to present a survey of the involvement of the $\sigma-\pi$ rearrangement reaction in catalysis. Because of the very large number of catalytic reactions presently known, complete coverage of this topic is impossible in a limited space. Instead we have attempted to present a representative sample of catalytic processes with the $\sigma-\pi$ rearrangement as a unifying thread.

For a more complete description of the principles of transition metal catalysis, a number of the excellent reviews published recently should be consulted. These include, in addition to those already cited, several reviews of rhodium (89) and palladium (90–93) chemistry, addition (94, 95) and insertion (96) reactions, homogeneous catalytic hydrogenation (97), hydride complex chemistry (98), nitrogen fixation (99), and organometallic complexes as potential synthetic reagents (100).

ACKNOWLEDGMENTS

The preparation of this manuscript was made possible by the generous support of the Robert A. Welch Foundation, Grant A-420. The authors are also indebted to their co-workers, whose names appear in the literature cited. We would also like to especially thank Dr. Ronald Dubois for his assistance in the preparation of the manuscript.

REFERENCES

1. H. H. Zeiss and M. Tsutsui, *J. Am. Chem. Soc.* **79,** 3062 (1957).
2. T. J. Kealy and P. L. Pauson, *Nature (London)* **168,** 1039 (1951).
3. E. O. Fischer and W. Hafner, *Z. Naturforsch., Teil B* **10,** 665 (1955).
4. E. O. Fischer and W. Hafner, *Z. Anorg. Chem.* **286,** 146 (1956).
5. R. B. Woodward, M. Rosenblum, and M. C. Whiting, *J. Am. Chem. Soc.* **74,** 3458 (1952).
6. M. Tsutsui, *Z. Chem.* **6,** 215 (1963).
7. M. Tsutsui, *Ann. N.Y. Acad. Sci.* **125,** 147 (1965).
8. M. Rosenblum, "Chemistry of the Iron Group Metallocenes." Wiley, New York, 1969.
9. M. Tsutsui and M. N. Levy, *Z. Naturforsch., Teil B* **21,** 823 (1966).
10. M. Tsutsui, *Trans. N.Y. Acad. Sci.* **30,** 658 (1968).
11. M. Tsutsui and G. Chang, *Can. J. Chem.* **41,** 1255 (1963).
12. M. Tsutsui, M. Hancock, J. Ariyoshi, and M. N. Levy, *Angew. Chem., Int. Ed. Engl.* **8,** 410 (1969).
13. J. P. Collman, *Acc. Chem. Res.* **1,** 136 (1968).
14. M. Hancok, M. N. Levy, and M. Tsutsui, *in* "Organometallic Reactions" (E. I. Becker and M. Tsutsui, eds.), Vol. 4, pp. 3–72. Wiley, New York, 1972.
15. M. Tsutsui, M. Hancock, J. Ariyoshi, and M. N. Levy, *J. Am. Chem. Soc.* **91,** 5233 (1969).
16. M. Tsutsui and C. E. Hudman, *Trans. N.Y. Acad. Sci.* **34,** 595 (1972).
17. M. Tsutsui and C. Hudman, *Chem. Lett.* p. 777 (1972).
18. H. Brintzinger, *J. Am. Chem. Soc.* **88,** 4305 (1966).

19. H. Brintzinger, *J. Am. Chem. Soc.* **88**, 4307 (1966).
20. H. Brintzinger, *J. Am. Chem. Soc.* **89**, 6871 (1967).
21. W. P. Long and D. S. Breslow, *J. Am. Chem. Soc.* **82**, 1953 (1960).
22. K. Causs and H. Bestian, *Justus Liebigs Ann. Chem.,* **654**, 8 (1962).
23. M. E. Vol'pin and V. B. Shur, *Nature (London)* **209**, 1236 (1966).
24. H. Brintzinger, *Biochemistry* **5**, 3947 (1966).
25. M. L. H. Green and P. L. I. Nagy, *J. Chem. Soc.* p. 189 (1963).
26. J. Kuratek and J. K. Seyler, *J. Organomet. Chem.* **3**, 421 (1965).
27. J. K. P. Ariyartne and M. L. H. Green, *J. Organomet. Chem.* **1**, 90 (1963).
28. M. L. H. Green and M. Cousins, *J. Chem. Soc.* p. 889 (1963).
29. P. Cossee, *J. Catal.* **3**, 80 (1964).
30. E. J. Arlman, *J. Catal.* **3**, 89 (1964).
31. E. J. Arlman and P. Cossee, *J. Catal.* **3**, 99 (1964).
32. M. Tsutsui, *Trans. N.Y. Acad. Sci.* **26**, 423 (1964).
33. M. Tsutsui, J. Aryoshi, T. Koyano, and M. N. Levy, *Adv. Chem. Ser.* No. 70, p. 266 (1968).
34. M. Tsutsui and J. Ariyoshi, *Trans. N.Y. Acad. Sci.* **26**, 431 (1964).
35. M. Tsutsui and T. Koyano, *J. Polym. Sci., Part A-1,* **5**, 681 (1967).
36. M. Tsutsui and T. Koyano, *J. Polym. Sci., Part A-1,* **5**, 683 (1967).
37. R. F. Heck, "Organotransition-Metal Chemistry," pp. 104–106. Academic Press, New York, 1974.
38. J. K. Stille and D. E. James, *J. Am. Chem. Soc.* **97**, 674 (1975).
39. J. K. Stille, L. F. Hines, R. W. Fries, P. K. Wong, D. E. James, and K. Lau, *Adv. Chem. Ser.* No. 132, p. 90 (1974).
40. J. K. P. Ariyaratne and M. L. H. Green, *J. Chem. Soc.* p. 1 (1964).
41. A. Cutler, S. Raghu, and M. Rosenblum, *J. Organomet. Chem.* **77**, 381 (1974).
42. Y. Wakatsuki, S. Nozakura, and S. Murahashi, *Bull. Chem. Soc. Jpn.* **42**, 273 (1969).
43. H. Thyret, *Angew. Chem., Int. Ed. Engl.* **11**, 520 (1972).
44. M. Tsutsui, M. Ori, and J. Francis, *J. Am. Chem. Soc.* **94**, 1414 (1972).
45. J. Hillis and M. Tsutsui, *J. Am. Chem. Soc.* **95**, 7907 (1973).
46. J. Hillis, J. Francis, M. Ori, and M. Tsutsui, *J. Am. Chem. Soc.* **96**, 4800 (1974).
47. F. A. Cotton, J. N. Francis, B. A. Frenz, and M. Tsutsui, *J. Am. Chem. Soc.* **95**, 2483 (1973).
48. M. Tsutsui and J. Francis, *Chem. Lett.* p. 663 (1973).
49. F. A. Cotton and G. Wilkinson, "Advanced Inorganic Chemistry," 2nd Ed., pp. 791–792. Wiley (Interscience), New York, 1966.
50. J. A. Osborn, F. H. Jardine, J. F. Young, and G. Wilkinson, *Chem. Commun.* p. 131 (1965).
51. J. A. Osborn, F. H. Jardine, J. F. Young, and G. Wilkinson, *Chem. Ind. (London)* p. 560 (1965).
52. W. O. Haag and D. D. Whitehurst, *Proc. Int. Congr. Catal., 5th, 1972* p. 465 (1973).
53. G. W. Parshall, *Acc. Chem. Res.* **8**, 113 (1975).
54. N. Calderon, *Acc. Chem. Res.* **5**, 127 (1972).
55. R. J. Haines and G. J. Leigh, *Chem. Soc. Rev.* **4**, 155 (1975).
56. R. L. Banks and G. C. Baily, *Ind. Eng. Chem., Prod. Res. Dev.* **3**, 170 (1964).
57. N. Calderon, H. V. Chen, and K. W. Scott, *Tetrahedron Lett.* p. 3327 (1967).
58. N. Calderon, E. A. Ofstead, W. A. Judy, and K. W. Scott, *J. Am. Chem. Soc.* **90**, 4133 (1968).
59. R. Hoffmann and R. B. Woodward, *Acc. Chem. Res.* **1**, 17 (1968).
60. E. L. Muetterties, *Inorg. Chem.* **14**, 951 (1975).

61. D. J. Cardin, B. Cetinkaya, M. J. Doyle, and M. F. Lappert, *Chem. Soc. Rev.* **2**, 99 (1973).
62. Wilke, B. Bogdanović, P. Hardt, P. Heimbach, W. Keim, M. Kröner, W. Oberkirch, K. Tanaka, E. Steinrücke, D. Walter, and H. Zimmerman, *Agnew. Chem., Int. Ed. Engl.* **5**, 151 (1966).
63. G. Wilke, *Proc. Robert A. Welch Found. Conf. Chem. Res., IX, Organomet. Compounds, R. A. Welch Found., Houston, Tex.* p. 165 (1965).
64. M. Tsutsui, *Ann. N.Y. Acad. Sci.* **93**, 133 (1961).
65. M. Tsutsui and J. Ariyoshi, *J. Polym. Sci., Part A*, **3**, 1729 (1965).
66. K. Tomao, Y. Kiso, K. Sumifani, and M. Kumada, *J. Am. Chem. Soc.* **94**, 9268 (1972).
67. M. Kumada, *in* "Organotransition-Metal Chemistry" (Y. Ishii and M. Tsutsui, eds.), pp. 211–222. Plenum, New York, 1975.
68. Y. Kiso, K. Tomao, N. Miyake, K. Yamamoto, and M. Kumada, *Tetrahedron Lett.* p. 3 (1974).
69. R. E. Harmon, S. K. Gupta, and D. J. Brown, *Chem. Rev.* **73**, 21 (1973).
70. A. Nakamura and S. Otsuka, *J. Am. Chem. Soc.* **95**, 7262 (1975).
71. A. Nakamura and S. Otsuka, *J. Am. Chem. Soc.*, **95**, 5091 (1973).
72. A. Nakamura and S. Otsuka, *Tetrahedron Lett.* p. 4529 (1973).
73. E. N. Frankel and R. O. Butterfield, *J. Org. Chem.* **34**, 3930 (1969).
74. E. N. Frankel, E. Selke, and C. A. Glass, *J. Org. Chem.* **34**, 3936 (1969).
75. J. Schwartz, D. W. Hart, and J. Holden, *J. Am. Chem. Soc.* **94**, 9269 (1972).
76. D. Neville Jones and S. D. Knox, *Chem. Commun.* p. 165 (1975).
77. D. Neville Jones and S. D. Knox, *Chem. Commun.* p. 166 (1975).
78. D. Crowfoot-Hodgkin, *Proc. R. Soc., Ser. A* **288**, 294 (1965).
79. G. N. Schrauzer, *Acc. Chem. Res.* **1**, 97 (1968).
80. H. A. O. Hill, *in* "Inorganic Chemistry" (G. L. Eichhorn, ed.), p. 1068. Elsevier, New York, 1973.
81. D. G. Brown, *Prog. Inorg. Chem.* **18**, 177 (1973).
82. T. C. Stadtman, *Science* **171**, 859 (1971).
83. R. B. Silverman, D. Dolphin, T. J. Carty, E. K. Krudel, and R. H. Abeles, *J. Am. Chem. Soc.* **96**, 7096 (1974).
84. R. B. Silverman and D. Dolphin, *J. Am. Chem. Soc.* **96**, 7094 (1974).
85. R. H. Prince and D. A. Stotter, *J. Inorg. Nucl. Chem.* **35**, 321 (1973).
86. B. M. Babior, *J. Biol. Chem.* **245**, 6131 (1970).
87. R. B. Silverman and D. Dolphin, *J. Am. Chem. Soc.* **95**, 1686 (1973).
88. K. L. Brown and L. L. Ingraham, *J. Am. Chem. Soc.* **96**, 7681 (1974).
89. R. Cramer, *Acc. Chem. Res.* **1**, 186 (1968).
90. P. M. Henry, *Acc. Chem. Res.* **6**, 16 (1973).
91. J. Tsuji, *Acc. Chem. Res.* **6**, 8 (1973).
92. J. Tsuji, *Acc. Chem. Res.* **2**, 144 (1969).
93. R. F. Heck, *Proc. Robert A. Welch Found. Conf. Chem. Res., XVII, Org. Inorg. Reagents Synth. Chem., R. A. Welch Found., Houston, Tex.* p. 53 (1974).
94. J. Halpern, *Acc. Chem. Res.* **3**, 386 (1970).
95. R. F. Heck, *Acc. Chem. Res.* **2**, 10 (1969).
96. G. P. Chinsoli, *Acc. Chem. Res.* **6**, 422 (1973).
97. G. Brieger and T. J. Nestrick, *Chem. Rev.* **74**, 567 (1975).
98. H. D. Kaesz and R. B. Saillant, *Chem. Rev.* **72**, 231 (1972).
99. A. D. Allen and F. Bottomley, *Acc. Chem. Res.* **1**, 360 (1968).
100. M. Rosenblum, *Acc. Chem. Res.* **7**, 122 (1974).

Characterization of Molybdena Catalysts

F. E. MASSOTH

Department of Mining and Fuels Engineering
University of Utah
Salt Lake City, Utah

I. Introduction 266
II. Background 267
III. Catalyst Preparation 268
IV. Characterization Techniques 269
    A. X-Ray Diffraction Analysis (XRD) 272
    B. Gravimetric and Volumetric Measurements 273
    C. Electron Spin Resonance (ESR) 276
    D. Diffuse Reflectance Spectra 278
    E. Magnetic Measurements 280
    F. Photoelectron Spectroscopy (ESCA) 281
    G. Infrared 283
    H. Acidity 284
    I. Adsorption 285
    J. Sulfur Analyses 286
    K. Miscellaneous 288
V. State of Molybdena Catalysts 289
    A. Oxidized State 289
    B. Reduced State 291
    C. Sulfided State 292
VI. Catalyst Activities 294
    A. Reduced Catalyst 294
    B. Sulfided Catalyst 295
VII. Proposed Models of the Sulfided Catalyst 298
    A. Monolayer Model 298
    B. Intercalation Model 300
    C. Contact Synergism Model 301
    D. Relevance of Models 301
VIII. Role of Cobalt 302
IX. Nickel–Molybdena Catalysts 303
X. Active Sites—Concluding Remarks 304
    A. Reduced Catalysts 304
    B. Sulfieded Catalysts 305
    References 306

I. Introduction

Molybdena catalysts have been with us for quite a long time. The term molybdena is used here to denote a composite catalyst consisting of molybdenum oxide supported on an activated support, commonly alumina. Early it was found that certain transition metals, notably cobalt and nickel, promote the molybdena catalyst for hydrodesulfurization (HDS) reactions.

The so-called *cobalt molybdate* catalyst has been used much in the petroleum industry for hydrotreating and hydrodesulfurization. More recently, these catalysts have been employed in coal liquefaction and synthoil upgrading. The latter probably accounts for the recent rash of publications on this very interesting catalyst system. Indeed, of the papers surveyed for this review, the majority have been published in the past 5 years with no letup in sight.

Part of this renewed interest may also be due to the development of newer and better techniques to characterize catalysts and the emergence of "catalyst characterization" as a legitimate endeavor in catalysis. Even some of the old classical studies on catalysts are receiving a second look with these new sophisticated techniques.

Interest in this catalyst is worldwide, as attested by the national origin of authors of published papers on the subject. In almost every major country and in many smaller ones work is going on in some aspect of this catalyst. Communist countries are not well represented in this review because of the difficulty in obtaining and/or translating the literature. However, considerable work is being done in these countries as evidenced in the recent book by Weisser and Landa (*1*) on sulfided catalysts. On the other hand, much of this work is of an applied nature and little basic information on catalyst states is reported.

The object of this review is threefold: (1) to discuss the various characterization techniques which have been applied to this catalyst system, (2) to relate what each technique reveals about the nature of the catalyst, and (3) to present an overall picture of the state of the catalyst as it now appears. We will not discuss the vast literature on catalyst activity testing, kinetics, or mechanisms here. These are subjects for review themselves. However, we will mention some selective catalyst activity tests which were designed to give some fundamental insight into the catalyst state or active sites present. Also, we will not discuss in detail the considerable work reported on pure compounds (unsupported) of molybdenum, cobalt, and/or aluminum; but we will have occasion to compare some of their properties to our catalyst systems to assess to what degree they may be present in the catalyst.

II. Background

Many reviews have been written on hydrodesulfurization. Of those dealing with the molybdena catalyst may be mentioned the articles by McKinley (2), Mitchell (3), Schuman and Shalit (4), Weisser and Landa (1), and Schuit and Gates (5). Other reviews, more specific to catalyst states, but predominantly expressing the authors' viewpoints, are given by Amberg (6), De Beer and Schuit (7), Massoth (8), and Delmon (9).

The molybdena catalyst probably represents one of the best examples of a catalyst which has been studied by a wide variety of techniques. This, in itself, is a good gauge of the complexity of the catalyst.

Standardized characterization techniques have been well developed for metallic catalysts and acidic-type catalysts. No such standard techniques have yet emerged for characterizing supported oxide or sulfide catalysts, with few exceptions; and especially how these may relate to catalyst activity. Thus, we are not at the stage where we can discuss turnover numbers or facile versus demanding reactions for these catalysts at the present time.

Table I lists the major characterization techniques which have been applied to the molybdena catalyst. They may be grouped into two broad categories: nonspectroscopic and spectroscopic methods. Space does not permit a full discussion of the theory, experimental techniques, or interpretation of results of these techniques—we give here only the author's interpretations of their results. The reader is referred to any number of standard texts or reviews on the specific technique for a more complete description.

The basis of most of these methods is comparative in nature, that is, results are compared to known pure compounds. The inference is that if the catalyst resembles some property of the known compound, then that compound is present in the catalyst. However, caution must be exercised in this interpretation because we may be dealing with indefinite and nonstoichiometric surface complexes which may have some properties in

TABLE I
Major Characterization Techniques Employed

Nonspectroscopic	Spectroscopic
X-Ray diffraction (XRD)	Optical (diffuse reflectance)
Gravimetric/volumetric	Infrared
Magnetic measurements	Photoelectron (ESCA)
Electron microscopy	Electron spin resonance (ESR)
Catalytic activity	

common with a bulk phase but others which are not. Hence, to characterize it on the basis of known bulk phases may be misleading.

Along this line, the limitations of the technique used must be recognized. Some measure predominantly bulk properties, e.g., X-ray diffraction and magnetic susceptibility; whereas, others are sensitive to surface composition, e.g., adsorption and ESCA. For example, in one reported study only cobalt in tetrahedral coordination was found on a catalyst by diffuse reflection spectroscopy, but magnetic measurements revealed that octahedral cobalt must also be present (*10*). Thus, it is dangerous to rely on any one method to characterize these catalysts.

III. Catalyst Preparation

Anyone who has had occasion to peruse the literature on these catalysts will be struck by the diversity and apparent contradictions of results and interpretations reported by different authors. This dismaying state of affairs may account for the tendency of workers to quote other workers' results which support theirs but to ignore other data which disagree. It almost seems as though everyone is working on a different catalyst—and this may not be as farfetched as it seems. Much of the disparity in results may be due to differences in catalyst preparation, which has generally not been fully appreciated.

Some factors which may be important in determining the final state of the catalyst are as follows:

1. *Support used:* The type of support, e.g., gamma-, eta-, boehmite-alumina, could greatly affect the disposition and/or degree of interaction with Mo and Co. Also, impurities in the support could change the acidic character of the support and consequent interactions. Some commercial aluminas contain sodium, chloride, sulfate, or silica impurities.

2. *Mode of preparation:* Differences can be expected between catalysts prepared by simple impregnation using incipient wetness, impregnation by immersion in a solution of the components, coprecipitation, kneading, or mixed-mull extrusion. The order of addition of Co and Mo when calcination is made between impregnations can also affect the outcome.

3. *Calcination:* The rate of heat schedule as well as the final temperature employed are important variables. A high enough temperature must be employed to ensure decomposition of the salts used in the preparation. At sufficiently high temperature, some solid-state reactions are engendered. Further, differences can be obtained whether using forced

dry air or a closed furnace. The latter can cause hydrothermal steaming of the catalyst from the water liberated in the calcination.

4. *Pretreatment:* In order to compare results of catalyst treatments, e.g., reduction and sulfidation, precise conditions should be given, such as temperature, time, reagents, and partial pressures. Also, whether the catalyst has been treated under static, recirculation, or flow conditions is important. Finally, the desorption conditions after treatment should be specified as some adsorbed species are only slowly removed and could contribute to misleading results if sufficient time is not allowed for their removal. This is particularly true of H_2S, H_2O, and other polar compounds as well as organic heterocyclics.

IV. Characterization Techniques

The individual techniques used to characterize molybdena catalysts are now considered. Table II presents a listing of articles concerning the characterization of molybdena catalysts. Unless otherwise specified, we implicitly refer to Mo and/or Co supported on an activated alumina, commonly γ-Al_2O_3. Most work has been done on the calcined (oxidized) state of the catalyst because of ease of sample handling. Reduced and sulfided catalysts are more difficult to work with since for meaningful results, exposure of these samples to air or moisture should be rigorously avoided. Therefore, sample transfer or special *in situ* treatment facilities must be provided.

TABLE II
Characterization of Molybdena Catalysts

Co[a]	Mo[a]	Preparation[b]	Temperature (°C)[c]	State[d]	Techniques[e]	Refs.
—	1–20	C, I	490	O, R	G, AD	30
5–75	—	P	700	O	MS	60
1–7.5	—	I	500	O	XR	116
0.5–4.6	7.5	I	540–816	O, S	MS, CS	59
—	2–50	I(?)	500	R	G	19
3	8	A, I[f]	500	O	XR, MS, RS, IR	10
—	5.5–7.5	I, C	500	R	V	28
1–5	4–8	I	550	R	ES	42
—	7–33	I, M	550	O, R	ES	38
—	5–63	P	400	O	XR, ES	117
0.5–3	3–15	A	500	O	MS, RS, IR	48
3	8	C	?	O	XR, MS, RS, IR	50
3	8	C	?	R	CS	79

(continued)

TABLE II (*Continued*)

	2–11	I	450–500	O	XR, RS	*118*
—	1–3	I	500	R	RS	*119*
—	2–17	I	500	O	XR, ES, PP	*36*
1	1.5, 13	I	500	O	RS, ES	*120*
—	4–16	I[f]	500	R	ES, CO	*39*
—	2–16	I	540	S	G, ES	*33*
2.5	3.5	I	500	O, R	AD, ES	*121*
0.5–3	3–19	A, C	?	S	XR, MS, RS	*89*
1	—	I	450	O	RS	*49*
5	10	?	360–450	R	G	*24*
2	8	A, M	550	O	G, CS, OO[g]	*106*
—	10	M	450–600	O	EC	*63*
—	1–7	I	450–600	R	ES, RS	*44*
0–5	8	I, C	500	R	CS	*114*
2	8	A	?	O, S	G, ES, CS, CO	*122*
0–6	10	M, I	500	R, S	XR, G	*16*
—	5–15	I	500	O	ES, AD	*123*
0–5	2–7	M	300[h]	S	IR, AD	*90*
0–5	2–12	I[i]	?	O	XR, PP, CS	*95*
0.3–5	3–8	I, A	550	O	ES, AD, CS	*47*
2–3	8	C	?	O, S	EC	*64*
—	2–20	I, P	500	O, R	XR, ES, G, RS, IR, PP	*51*
0.2–3	1–25	A	500	R	AD	*124*
2	2–8	I	500	O	IR	*80*
0.5–5	8	I	500–600	O, S	XR, RS, MS	*53*
2–5	8	I	500–600	O, R, S	ES	*37*
—	2–25	I	540	R	G	*23*
—	12	A, O[j]	500	O, R	XR, OO[g]	*75*
—	10	I	500	R	ES	*40*
—	2–6	I	25	O	OO[k]	*125*
4.3–5.3	6	I, M	?	S	PP, CS	*100*
3	8	C	?	O, R, S	EC	*74*
—	8	I	540	R	V	*20*
2.5	7.5	C	?	S	RS, MS, IR	*57*
2	8	I	550	O	OO[l]	*81*
3	8	I, C	700–900	O	XR, OO[k]	*11*
—	2–15	I	500	O	XR, OO[m]	*13*
1.5	12	I	480	R	V	*31*
0.5	8	I	650	O, S	MS, AD, CS	*61*
1.5–4.5	8	I	400–600	O, R	EC	*62*
3	8	C	?	O	AD	*82*
—	2–20	I, P	500	O	XR, G, RS, PP, OO[g,k]	*83*
—	2–20	I, P	500	R	G, RS, ES	*25*
—	2–20	I, P	500	O, R	RS, ES, CO	*102*
—	8	I	540	S	G, AD, OO[n]	*21*
1–3	8	I	650	O	RS, IR	*15*
5–10	—	I	500, 1000	O	EC	*66*

TABLE II (Continued)

2	8	I	550	O, R, S	G, IR, CS, OO[l]	27
2	8	I	500	O	EC	67
3	2–12	I, C	600	S	XR, CS	92
4	1–10	I	600	S	XR, RS, CS	17
—	7	A, O[j]	600	O, R	IR	76
0–15	0–15	A	500	O, S	XR, EC, CS	18
0–1	0–9	I, A(?)	500	O	EC	65
0–1	0–9	A	500	R	EC	70
—	8	I	540	R	V, ES, MS	29
2–3	3–8	I	500–900	O	PP, CS	97
0.4–12	—	I	600	O	XR, RS, MS	54
0–4	6–9	I[f]	600(?)	O	RS, MS, AS, OO[g]	56
3	10	I	400–700	O, S	IR, CS	55
2	10	C	?	O, R, S	EC	71
—	2–12	I, A	500	O, R	RS, AO	52
4	9	I, M	450–600	S	AS	91
2	8	I	550	O, R	ES, MS	43
3	8	C	?	O, S	EC, AD, AS	73
2.4	10	C[o]	?	O, S	IR[p]	77
—	8	I	540	R	V, AO	84
—	8	I	540	R, S	G, AS	94
2.4	10	I, C	550	R	AD	69
0.2	8	A	500	O, S	CS, OO[l]	98
4	8	I	?	O, S	RS	58
0, 3.5	5–20	I	500	S	EC	74a
?	?	C	?	S	AD	85a

[a] Values are ranges covered in wt% metal.
[b] Preparation: A, adsorption from solution; C, commercial catalyst; I, impregnation; P, precipitation; M, mixture; O, other.
[c] Calcination temperature.
[d] Catalyst states studied: O, oxidized; R, reduced; S, sulfided.
[e] Symbols: AD, adsorption; CO, catalytic activity for reactions other than desulfurization; CS, catalytic activity for desulfurization; EC, ESCA; ES, ESR; G, gravimetric; IR, infrared; MS, magnetic susceptibility; OO, other; PP, physical properties; RS, reflectance spectra; V, volumetric; XR, X-ray diffraction.
[f] Hydrous alumina gel used.
[g] Solution extraction.
[h] Treated in H_2S/H_2 mixture.
[i] η-Al_2O_3 used.
[j] Vapor phase deposition.
[k] DTA and TGA.
[l] Acidity by n-butylamine with pK indicators.
[m] Luminescence spectra.
[n] D_2 exchange.
[o] SiO_2–Al_2O_3 support.
[p] Raman spectra.

A. X-Ray Diffraction Analysis (XRD)

The most obvious choice to determine phases that may be present in the molybdena catalyst is XRD. Matching of diffraction lines obtained for the catalyst with those of pure bulk compounds gives unequivocal identification of phases present. This is one of the few techniques that yields positive results. The absence of matching diffraction lines, however, is not proof that the phase in question is not present in the catalyst. The XRD technique is limited to particle sizes of above approximately 40 Å for oxides or sulfides, lower sized particles giving no discernible pattern over that of the broad alumina pattern. Thus, the presence of a highly dispersed phase, either as small crystallites or as a surface compound of several layers thickness will not be detected. Also, if the phase is highly disordered (amorphous), a sharp pattern will not be obtained, although some broad structure above that of the alumina may be detected. It is a moot point as to whether such a case is considered as a separate phase or a perturbation of the alumina structure. Ratnasamy et al. (11) have examined their CoMo/Al catalyst from the latter point of view, with particular emphasis on the effect of calcination temperature.

A summary of the findings by XRD is presented in Table III. The oxidized catalyst (normal calcination temperature 500°–550°) has been examined by many workers with generally negative results, i.e., no discernible diffraction pattern attributable to Mo or Co phases was found if the respective levels on the catalyst were below certain values. For Mo/Al this was approximately 15% (weight). Above this level some workers report MoO_3, while others find $Al_2(MoO_4)_3$. This value is a rough average, some workers finding MoO_3 at somewhat lower concentrations and others failing to detect Mo phases at considerably higher levels. At higher calcination temperature (750°), $Al_2(MoO_4)_3$ is formed from a mixture of MoO_3 and Al_2O_3 (12). Stork et al. (13) reported that $Al_2(MoO_4)_3$ was present on their impregnated catalysts at concentrations as low as 3% Mo by a luminescence spectroscopic technique using Cr^{3+} tracer ions. Of

TABLE III
X-Ray Diffraction Results

State	Mo/Al	CoMo/Al	Co/Al
Oxide	>15% MoO_3 or $Al_2(MoO_4)_3$	Same >4% Co–Co_3O_4	>2% Co_3O_4
Sulfided	>15% MoS_2	Same >4% Co_9S_8	—

interest, these authors (14) found no evidence for $Al_2(MoO_4)_3$ on a catalyst prepared by the adsorption method (Mo level not specified).

For Co/Al, the detection level was about 2%, above this value bulk Co_3O_4 being detected by XRD. However, addition of 3% Co in separate 1% impregnation steps with calcination between steps resulted in a catalyst having no detectable Co_3O_4 phase present (15). Interestingly, when both Mo and Co were present together, higher levels of Co could be accommodated ($\sim 4\%$) before Co_3O_4 was detected. But, in one study with 8% Mo, MoO_3 was observed to appear when the Co content exceeded 3% (16). It is significant that $CoMoO_4$ has not been detected in any of these catalysts.

On a silica support, results were quite different. Now, at practically all levels, MoO_3 and Co_3O_4 phases were found on Mo/Si and Co/Si catalysts, respectively. In the CoMo/Si catalyst, these phases and/or $CoMoO_4$ were found, the presence of the latter depending upon the Co/Mo ratio employed (17).

Colors are revealing. The bulk compounds have the following colors: MoO_3, yellow; $Al_2(MoO_4)_3$, white; CoO, brown; Co_3O_4, black; $CoAl_2O_4$, blue; $CoMoO_4$, purple (or olive green). The Mo/Al catalyst is white at low concentrations (no XRD pattern) and yellow at high concentrations (MoO_3 detected). The Co/Al catalyst is blue or black, depending on concentration. The CoMo/Al catalyst is blue to purple to black, depending on concentrations. Too much stock should not be placed on assignment of phases from color observations, however, since color is sensitive to coordination environment and not necessarily to a specific phase structure. Thus, the blue Co/Al catalyst could be due to a surface complex or interdiffusion compound with alumina of stoichiometry different from that of bulk phase $CoAl_2O_4$.

Surprisingly, no XRD results have been reported for reduced catalysts. Sulfided catalysts generally gave negative results for Mo and Co levels about the same as for the oxidized catalysts. At higher levels, MoS_2 or Co_9S_8 are observed. However, under more severe sulfiding conditions (higher temperature or pure H_2S), these phases were detected even at lower levels of Mo and Co (18).

In summary, no bulk Mo or Co phases are observed on catalysts having low to moderate Mo and Co concentrations (encompassing commercial compositions) in either oxidized or mildly sulfided forms.

B. Gravimetric and Volumetric Measurements

If the molybdenum is not observed by XRD, in what state is it? The first hint of an interaction between molybdena and the alumina support came from studies of Sontag et al. (19). Wishing to increase the reducibility of bulk MoO_3 in H_2, they supported the MoO_3 on an activated alumina with the thought that dispersion on the support would provide smaller particles and greater reactivity. Surprisingly, they found the reduction to be markedly slower than that for bulk MoO_3. They explained these results as being due to either (1) an interaction between the MoO_3 and Al_2O_3 yielding a less reactive specie or (2) the water from reduction being partially retained by the Al_2O_3 in close proximity to the MoO_3, thereby retarding reduction of the latter (water vapor is known to retard reduction of bulk MoO_3). Their work involved determination of the oxygen left by a chemical method.

Subsequent work on catalyst characterization via reduction studies has involved gravimetric or volumetric techniques. The gravimetric method measures weight loss (due to O loss), while the volumetric method measures amount of H_2 consumed [and amount of water formed (O loss)]. In the latter case, the amount of H retained by the catalyst can also be determined (20). The gravimetric technique has also been applied to sulfided catalysts (21). From these measurements, the stoichiometric state of the catalyst can be determined.

Table IV summarizes the findings of such studies. The results of Sontag et al. have been confirmed many times, viz., the Mo/Al catalyst reduces slower than bulk MoO_3. Whereas, bulk MoO_3 reduction proceeds rapidly to MoO_2, then more slowly to Mo metal (22), no such sequence is observed for the catalyst (23). Further, despite occasional claims in the literature, reduction does not stop at the MoO_2 state—more than one O/Mo is removed at high temperature, and considerably less than one O/Mo is removed at low temperature. In one case, it was reported that reduction continued even after 2 days (24). Fractional reduction increased with increase in the Mo content of the catalyst (16, 23, 25). Reduction rates have generally followed the Elovich law, indicative of a surface

TABLE IV
Gravimetric–Volumetric Results

State	Mo/Al	CoMo/Al	Co/Al
Reduced	MoO_x, $1^a < x < 3$	Same	CoO_z, $0^a < z < 1$
	H_1	—	—
Sulfided	MoO_xS_y, $0^a < x < 3$, $0 < y < 3^a$	Same	CoO_z, Co_9S_8
	H_1	—	—

[a] High level or severe treatment conditions.

reaction (*23, 24, 26*), although the Avrami–Erofeev law has been evoked in one case (*27*). A variable activation energy for reduction has been noted, the activation energy increasing with extent of reduction. The above facts have been interpreted as signifying a monolayer of MoO_3 attached to the alumina surface with a varying strength of bonding. Reducibility decreased with increasing calcination temperature, and was greater for a silica–alumina support (*28*).

Volumetric measurements have revealed that a significant amount of strongly adsorbed hydrogen (irreversible H_I) remains on the reduced Mo/Al catalyst (*20*). This hydrogen could only be removed by heating to higher temperature or by oxidization with air, both cases producing water. Mild reduction has been characterized in terms of anion vacancies. On this basis, exactly $2H_I$ per vacancy has been found on low to moderately reduced catalysts (*29*). H_2–D_2 exchange studies on reduced Mo/Al catalysts showed that H associated with Al_2O_3 decreased as the Mo content increased up to 8% (*30*). At the same time, the H associated with the reduced Mo increased due to H_I (*21*).

Reduction of Co/Al catalysts has been less extensively studied. At low levels (~1%), no reduction occurs up to 500° and the sample remains blue in color. At higher Co contents, appreciable reduction to Co metal occurs.

The situation with respect to reduction of the CoMo/Al catalyst is more confusing. Various authors claim that the presence of cobalt at a low level accelerates (*16*), retards (*27*), or has no effect on (*31*) the reduction of the molybdena. Of course, at high Co loadings, more reduction is obtained than for the Mo/Al alone, due to reduction of the Co_3O_4 phase present, but it is difficult to assess whether the molybdena is itself affected by the cobalt reduction. It is well known that transition metals can catalyze reduction of oxides (*32*). It is probable that the different results obtained could be due in large part to differences in preparation or calcination temperature as pointed out earlier.

In spite of the fact that these catalysts are presulfided or become sulfided in hydrodesulfurization processes, surprisingly little work has been done on determining the stoichiometry of the sulfided catalyst state. Where sulfur levels have been analyzed, the catalyst has usually been found to be incompletely sulfided to the respective sulfides, MoS_2 and Co_9S_8 (see Section IV.J). The basic question here is whether only part of the molybdena is converted to MoS_2 or a molybdena oxysulfide complex is formed. In sulfiding bulk MoO_3 with an H_2S/H_2 mixture, the reaction proceeds through the topochemical path

$$MoO_3 \rightarrow MoO_2 \rightarrow MoS_2$$

the first step being rapid and the second slower (33), with no evidence for any mixed molybdenum oxysulfide phases (34, 35, 35a).

Massoth (21) was the first to address the question using combined gravimetric and sulfur analysis. Sulfiding was more rapid than reduction and appeared to reach a steady-state value within 2 hr, which value increased with temperature. Incomplete sulfiding was obtained, except at high sulfiding temperatures. The main reaction was exchange of catalyst O for S, with some additional O loss (vacancies) due to reduction. But the total O lost (O_L) was considerably less than $O_L/S = \frac{3}{2}$ required for conversion of MoO_3 to MoS_2. Therefore, either an Mo–O–S complex is formed without a discrete MoS_2 phase being present, or if MoS_2 is formed, only part of the sulfur is in this phase, the rest residing in an Mo–O–S complex. Prereduced catalysts sulfided to a lesser extent.

At low concentrations, Co/Al apparently does not sulfide, but at higher Co levels, Co_9S_8 is obtained, as can be observed by XRD. It is likely that only the reducible Co sulfides, so that somewhat less than stoichiometric S/Co ratios are obtained at intermediate Co levels.

Little definitive data are available on the sulfided state of CoMo/Al catalysts. Generally, higher sulfide levels are obtained than with comparable Mo/Al catalysts, but it is not certain whether this is due to sulfiding of the cobalt or additional sulfiding of the molybdena.

In summary, the evidence strongly points to a model of the oxidized catalyst consisting of a high dispersion of molybdena at the alumina surface, possibly as an epitaxial monolayer. This layer interacts strongly with the alumina substrate, exhibiting a range of bonding strengths. Reduction does not stop at MoO_2, the extent of reduction depending upon the Mo content, temperature, and time of reduction. The true effect of cobalt on reduction is uncertain at this time. Sulfiding is incomplete with respect to MoS_2 and Co_9S_8 formation under usual sulfiding conditions employed in industrial practice. The sulfur level is predominantly determined by the sulfiding temperature employed. The sulfided catalyst either consists of a mixed surface oxysulfide specie or some bulk MoS_2 and another oxysulfide specie. The state of Co in the sulfided catalyst is uncertain.

C. Electron Spin Resonance (ESR)

It was discovered early that reduced Mo/Al catalysts produce a rather large ESR signal, which was attributed to an Mo^{5+} specie. An ESR signal arises when unpaired electrons are present. The structure of the signal gives information about the coordination of the specie. It is very sensitive to small concentrations.

Some results of ESR studies are depicted in Table V. The oxidized

TABLE V
ESR Results

State	Mo/Al	CoMo/Al	Co/Al
Oxide	(Mo^{5+})	Co^{2+}[T], Co^{2+}[O]	Co^{2+}[T], Co^{2+}[O] Co^{3+}[O](?)
Reduced	Mo^{5+} ∧∨ >	Mo^{5+} Co^{2+}[T], Co	Co^{2+}[T], Co
Sulfided	Mo^{5+}	Mo^{5+} Co^{2+}[T], Co_9S_8	Co[T], Co_9S_8

$$\diagdown_{}\!\!\!\!\!\overset{O}{\underset{\diagup}{Mo}}\!\!\!\!\!\overset{OH}{\diagup}\,,\ MoO_4^{3-},\ HMoO_4^{2-},\ MoO(OH)^{2+},\ MoO_3^{+}$$

Mo/Al catalyst gives a weak Mo^{5+} signal, more so when supported on η-than γ-Al_2O_3 (36), but the signal is absent when Co is present (37). Upon reduction, the ESR signal for Mo^{5+} is enhanced some 100-fold and increases with the Mo content (38, 39). Nevertheless, the fraction of Mo^{5+} is rather small (1–10%). (Such enhancement is not observed on an SiO_2 catalyst.) During reduction, the Mo^{5+} signal increases with time, goes through a maximum and then decreases to a low level (40). It is lower at higher reduction temperatures. The Mo^{5+} state is believed to be an intermediate in the reduction of Mo^{6+} to Mo^{4+} or lower valences. No authors have reported observing an Mo^{3+} signal in their ESR spectra of the reduced Mo/Al catalyst, although a spectra given in a recent publication (41) purports to show Mo^{3+} on a reduced Mo/Al catalyst. The presence of Co lowers the Mo^{5+} signal on the reduced catalyst; this has been interpreted in terms of the Co increasing the reducibility of Mo to Mo^{4+} (42), but it could also be due to a decreased reducibility to Mo^{5+} (43) or other factors. The Mo^{5+} environment appears to be the same with Co present (43).

It is clear that the Al_2O_3 support contributes to the stability of Mo^{5+}. Various authors have attributed the Mo^{5+} to different species. Some of the proposed species are shown in Table V. The H-containing species are particularly attractive in view of the irreversible H_I found on these reduced catalysts. Asmolov and Krylov (44) consider Mo^{5+} to be in a tetragonally compressed C_{4v} complex. Abdo and co-workers (45) have recently proposed the structure

$$\diagdown_{}\!\!\!\!\!\overset{}{\underset{\diagup}{Mo}}\!\!\!\!\!\overset{OH}{\underset{O}{\diagup}}$$

in which the Mo is in tetragonal coordination.

The sulfided Mo/Al catalyst also exhibits an Mo^{5+} signal but it is weaker than for the reduced catalyst (33).* This could be due to less reduction of the Mo as sulfiding is usually done at lower temperature than reduction, although other explanations are possible. A sulfur species, assigned to free sulfur has also been noted on the sulfided catalyst after exposure to air (33, 46).

Cobalt species have also been detected by ESR. For oxidized Co/Al, the Co environment has been ascribed to Co^{2+} in tetrahedral coordination, Co[T] (37). It is not affected by reduction (or sulfidation) except at higher Co contents, whence a second broad signal appears, probably due to metallic Co. In another study (47), it is claimed that Co^{2+} in Co/Al is in octahedral coordination, Co[O], but when Mo is present, both Co^{3+}[O] and Co^{2+}[T] exist.

In summary, the presence of a relatively strong Mo^{5+} signal found on Al_2O_3 catalysts but not on SiO_2 catalysts can be considered as additional evidence for an interaction between the molybdena and the alumina, which permits stabilization of the Mo^{5+} state in the reduced and sulfided catalyst. An H-containing specie seems a reasonable hypothesis for this state. The Co may be in tetrahedral or octahedral environment or both.

D. Diffuse Reflectance Spectra

This technique measures the optical spectrum of light which is diffusely scattered off the catalyst sample. Absorption frequencies are characteristic of certain arrangements of molecules and their environment. Even pure compounds give rather broad diffuse spectra, and catalysts show even broader spectra. Hence, results are only semiquantitative at best. In regards to molybdena catalysts, the information derived with this technique is the coordination environments of Mo and Co in the catalyst.

Table VI summarizes the findings by this technique. Although early work on the Mo/Al catalyst showed contradictory results, i.e., Mo^{6+} present as Mo[T] (48, 49) versus Mo[O] (50), later studies point to the presence of both species in the oxidized catalyst. At low Mo concentrations (and for coprecipitated catalysts), Mo[T] dominates (17), whereas at higher Mo levels, the ratio of Mo[O]/Mo[T] increases (51). Because of similarities with pure compounds, the Mo[T] has been ascribed to a surface $Al_2(MoO_4)_3$ phase and the Mo[O] to a free MoO_3 or polymeric Mo-oxide phase on the surface. As noted earlier, one must be cautious in

---

* The absolute values of Mo^{5+} given in this article are erroneously high based on later results (40).

TABLE VI
Reflectance Spectroscopy Results

State	Mo/Al	CoMo/Al	Co/Al
Oxide	$Mo^{6+}[T]$, $Mo^{6+}[O]$	Same	$Co^{2+}[T]$, $Co^{2+}[O]$
	$MoO_3{}^a$		$CoAl_2O_4(?)$
	Mo dimers		$Co_3O_4{}^a$
Reduced	Bis–Mo–like	—	$Co^{2+}[T]$
Sulfided	—	MoO_xS_y	$Co^{2+}[T]$
		MoS_2	Co_9S_8

[a] High level

interpreting these as distinct phases. The reduced catalyst was reported to contain bis-Mo-like structures (25), while another study claimed $Mo^{5+}[T]$ and $Mo^{5+}[O]$ species to be present (52).

Almost total agreement exists that on the Co/Al catalyst, Co is as Co^{2+} both in Co[T] and Co[O] coordination. At low concentrations, the Co is mainly in tetrahedral symmetry (53), but the Co[O]/Co[T] ratio increases with increasing Co (48). At higher than 2–2.5% Co, a separate Co_3O_4 phase is present as evidenced by the appearance of Co^{3+} in the spectra (54). Higher calcination temperature favors interaction of the Co with the Al_2O_3 structure (55). Co[T] has been ascribed to a surface $CoAl_2O_4$ phase or to Co substitution in the Al_2O_3 spinel lattice, whereas Co[O] has generally been associated with bulk Co_3O_4.

The CoMo/Al catalyst generally shows similar properties to the individual, separate catalysts except that higher levels of each component can apparently be tolerated before separate phases are detected (53). Martinez et al. (56) have reported a broadening of the Mo peak with increasing Co level, which they ascribe to greater spreading of the Mo over the Al_2O_3 surface caused by the Co. The sulfided catalyst has been described as similar to thiomolybdate $[MoO_xS_{4-x}]^{2-}$ and unlike MoS_2 or CoS (57). However, there is some question about the validity of these results as the sulfided catalyst apparently was exposed to air, which results in appreciable oxygen uptake. *In situ* spectra were reported to reveal the presence of MoS_2 on the sulfided catalyst (58). Reoxidation of the sulfided catalyst returned the spectra to that of the original oxidized catalyst, without evidence of MoO_3 or Co_3O_4 being present (56).

In summary, it may be said that the oxidized catalyst most probably contains Mo^{6+} and Co^{2+} in both tetrahedral and octahedral configurations, the relative amounts of each depending on concentration and calcination temperature. Generally, no strong evidence has been found for definite compounds on the oxidized or sulfided catalyst.

E. MAGNETIC MEASUREMENTS

One of the earliest studies on the CoMo/Al catalyst was done by Richardson (59) employing magnetic measurements. He characterized the oxidized catalyst in terms of bulk compounds, which is clearly incorrect in view of later work. On this basis, and the known sulfidibility of the compounds, he deduced the active catalyst consisted of an MoS_2 phase containing Co of an unknown stoichiometry.

The measurement of magnetic moments is a companion technique to reflectance spectra in that coordination environments are discernible. Magnetic moment depends on unpaired electrons and has been used to characterize Co in these catalysts. It should be noted that an average magnetic moment is obtained from these measurements. Thus, a magnetic moment falling between that for Co[T] and that for Co[O] has been ascribed to the presence of both species on the catalyst.

The results of findings of various investigators are summarized in Table VII. At low levels, Co is predominantly as Co[T] (60), although some Co[O] has also been noted (10). The Co[O]/Co[T] ratio increases with Co concentration (48). At Co greater than 2%, a Co_3O_4 phase segregates (54). Co[T] apparently neither reduces nor sulfides (61), although Ramaswamy et al. (43) report Co[O] as well as Co[T] present in the reduced catalyst. At higher levels, part of the Co undergoes sulfidation, presumably that part due to Co_3O_4 present. Similar results were found for the CoMo/Al catalyst.

Hall and Lo Jacono (29) measured the paramagnetism of a reduced Mo/Al catalyst. Ascribing this to Mo^{5+}, they showed it agreed with volumetric measurements of irreversibly retained H_I.

In summary, magnetic measurements agree with other studies in that

TABLE VII
Results of Magnetic Measurements

State	Co/Al	CoMo/Al
Oxide	Co^{2+}[T], Co^{2+}[O]	Same
Reduced	Co^{2+}[T], Co, Co[O]	Same
Sulfided	—	Co^{2+}[T]

both Co[T] and Co[O] species appear to be present on the catalyst, the former predominating at low concentration and the latter at high.

F. Photoelectron Spectroscopy (ESCA)

A relatively new entry into the catalyst characterization field is ESCA, which stands for electron spectroscopy for chemical analysis; it is also referred to as photoelectron spectroscopy. This technique has only recently been applied to the CoMo catalyst, but studies have rapidly proliferated. The basis of the technique involves the bombardment of the sample with X-rays, measuring the binding energy of the emitted electrons. The binding energy is not only characteristic of the element, but more importantly its valence state. Binding energy shifts due to environment are small. Measurements are confined to surface layers, penetration being some 10 to 20 atomic layers deep.

Some results obtained by this technique are shown in Table VIII. The oxidized Mo/Al catalyst exhibits appreciable peak broadening compared with bulk MoO_3. This has been interpreted as due to several phases present or to differences in environment, i.e., binding energies of Mo. The binding energy (*BE*) of the catalyst has been reported as the same (*62*) or higher than that for bulk MoO_3 (*63, 64*). The latter has been taken as evidence for a strong interaction of the molybdena with the alumina surface, variously explained as due to partial transfer of electrons from O to Al (*63*) or displacement of Mo on different Al sites (*18*). No difference in the nature of the ESCA signal was reported for catalysts up to 5% Mo, the Mo occupying both [T] and [O] sites (*65*). Results with the oxidized Co/Al catalyst are less clear. There appears to be at least two types of Co^{2+} on the catalyst, depending on the Co level. At low Co, only Co[T] is present, while at high Co, a Co_3O_4 phase is present (*65*). Co^{3+} has been reported at high Co concentrations (*66*). Results for the CoMo/Al catalyst were simi-

TABLE VIII
ESCA Results

State	Mo/Al	CoMo/Al	Co/Al
Oxide	B.E. MoO_3	Same Co^{2+}(two types)	Co^{2+}, Co^{3+}
Reduced	—	Mo^{6+}, Mo^{5+}, Mo^{4+}	Co^{2+}
Sulfided	—	MoS_2, Mo^{4+}–S Mo^{6+}–O, Mo^{4+}–O(?) Co–O, Co–S(?)	—

lar to the Mo catalyst in most cases. In one, however, it was reported that the BE for the CoMo/Al catalyst was the same as for MoO_3 but the Mo/Al catalyst was higher (67). This same study claimed three Co phases to be present, viz., (1) Co^{2+} in surface $CoAl_2O_4$, (2) Co^{2+} in Al_2O_3 phase, and (3) Co^{2+} in coordination with strongly bound water. Walton (68) has recently questioned the latter phase. Curiously, when Co was impregnated on a previously calcined Mo/Al catalyst and calcined at 550°, no Co was detected by ESCA (69).

Studies on reduced and sulfided catalysts were predominatly done on the CoMo/Al catalyst. In a detailed study of catalysts reduced at 600°, Grimblot and Bonnelle (70) concluded that a Mo_4Co phase was present, which they denoted as

$$Mo_4^{6+}[O]Co^{2+}[T]Al_2O_3$$

its amount varying with the Co and Mo level. At low Co concentration, the Co and Mo are in separate and independent states, with no evidence of interaction between them. Patterson et al. (71) made a detailed study on a commercial CoMo/Al catalyst. They found that the Co did not reduce up to 500° and only slightly sulfided at 400°. By deconvolution techniques, they described the reduced catalyst in terms of Mo^{6+}, Mo^{5+}, and Mo^{4+} states, the average state of reduction decreasing with time of reduction. Their analysis showed that 50% of the Mo^{6+} was converted to Mo^{5+} before Mo^{4+} began to appear. These results are in qualitative agreement with stoichiometric measurements (23). They described the mildly sulfided catalyst in terms of MoS_2 and MoO states. The more severely sulfided catalyst gave more of the former than the latter, but in no case was complete sulfiding achieved. Of interest, reoxidation of the sulfided catalyst gave a spectrum identical to the original oxidized catalyst. Brinen (72) also reported both Mo sulfide and oxide present on a used commercial catalyst. The BE for the sulfided catalyst was close to that of MoS_2 (73). This has been taken as evidence for destruction of the Mo oxide–Al_2O_3 interaction with formation of a separate MoS_2 phase; however, Walton (68) has pointed out that this interpretation may not be unequivocal. Friedman et al. (74) found two Co phases in the sulfided catalyst which they interpreted as due to CoS and $CoAl_2O_4$. On the other hand, in a study of molybdena catalysts sulfided with a thiophene/H_2 mixture, Okamoto et al. (74a) conclude that the monolayer essentially remains intact except for high Mo catalyst levels. In contradiction to other workers, they found no evidence for Co sulfides but rather the appearance of Co metal under their in situ reaction conditions.

Although not entirely unequivocal, ESCA results appear to be consistent with the monolayer model of the oxidized catalyst. More uncertainty exists for the sulfided catalyst: either a MoS_2 phase separates from the support or if the monolayer becomes partially sulfided, its interaction with the support is apparently weakened.

G. INFRARED

Infrared (IR) has been used in two ways: for structural characterization and for acidity characterization. The latter will be taken up in the next section. Vibrations between atoms produce characteristic frequencies in the IR region; these are compared to known compounds to infer structural aspects of the catalyst.

Lipsch and Schuit (50) reported Mo to be predominantly as Mo[O] on an oxidized commercial catalyst. Giordano et al. (51) report Mo[T] exclusively on coprecipitated Mo/Al catalysts and both Mo[T] and Mo[O] on impregnated catalysts. In the latter case, the Mo[O]/Mo[T] ratio increased with increase in Mo concentration. They attributed Mo[O] to polyoxyanion Mo forms. Mitchell and Trifiro (57) interpreted their results in terms of oxomolybdenum(VI) species with bridge and terminal oxygens.

Addition of Mo to Al_2O_3 has been reported to lower and shift the frequency of the Al–O–H band (57, 75). This has been interpreted as due to covering of the Al_2O_3 surface by a molybdena layer, or actual loss of OH by interaction during preparation or calcination of the catalyst. The latter view agrees with a suggested reaction put forth by Dufaux et al. (36),

$$\underset{HO}{\overset{O}{\diagdown}}\underset{}{\overset{}{Mo}}\underset{OH}{\overset{O}{\diagup}} + -O-\underset{\underset{}{|}}{\overset{OH}{|}}Al-O-\underset{\underset{}{|}}{\overset{OH}{|}}Al- \xrightarrow{-2H_2O} \underset{O-Al-O-Al-}{\overset{O\diagdown Mo\diagup O}{\underset{|}{O}\underset{|}{O}}} \quad (1)$$

Massoth (21) also found less H on an Mo/Al catalyst than on the Al_2O_3 support alone by an H_2-D_2 exchange technique.

Ratnasamy et al. (27) reported that on reduction, terminal oxygen is lost first, and Fransen et al. (76) found the OH band to reappear. The latter authors speculated that on reduction, breaks in the monolayer appear caused by migration of Mo^{4+} ions to octahedral sites. An MoO_2 phase did not appear and oxidation gave back the same spectra as the freshly oxidized catalyst.

Mitchell and Trifiro (57) reported that the sulfided CoMo/Al catalyst did not resemble any known sulfides of Co or Mo. Terminal oxygen was still present on the sulfided catalyst. They proposed a surface $O-Mo^{6+}-S$

species to be present, but their results must be considered inconclusive since their sulfided sample was apparently exposed to air before running the IR spectrum. Reoxidation of the sulfided catalyst restored the IR spectrum to that of the fresh catalyst.

Brown and Makovsky (77) have studied the Raman spectra of CoMo/Al commercial catalyst containing 5% SiO_2. The oxidized catalyst showed bridged Mo—O—Mo and terminal Mo=O structures with no evidence for bulk MoO_3. The sulfided catalyst showed spectra similar to MoS_2 (also detected by XRD). There is reason to believe that Mo interaction with silica–alumina is weaker than with Al_2O_3 (28); thus, sulfiding may more easily destroy the surface interaction complex present in the oxidized catalyst.

The IR spectra of thiophene adsorbed on MoS_2 and sulfided catalyst were determined by Nicholson (78), who interpreted the results in terms of 1-, 2-, and 4-point adsorption. However, Lipsch and Schuit (79) questioned these assignments, ascribing the adsorption to a 1-point mode only.

In summary, the IR results are in agreement with other techniques and in particular provide evidence for Mo–Al surface interactions. Also, the finding of terminal and bridged Mo–O species agrees with the picture of a monolayer of molybdena on the alumina surface.

H. Acidity

Infrared has also been used to assess the type of acidity present on a catalyst. The method involves measuring the spectrum of adsorbed pyridine on the catalyst; certain characteristic absorption frequencies are assigned to Lewis acid centers (coordinately bound pyridine) and others to Brønsted sites (pyridine adsorbed as pyridinium ion).

Kiviat and Petrakis (80) were the first to show that Mo on Al_2O_3 developed Brønsted acidity. The Al_2O_3 support and Co/Al showed only Lewis acidity. The CoMo/Al catalyst gave decreased Brønsted/Lewis ratio over the Mo catalyst, which led them to conclude that the Co interacts with the Mo. Reduced catalysts gave essentially identical results to oxidized catalysts; thus, the irreversibly retained hydrogen on the reduced catalyst does not appear to partake of Brønsted acid character. Mone and Moscou (15) reported slightly different results; they only found Brønsted acidity after rehydrating and recalcining their catalyst. Further, they found little Brønsted acidity for the CoMo/Al catalyst calcined at 500°, but increased Brønsted acidity when calcined at 650°. They explained these results on the basis that Co ions on the surface neutralize the Brønsted sites of the Mo/Al at low calcination temperature, but at high temperature, the Co ions move into the Al_2O_3 lattice, regenerating Brønsted sites. Mone (55) also quotes evidence for interaction of Co ions with the Mo–Al surface by the appearance of a new Lewis band. Of importance, he found no Brønsted acidity on a sulfided catalyst. So the relevance of Brønsted acidity to catalytic activity may be moot for catalysts operating under HDS conditions.

Another way of characterizing acidity of catalysts is by acid–base (Hammett) indicators. Amounts and strengths of acid sites can be obtained by n-butylamine titration coupled with indicators of definite pK_a values. The only published work on this is by Ratnasamy et al. (81). They showed that Mo/Al and CoMo/Al had strong acid sites ($pK_a < -5.6$) whereas Co/Al and Al_2O_3 exhibited weaker acidity. The order of addition apparently affected the acidity: Mo added to Co/Al was about the same as Mo/Al, but Co added to Mo/Al or both added together exhibited higher acidity. They also demonstrated that use of an Al_2O_3 containing small amounts of sodium suppressed formation of strong acid sites in the CoMo/NaAl catalyst.

The findings of enhanced acidity on addition of Mo to Al_2O_3, and particularly generation of Brønsted acid sites are very interesting findings. Concerning the latter, it is uncertain whether the Brønsted sites are directly associated with Mo or Al atoms. Once again, the results obtained are indicative of an interaction between the Mo phase and Al_2O_3.

I. Adsorption

Numerous adsorption measurements have been made on molybdena catalysts to ascertain the nature of the active sites that may be present. These generally involved volumetric or gravimetric determinations.

Adsorption measurements on oxidized catalysts have been done at low temperatures to avoid reduction or reaction with the molybdena present. Adsorption of H_2 appears to be negligible at room temperature (73). Dollimore et al. (82) reported an adsorption ratio of $1.8H_2S/Mo$, indicative of a high Mo dispersion on their catalyst. In one study, thiophene was reported to be reversibly adsorbed (82) whereas another found appreciable irreversible adsorption (73). Chemisorption of 1-butene at 100° was found to be the same for a series of catalysts of varying Co levels with the same Mo content (61). One butene molecule was adsorbed per two Mo atoms. A calcined Mo catalyst gave $2H_2O/Mo$ (83) adsorbed at room temperature; but this value may be incorrect as Al_2O_3 itself can adsorb considerable water.

Appreciable reversible H_2 adsorption was found on reduced Mo catalysts. This is reported to be of an activated type (79) as the amount increased with temperature. Lo Jacono and Hall (84) claimed the reversible H_2 to be associated with anion vacancies. Since increased temperature results in an increase in the vacancy concentration, this could account for the activated hydrogen referred to above. Reversible NH_3 adsorption was also found to be proportional to vacancies (21). Interestingly, both reversible H_2 and NH_3 adsorption gave similar results, about one adsorbed molecule per three vacancies present. Both thiophene and pyridine were partly irreversibly adsorbed on a reduced CoMo/Al catalyst (79). Low-temperature adsorption of O_2 was used to measure the area of MoO_2 on reduced catalysts (69). However, it is doubtful that a separate MoO_2 phase exists on these catalysts as the monolayer was found to remain intact during reduction (85); so the interpretation of these results is questionable.

On sulfided catalysts, adsorption of H_2 was reported to be activated and predominately reversible (73). The kinetics of adsorption has been studied between 150°–280° (85a). The reported activation energy of around 50 kcal/mole is indicative of dissociative chemisorption. Also reversibly adsorbed is H_2S (21, 27), NH_3 (21) and to a large degree thiophene (86), although some irreversibility has been reported for the latter (73). In a temperature-programmed desorption study of thiophene adsorbed on sulfided catalysts at room temperature, two desorption peaks were obtained, one being attributed to an Mo site and the other to an Al site (61). On the basis of relative differences in the two peak areas, it was concluded that, in sulfiding, the Mo/Al catalyst retained a monolayer structure whereas the CoMo/Al catalyst underwent breakup of the monolayer, forming a separate $CoMoS_2$ bulk phase. Pyridine has been reported to be strongly adsorbed on the sulfided catalyst (87, 87a).

The salient findings from adsorption studies are that reversible H_2 of an activated type exists on reduced and sulfided catalysts and that polar-type compounds are strongly adsorbed on these catalysts.

J. Sulfur Analyses

Many workers (16, 21, 33, 71, 72, 88–91) report that sulfided catalysts are incompletely sulfided with respect to MoS_2 and Co_9S_8. A notable exception is the data of De Beer et al. (92), who found unusually high sulfur values. They concluded that the Mo was essentially completely sulfided to MoS_2 (except at low Mo levels) and the Co partially sulfided to Co_9S_8. In a later publication, De Beer and Schuit (7) claim that the lower values obtained by others are in error due to exposure of the samples to air. This was based on their earlier observation (92) that the sulfided catalyst upon exposure to air underwent a strong exothermic reaction with liberation of SO_2. It should be pointed out that their sulfiding treatment involved cooling down the sample in the sulfiding mixture followed by an N_2 flush for 10 min. Under these conditions, appreciable adsorbed H_2S can be expected to be present on the samples.

We have investigated this question in our laboratory in the following experiment:

A 1% Co 8% Mo/γ-Al_2O_3 catalyst was sulfided in a 10% H_2S/H_2 flow at 400° for 4 hr conducted in a glass vacuum system at atmospheric pressure. After purging in a flow of N_2 for 2 hr, the catalyst was evacuated and the temperature lowered to 0°. The catalyst was isolated and air was admitted into the system to 600 torr pressure. The air was cycled in the closed system using a glass–Teflon circulation pump. A small portion of the gas stream was continually drawn into a mass spectrometer. The catalyst was then exposed to the air by circulation through the catalyst bed. An immediate and rapid uptake of O_2 by the catalyst sample was observed, but no volatile sulfur species were detected. On gradually raising the temperature of the catalyst, SO_2 was observed only when the temperature exceeded about 100°. A similar experiment was done with an 8% Mo/γ-Al_2O_3 catalyst with the same results.

These results demonstrate that no volatile sulfur compounds are liberated on exposing a well-purged sulfided catalyst to air. It seems likely, therefore, that the sulfur loss referred to by De Beer *et al.* was due to some residual adsorbed H_2S and not the sulfide sulfur. Owns and Amberg (93) report that H_2S is strongly adsorbed on the sulfided catalyst, and Massoth (21) showed that at least a 1-hr purging at temperature was needed to remove it. A 10-min purge at room temperature could hardly be expected to remove much adsorbed H_2S. We conclude that the high sulfur values reported by De Beer *et al.* do not accurately reflect the stoichiometric sulfide content of the catalyst, and that catalysts need to be well purged of adsorbed H_2S prior to a sulfur determination to obtain true stoichiometric values.

The same authors (92) also report a substantial loss of sulfur in an H_2 posttreatment of the sulfided catalyst. This is probably due to the purging action of H_2 removing adsorbed H_2S rather than a true loss of sulfide sulfur as they assumed. Owens and Amberg (93) mention an appreciable loss of adsorbed H_2S but very little sulfide sulfur loss in a similar treatment. Massoth and Kibby (94) showed this treatment did not result in loss of any appreciable sulfur on well-purged samples.

De Beer *et al.* (7) also mention that disulfide species found on exposure of sulfided catalysts to air (46) can cause incorrect assessment of sulfur stoichiometry. Again, however, this can be attributed to adsorbed H_2S on their catalyst and its overall contribution to the sulfur content of well-purged samples was found to be small (33).

K. Miscellaneous

Surface areas of calcined Mo/Al catalysts of varying Mo content have been determined by Giordano *et al.* (*83*), who claim the results support a monolayer coverage of Mo up to about 15% Mo. Todo *et al.* (*95*) claim no change in surface area or pore volume upon sulfiding CoMo/Al catalysts. De Beer *et al.* (*92*) present surface area data for a number of Mo/Al and CoMo/Al catalysts, both oxidized and sulfided. Massoth (*96*) has recently analyzed these data and showed them to be consistent with the monolayer model for the sulfided as well as the oxidized state.

Electron microprobe studies have been made to determine the elemental distribution of Mo and Co throughout the catalyst particle. This is a relative macroscale, not microscale measurement; thus, intimate mixing at the atomic scale cannot be determined. Some workers reported rather even distributions (*51, 95*), but one group found higher concentrations of Co near the catalyst edge in their catalysts (*97*). Stevens and Edmonds (*73*) report no difference in ESCA spectra between the catalyst particle exterior and crushed particles, demonstrating homogeneity throughout the particle.

Alumina adsorbs molybdenum and cobalt from aqueous solution at room temperature. Martinez *et al.* (*56*) reported maximum uptakes for Mo and Co from their separate solutions equivalent to 4% Mo and 2% Co, respectively. With both present, 7% Mo and 3% Co were obtained. This was taken as evidence of interaction between Mo and Co on the alumina surface with Co causing a greater fractional surface coverage of Mo. Using a solution flow technique over a bed of alumina particles, Sonneman and Mars (*75*) showed that up to 15% to 20% Mo could be added to the alumina in essentially a monolayer. The lower Mo value obtained by Mitchell was explained by Fransen *et al.* (*85*) as due to a difference in solution pH, lower pH favoring more Mo adsorption. The higher Mo addition obtained by Mitchell in the presence of Co may then have been due to a lower pH for the mixture as compared to the Mo solution, rather than a true Mo–Co interaction. For impregnated catalysts, adsorption of Mo would then be determined by the solution pH, and excess Mo would presumably form bulk MoO_3 on calcination.

Yamagata *et al.* (*98*) report a direct correlation between catalyst support OH groups and Mo added by an immersion technique. They studied silica–aluminas having various Si/Al ratios and various aluminas, these supports having differing amounts of OH. The OH content of the supports was measured by titration with ammonium fluoride.

Molybdenum and cobalt on calcined alumina catalysts are partly extractable by washing. The amount of Mo removed by extraction with water increased with the Mo content, showing a break between 6% and 7% Mo, indicative of two species present; the amount of Co removed increased with Co content but leveled out at a saturation level corresponding to 1.8% Co monolayer (56). With ammonia solution, about two-thirds of the Mo was extracted, octahedral Mo being preferentially removed (83).

V. State of Molybdena Catalysts

It is obvious that there are many differences in results found and interpretations made on the state of molybdena–alumina catalysts, especially the sulfided state. Nevertheless, some common ground exists in these investigations, and we shall try to present a general picture of the state of these catalysts, pointing out areas of major disagreement that still persist. We shall limit our discussion to catalysts below 15% Mo and 4% Co, which encompasses the formulations of most commercial catalysts.

A. OXIDIZED STATE

There can be no doubt that Mo interacts with the alumina support in properly prepared catalysts and most commercial catalysts. A reaction path involving Al—OH groups according to reaction (1) seems most plausible in principle if not in specific detail. Loss of OH from Al—OH (57, 75) and loss in H content (21) are consistent with this scheme. A conceptual reaction scheme is illustrated in Fig. 1, which attempts to summarize the major findings.

In the Mo/Al catalyst, the precursor is normally ammonium paramolybdate, which has the formula $(NH_4)_6Mo_7O_{24}$. The Mo

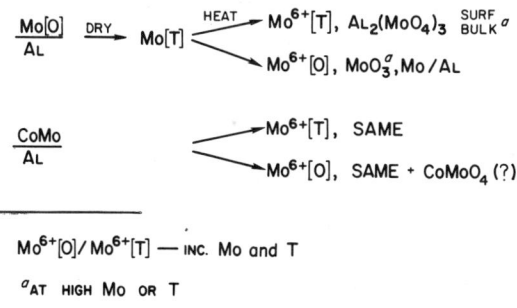

FIG. 1. Mo–support interactions.

coordination in the paramolybdate ion is octahedral, but already in the oven-dried catalyst this becomes tetrahedral. This has been explained as due to the reaction (48)

$$Mo_7O_{24}^{6-} + 4H_2O \rightarrow 7MoO_4^{2-} + 8H^+ \quad (2)$$

The MoO_4^{2-}[T] is preferentially adsorbed, thus shifting the equilibrium to the right. Upon calcination, the evidence suggests that both Mo[T] and Mo[O] species are present, the latter increasing with Mo content of the catalyst. Mo[T] has been variously attributed to $Al_2(MoO_4)_3$, either as a surface complex or bulk phase; whereas Mo[O] has been assigned to a surface complex or bulk MoO_3. However, most authors now agree that the Mo is essentially as a monolayer on the alumina surface, so any bulk MoO_3 or $Al_2(MoO_4)_3$ present must be rather small. It is unlikely that Mo penetrates into the Al_2O_3 subsurface to any degree.

Much evidence also exists for interaction of cobalt with the alumina. Suggested reaction pathways are shown in Fig. 2. Cobalt is apparently in an octahedral form on the oven-dried Co/Al catalyst, but on calcination both Co[T] and Co[O] species are present. Above 1–2% Co, a separate Co_3O_4 phase appears. Co[T] has been ascribed to surface or bulk $CoAl_2O_4$ and Co[O] to a surface complex, bulk Co_3O_4 or bulk CoO. Since Co_3O_4 contains Co^{2+}[O], the Co^{2+}[O]/Co^{2+}[T] ratio increases with increasing Co content. Also, Co may penetrate into the alumina subsurface some distance, occupying tetrahedral or octahedral sites in the defective alumina spinel structure. Such penetration must either displace Al to the surface or be accompanied by an oxygen anion residing on the surface to maintain charge neutrality. When Mo is also present, the level of Co needed before Co_3O_4 forms is considerably raised (3–4% Co). This

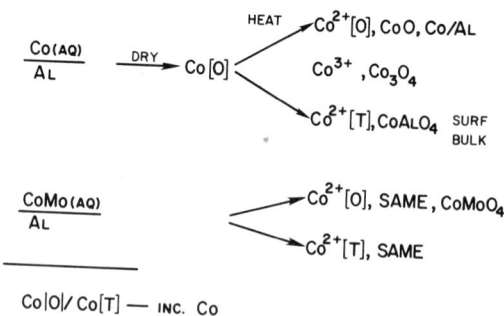

FIG. 2. Co–support interactions.

depends somewhat on the preparation and mode of addition of the separate elements, but would seem to indicate an interaction between Co and Mo on the catalyst surface. This may be a precursor state of $CoMoO_4$, which is found on silica-supported catalysts as a distinct phase (17).

B. REDUCED STATE

Molybdena catalysts generally need to be activated by reduction or sulfidation in order to obtain an active catalyst for most reactions in which they are employed (except for oxidation-type reactions). Therefore, it is important to determine what changes occur in the state of the oxidized catalyst when it is subjected to these activation pretreatments.

Reduction in hydrogen causes a loss in oxygen and a lowering in the Mo valence, but no known stoichiometric Mo compounds appear to be formed. Rather, reduction proceeds gradually and smoothly from the Mo^{6+} state to lower valence states without indication of definite intermediate states in the reduction process. Thus, depending on reduction temperature and time, species of Mo^{6+}, Mo^{5+}, Mo^{4+}, and even lower valence states may exist. It is probable that at any one state of reduction, more than one valence of Mo is present, e.g., Mo^{6+}, Mo^{5+}, and Mo^{4+} were simultaneously observed by ESCA on a reduced CoMo/Al catalyst (71). This has been interpreted in terms of differing reducibilities of various Mo species on the oxidized catalyst, as shown in Fig. 3. For example, Ratnasamy et al. (27), claimed three different Mo species to be present: a nonreducible phase (A) of Mo[T], which is Al_2MoO_4-like; an easily reducible phase (B) of bulk MoO_3; and an intermediate reducible phase (C), probably Mo[O], whose reducibility is increased with Co.

Another interpretation, based on a molybdena monolayer, invokes differences in the Mo–O bond strength due to Mo interactions with Al sites of varying strength (23). Reduction then involves removal of the

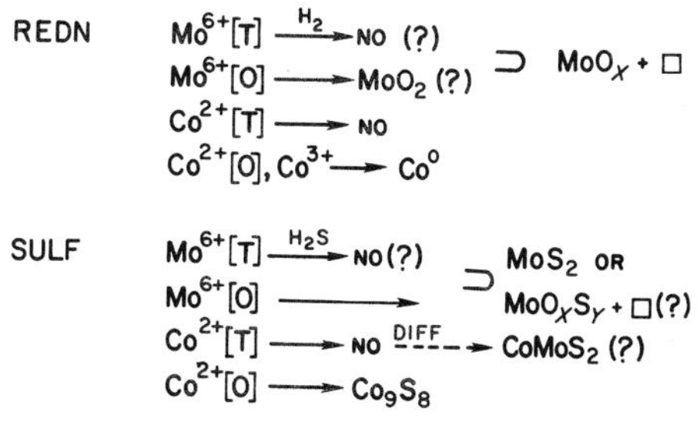

REDN
$Mo^{6+}[T] \xrightarrow{H_2} NO$ (?)
$Mo^{6+}[O] \longrightarrow MoO_2$ (?) $\supset MoO_x + \square$
$Co^{2+}[T] \longrightarrow NO$
$Co^{2+}[O], Co^{3+} \longrightarrow Co^0$

SULF
$Mo^{6+}[T] \xrightarrow{H_2S} NO(?)$
$Mo^{6+}[O] \longrightarrow$ $\supset$ MoS_2 OR $MoO_xS_y + \square(?)$
$Co^{2+}[T] \longrightarrow NO \xrightarrow{DIFF} CoMoS_2$ (?)
$Co^{2+}[O] \longrightarrow Co_9S_8$

FIG. 3. Catalyst reduction–sulfidation reactions.

weaker oxygen anions first, the stronger ones requiring more severe reduction conditions. In this concept, removal of oxygen anions creates anion vacancies, i.e., coordinated unsaturated sites, similar to that proposed by Burwell et al. (99) for chromia catalysts. This idea has been amplified by Hall and co-workers (20, 29), who have accounted for irreversible hydrogen as well as vacancies. At low extents of reduction, exactly $2H_I/Mo$ are obtained, implying a fundamental relationship between vacancies and retained hydrogen, which may be represented as

$$3\left[\begin{array}{c}\diagdown_{6+}\diagup O\\Mo\\\diagup\diagdown O\end{array}\right]\xrightarrow[-H_2O]{2H_2}2\left[\begin{array}{c}\diagdown_{5+}\diagup OH\\Mo\\\diagup\diagdown O\end{array}\right]+\begin{array}{c}\diagdown_{4+}\diagup O\\Mo\\\diagup\diagdown\square\end{array}\quad(3)$$

$$2\left[\begin{array}{c}\diagdown_{6+}\diagup O\\Mo\\\diagup\diagdown O\end{array}\right]\xrightarrow[-H_2O]{2H_2}\left[\begin{array}{c}\diagdown_{5+}\diagup OH\\Mo\\\diagup\diagdown O\end{array}\right]+\begin{array}{c}\diagdown_{3+}\diagup OH\\Mo\\\diagup\diagdown\square\end{array}\quad(4)$$

Evidence for Mo^{5+} and vacancies has already been discussed earlier in this article. Fransen et al. (85), have shown that the monolayer is not destroyed by reduction.

The situation with respect to Co/Al catalysts is less certain. At low levels of Co, no reduction occurs, whereas at higher levels, some reduction occurs. It is likely that reduction in the latter case is due to the ready reducibility of Co_3O_4. One may surmise that Co[T] is not reducible in view of the resistance of $CoAl_2O_4$ to reduction. However, whether Co[O] in other forms besides bulk Co_3O_4 (or CoO) reduces (if indeed it is present) is not known.

When Co and Mo co-exist on the oxidized catalyst, total disagreement exists between various investigators on ease of reducibility. Some claim Co enhances Mo reduction, others that it impedes and still others that it has no effect. Here again, catalyst preparation and mode of addition of the Co and Mo could have a bearing on these results. Clearly, more research on this point is needed to get a better picture of the Co–Mo interactions which may be present.

C. Sulfided State

One of the more important uses of molybdena catalysts is in hydrodesulfurization processes. In operation, the catalyst is usually presulfided with hydrogen sulfide/hydrogen or other suitable sulfiding agents. Even when presulfiding is not employed, the catalysts become sulfided during processing due to the hydrogen sulfide liberated in the reaction. Therefore, it is most important to establish what happens to the catalyst state during sulfiding, and especially the nature of the sulfided catalyst state.

The bulk of the evidence suggests that the molybdena is incompletely sulfided to MoS_2, under most processing conditions employed (mild sulfiding conditions). Exceptions occur when the sulfiding conditions are more severe or when the molybdena–alumina interaction is weak, in which cases close to stoichiometric MoS_2 is apparently achieved. Sulfiding occurs more rapidly than reduction, with the result that catalyst oxide is replaced by sulfide on a one-to-one basis, at least initially. Some additional oxygen is also lost in the process due to catalyst reduction by hydrogen present in the sulfiding stream. These results would appear to indicate that oxide replacement is kinetically (and perhaps thermodynamically) more favorable than oxide removal. One might rationalize this on the basis that the former process would not involve any formal change in the Mo valence whereas the latter requires a reduction of valence from Mo^{6+} to Mo^{4+}. However, this is not entirely true owing to the complicating factor of irreversible hydrogen attending these processes. Thus, for example, the average reduction need only proceed to a state between $4+$ and $5+$ if Eq. (3) applies. Similarly, irreversible hydrogen on the sulfided catalysts could lower the effective Mo valence, even when no vacancies are present. So, it is not at all clear why sulfiding occurs more readily than reduction.

Various interpretations have been given to the reactions involved in sulfiding. Figure 3 depicts some of the more recurrent of these. In one view, $Mo^{6+}[T]$ does not sulfide while $Mo^{6+}[O]$ goes completely to MoS_2. This would account for less than complete sulfiding on the average, but would predict higher S/Mo ratios at higher Mo levels (where Mo[O]/Mo[T] is higher) contrary to some experimental results. Furthermore, it is dificult to see why Mo[O] should sulfide faster than Mo[T] in view of the fact that bulk MoO_3 (Mo[O]) sulfides only slowly. In fact, bulk MoO_3 reduces much faster than its sulfides, completely opposite to that of the Mo/Al catalyst (which incidentally is added support for the absence of bulk MoO_3 on the oxidized catalyst). Thus, there is no reason to believe that Mo[O] would sulfide in preference to Mo[T]. A companion viewpoint is that one or both forms of Mo sulfide incompletely, but that which does forms MoS_2. The balance presumably remains unreacted or may be reduced.

An alternate viewpoint (21) suggests that the incompletely sulfided catalyst consists of a surface layer containing both oxide and sulfide anions, sulfiding occurring preferentially on the more weakly bound oxide ions first. As sulfiding proceeds, the more tightly bound oxide anions also sulfide and some anion vacancies are formed. This interpretation has elements in common with a similar proposed scheme for reduction discussed above.

The fate of cobalt during sulfiding is also somewhat uncertain. It is clear that when a Co_3O_4 phase is present, it sulfides readily to Co_9S_8. Most authors believe Co[O] also sulfides readily to give Co_9S_8, and Co[T] sulfides only slowly or not at all. De Beer *et al.* (*100*) have put forth the interesting proposal that Co incorporated in the Al_2O_3 lattice diffuses to the surface during sulfiding and interacts with small MoS_2 crystallites on on the surface to form an intercalated sulfide compound. On the other hand, Mone (*55*) has presented data in conflict with this idea. Delmon (*9*) believes separate, bulk phases of MoS_2 and Co_9S_8 are present on the sulfided catalyst.

Further discussion of proposed models of the sulfided state will be deferred until after considering results of catalytic activity studies.

VI. Catalyst Activities

Molybdena catalysts have been used for a large number of catalytic reactions—the literature is extensive in its use. We will limit our discussion to only the most common reactions occurring over reduced or sulfided catalyst. Furthermore, only those studies which attempt to relate catalyst activity to catalyst properties will be covered here.

A. Reduced Catalysts

Few definitive studies have been reported for the reduced catalyst. One of the earliest studies which was made attempted to relate activity for aldol condensation of *n*-butylaldehyde to Mo^{5+} concentration on reduced Mo/Al catalysts (*39*). The catalysts contained varying amounts of Mo and were prereduced under identical conditions. Aldol conversions were approximately linear with Mo^{5+} signals (by ESR) up to about 10% Mo. Morris *et al.* (*101*) also report a correlation between Mo^{5+} and catalytic activity for ethylene polymerization over reduced Mo/Al catalysts. Unfortunately, in both studies, tests were not done on a catalyst having the same Mo content but prereduced under different conditions (to give different Mo^{5+} values). Therefore, the correlation obtained was not unequivocal since other parameters besides Mo^{5+} might also be proportional to Mo level.

Propylene disproportionation over Mo/Al was studied by Giordano *et al.* (*102*), who reported that the active site contained Mo^{5+} dimers (not observed by ESR); monomeric Mo^{5+} species, detected by ESR, did not correlate with activity. Lo Jacono *et al.* (*84*) studied the same reaction over a Mo/Al catalyst which was prereduced to different extents of reduction and concluded that activity was proportional to vacancy concentrations as determined by volumetric measurements. The nature of the active site was such that water was a poison for the reaction. A side reaction involving polymerization (*coking*) was also shown to be proportional to vacancies.

Isomerization of cyclopropane was studied by Hall and co-workers (*84, 103*), who reported catalyst activity to increase with vacancy concentration. They postulated that the active site consisted of a vacancy and neighboring Al—OH group. Hydrogenation of ethylene at room temperature was found to increase with increase in catalyst reduction and irreversible hydrogen was apparently not involved in the reaction (*104*).

B. Sulfided Catalysts

By far, the majority of definitive studies which have been made on this catalyst have used thiophene as a model compound for hydrogenolysis. Table IX summarizes some of the more important findings. We will not discuss all the individual papers because few of them attempt any quantiative (or even qualitative) correlations with catalyst properties; and there is considerable disagreement between workers on the effects obtained. The more significant features to be pointed out are the following:

1. Catalytic activity increases with Mo level.
2. The presence of Co enhances catalytic activity, the optimum Co/Mo ratio depending on the Mo level.
3. Appreciable hydrogenation accompanies desulfurization.
4. Polar-type molecules, especially nitrogen compounds, inhibit both hydrogenolysis and hydrogenation, but to differing extents.
5. Two different types of sites appear to be responsible for hydrogenolysis and hydrogenation.
6. Bulk sulfides and silica-supported catalysts give appreciably less hydrogenation than the alumina-supported catalysts.

Few studies have attempted to relate catalytic activity to catalyst parameters. Ueda and Todo (*47, 105*) have developed a complex correlation between hydrodesulfurization of thio-β-naphthol and paramagnetic species present on the catalyst. Their correlation involves Mo^{5+}, Co^{2+}, and a surface complex containing an organic species.

TABLE IX
Effect of Variables on Thiophene Activity over CoMo/Al Catalyst

Variable	Effect on activity
Mo level	Increases up to ~10% Mo
Calcination	500° > 650° (?)
Co promotion	Maximum 0.2–0.5 Co/Mo
Preparation order	Co impreg. > Mo 1st > Co 1st (?)
Pretreatment	Sulfided > reduced
Sulfiding	Increase initially, same at steady state (?)
Inhibition	Pyridine, H_2S, H_2O
Hydrogenation	Increases with Co
Isomerization	Decreases with Co

Massoth and Kibby (94) have studied hydrogenolysis activity of thiophene over a series of sulfided Mo/Al catalysts using combined conversions and stoichiometric analyses on catalysts under reaction conditions. They showed that the intrinsic rate constant could be expressed as

$$k \propto [(\Box)_E S_s + (\Box)_E O_s] \tag{5}$$

where k is the rate constant, $(\Box)_E$ is the effective vacancy concentration corrected for coke deposits, and S_s and O_s are surface sulfide and oxide anions, which they assumed were capable of adsorbing hydrogen. The sulfided monolayer model was used for these correlations. Catalysts having some O_s left had higher activity in the steady state than those having little or no O_s. In a follow-up work, Massoth (86) showed that the number of sites calculated from gravimetric adsorption measurements during reaction agreed well with vacancies obtained in the previous study, strongly suggesting that vancancies are indeed the active sites for thiophene adsorption. A summary of this work has been recently reported (8). Subsequent studies using pyridine as a specific site poison for benzothiophene hydrodesulfurization also showed agreement with these results (87a).

In a more qualitative vein, Kotera et al. (106) found that an ammonia extracted CoMo/Al catalyst, in which 50% of the Mo was removed, was as active as the original catalyst for residual oil HDS. They concluded that combined, not free, MoO_3 was the active species. Ratnasamy et al. (27) studied the HDS of kerosine containing 0.6 wt% sulfur. They found catalytic activity increased then decreased with increasing extent of reduction of the catalyst. From characterization studies, three phases were proposed to be present on the catalyst (see discussion under Section V.B), catalytic activity being ascribed only to phase C, the other phases being inactive. This phase undergoes sulfiding via coordination with surface vacancies (associated with Mo) created from partial reduction. In their view, Co enhances the number of vacancies (presumably the active sites) ensuring a "proper balance" between oxygen and sulfur atoms. This model has some features in common with the sulfided monolayer model. Okamoto et al. (74a) propose that the active site for thiophene hydrogenolysis is a sulfided monolayer having a composition of S/Mo(IV) = 1. This can be considered to be analogous to the $(\Box)_E S_S$ site of Eq. (5).

Martinez et al. (56) have ascribed thiophene hydrogenolysis activity to the amount of monolayer molybdena available in the oxidized catalyst state. According to them, the presence of Co causes a greater dispersion of the Mo than when it is absent. In sulfiding, the Mo monolayer forms small crystallites of MoS_2 into which the Co becomes incorporated. Their view of the active catalyst is thus identical to that of De Beer and Schuit (7) already discussed. However, an important difference is that the former authors claim only surface cobalt is available to intercalate whereas the latter authors believe lattice cobalt also participates in intercalation via diffusion from the lattice during sulfiding. Furthermore, Martinez et al. show that cobalt promotes hydrogenation (to butane) over the sulfided catalyst and claim that hydrogenation activity correlates with surface cobalt, not total cobalt. Gajardo et al. (18) also found two forms of cobalt present on their sulfided CoMo/Al catalyst, one corresponding to a "pseudo aluminate" and the other to sulfided cobalt. The former is claimed to have no catalytic role, whereas the latter to correlate with the synergistic promoter behavior. Delmon (9) has recently reviewed these findings.

A brief mention should be made concerning catalyst selectivities. Appreciable hydrogenation of the hydrocarbon C_4 fraction occurs during thiophene HDS. Based on pyridine adsorption studies and other tests, Desikan and Amberg (87) postulate two active sites to be present: (1) strong acid site, responsible for hydrogenation and with weak desulfurization activity, and (2) weak acid site, primarily active for desulfurization. The hydrogenation sites are evidently associated with the alumina support in some way as hydrogenation over bulk sulfides or silica-supported catalysts is appreciably lower than over alumina catalysts (17, 18). De Beer and Schuit (7) have ascribed the hydrogenation sites to a monolayer oxide residue which is very strongly bonded to the Al_2O_3. Finally, the addition of Co was found to decrease isomerization activity (*88, 107*) in contrast to the enhancement obtained in hydrogenolysis and hydrogenation activity. Since isomerization readily occurs over Al_2O_3 itself, it may be surmised that Co interacts with the Al_2O_3 isomerization sites, rendering them less active.

VII. Proposed Models of the Sulfided Catalyst

There are three current theories on sulfiding the CoMo/Al catalyst: monolayer, intercalation, and contact synergism. These are schematically depicted in Fig. 4. All start with the assumption of a monolayer model for the oxidized catalyst, differences appearing in the subsequent effect that sulfiding has on the monolayer structure. Space limitations preclude a full description of these models; the original references should be consulted for more detail.

A. Monolayer Model

In this model, proposed by Massoth (*21*), sulfiding effects replacement of some terminal oxide anions by sulfide anions and removal of some by reduction forming vacancies, the surface layer remaining essentially intact. It has been developed mostly from analogy with reduction characteristics of Mo/Al catalyst with interpretation of sulfiding studies along similar lines. In this respect, it is a modification of the model first proposed by Lipsch and Schuit (*79*) to explain their thiophene hydrogenolysis results over reduced molybdena catalysts and later extended by Schuit and Gates (*5*). However, an important difference is that Massoth envisions the oxidized catalyst as one-dimensional molyb-

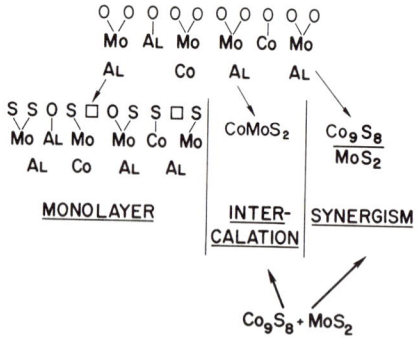

FIG. 4. Theories of the sulfided state.

dena chains attached to the alumina surface rather than the two-dimensional layer of Schuit. The latter does not permit addition of more than one S/Mo without disruption of the monolayer, whereas the chain model can accommodate considerably more S without breakup of the surface configuration. Vacancies generated by reduction or sulfidation are considered the active sites for adsorption of sulfur compounds, and surface sulfide (and oxide, if any left) for adsorption of hydrogen, hydrogenolysis occurring by a surface reaction between these surface species (86, 94). The location of cobalt has not been established but may be considered partly at the surface and partly in the Al_2O_3 lattice, that at the surface presumably undergoing sulfiding. The promotional role of cobalt on catalytic activity is uncertain in this model. It may have a kinetic effect on surface reactivities through changes in bonding strengths and/or may affect vacancy concentrations (27).

Support for this model is based mainly on the Mo/Al catalyst and includes the following relative to the sulfided catalyst:

1. Sulfur stoichiometry did not agree with simple conversion of MoO_3 to MoS_2 and initial results were closer to a one-to-one replacement of O for S (see Section IV.B).
2. Surface area measurements were consistent with a monolayer (96).
3. High H_I/S values (0.7–1.1) were measured (21).
4. NH_3 chemisorption was found to be proportional to vacancies and the same as equivalently reduced catalyst (21).
5. No MoS_2 crystallites were observbed by electron microscopy (96).
6. An appreciable Mo^{5+} signal was detected by ESR, but none on MoS_2 (33).
7. Reoxidation of the sulfided catalyst gave the same state as the original oxidized catalyst by reflectance spectroscopy, IR, and ESCA.
8. Thiophene hydrogenolysis data over a series of catalysts having different sulfided states correlated well with the model.

Some of these points have been taken up more fully in a recent note (96). Relative to point 3, lower H_I/S values would be expected if a bulk MoS_2 phase is present and chemisorption occurs on the surface of the MoS_2. Point 4 shows that vacancies on a sulfided catalyst are equivalent to those on a reduced catalyst, the latter being essentially as a monolayer. Point 7 shows that the sulfided catalyst returns to the original oxidized state on oxidation, which would not be expected if bulk MoS_2 were present, forming bulk MoO_3 instead.

De Beer and Schuit (7) have argued against a sulfided monolayer being present, holding the opinion that a separate MoS_2 phase is formed. Their arguments encompass similar reaction patterns for the catalyst and bulk mixed sulfide phases, high sulfur contents of the catalysts, and identification of MoS_2 by reflectance spectra. It hardly need be pointed out that HDS activity for bulk MoS_2 and promotional effects in mixed sulfides does not prove that these phases are present as such on the supported catalyst. The question of the high sulfur analyses has already been discussed (Section IV.J), it being this author's opinion that their results are incorrect with regard to stoichiometric sulfide sulfur on the catalyst. The interpretation of the reflectance spectra to prove the presence of a bulk phase seems somewhat tenuous to this author. For example, how closely would the catalyst spectra resemble that of MoS_2 for a Mo–S surface monolayer? Also, what effect does adsorbed H_2S on their catalyst samples (see Section IV.J) have on their spectra? Further, this technique is a qualitative one and may actually represent only a small fraction of the Mo in this form. Other evidence cited for support of a separate MoS_2 phase comes from ESCA and electron microscopy results. The former has been previously discussed (Section IV.F). As regards the latter, which purports to show small crystals of MoS_2 on a sulfided NiMo/Al catalyst, it would be necessary to know the catalyst composition, preparation, sulfiding conditions, fraction of Mo represented by the crystals and whether identical results would be obtained for sulfided Mo/Al and CoMo/Al catalysts. A recent report (96) cited unpublished work which failed to detect MoS_2 by electron microscopy on sulfided Mo/Al catalysts.

B. INTERCALATION MODEL

The intercalation model has been championed by De Beer and Schuit (7). In this model, sulfiding causes a breakup of the oxide monolayer with formation of discrete MoS_2 crystallites of small size. The cobalt can react with the MoS_2 forming intercalated $CoMoS_2$. Another important role of the Co is to produce smaller sized MoS_2 crystallites during sulfiding. Cobalt takes up edge positions in the MoS_2 crystal, resulting in formation of an Mo^{3+} cation, which is regarded as the active catalytic site. Presumably without Co present, the concentration of Mo^{3+} is low and thus, catalytic activity is low. Hydrogenation activity is assigned to a small amount of residual molybdena monolayer which neither reduces nor sulfides.

The evidence for a separate MoS_2 phase being present on the catalyst has already been discussed under A above. There is scant evidence for the existence of an intercalation phase on the supported catalyst. Much of the argument is based on the intercalation model proposed for the bulk Ni–WS_2 catalyst (108). Canesson and Grange (109) and Mone (55) have brought forth evidence contradictory to formation of an intercalation compound. Delmon (9) has recently presented arguments against the intercalation model.

One of the important tenets of the intercalation model is the role of Co in increasing the Mo^{3+} population, the presumed active site for HDS. Most workers have failed to detect Mo^{3+} by ESR or ESCA, but De Beer (41) cites ESR evidence for its presence on a reduced Mo/Al sample. Notwithstanding, its concentration must be extremely low not to be detected by others. However, even if this is the active site, an intercalation model need not be invoked to explain Co promotion. For example, the monolayer model could produce Mo^{3+} in reduction or sulfidation according to the structures,

$$\begin{array}{cc} HO\diagdown\quad\square & HS\diagdown\quad\square \\ \diagup Mo\diagdown & \diagup Mo\diagdown \end{array}$$

and the presence of Co could conceivably promote more of these sites by increasing catalyst reducibility.

C. Contact Synergism Model

Recently, Delmon (9) has advocated an alternate model which involves a contact synergism between Mo and Co sulfide phases. In this model, the oxide monolayer is also destroyed during sulfiding resulting in formation of separate phases of MoS_2 and Co_9S_8. The special promoting role of cobalt occurs only at the interfaces between the two separate phases; the nature of this enhancement is not known at present. Much of the arguments come from analogy with mixed bulk sulfide catalyst and supported catalysts in which bulk MoS_2 has been shown to be present by XRD analysis.

An objection to this model is that it fails to account for the effect of the support, which is known to increase the hydrogenation/hydrogenolysis ratio over that of bulk sulfides. De Beer and Schuit (7) concede that the model may have some merit at higher Co/Mo ratios, but that the major contribution to HDS activity is caused by intercalated systems.

D. Relevance of Models

Although each of the models have their advocates, none can be considered definitely proved at present. It is possible that more than one model may apply or that none are correct. Perhaps the monolayer model is more correct for mildly sulfided catalysts and either or both of the separate phase models may apply under more severe sulfiding conditions, or that the former may gradually revert to the latter under some processing conditions.

It should be appreciated that in some instances, bulk MoS_2 is unquestionably present on the sulfided catalyst, as proved by XRD analysis. Also, such catalysts are quite active and show Co promotional effects. In such cases, the monolayer model cannot obviously exclusively apply and either of the other models would be favored. However, even in this case, it is possible that an important fraction of the Mo could remain as a monolayer and exhibit catalytic activity.

VIII. Role of Cobalt

One of the more important and intriguing questions posed is the role which cobalt plays in catalysis over molybdena catalysts. Its promoting effect is unquestioned in HDS reactions and is a large reason for the commercial success of these catalysts. Also, it is probably the subject which has received the greatest attention in the study of these catalysts. Despite this effort, controversy still attends the primary function of cobalt. Many theories have been proposed based on limited results, many of these being in disagreement with other results.

Some of the more important theories are presented in Table X together with pertinent references. These depend on either alteration of catalyst phases or specific kinetic effects of the cobalt. The merits of each are briefly discussed below.

1. Increases Mo monolayer area. The addition of Co causes the Mo to better disperse on the support. Activity is then proportional to Mo areas. Ben-Yaacov's (61) findings disagree, showing no effect of Co on Mo dispersion.

2. Increases Mo reduction. The molybdena catalyst reduces more in the presence of Co and activity depends on extent of reduction. As discussed earlier, there is considerable disagreement on the former point and the latter is unproved, especially for sulfided catalysts.

3. Increases H_2 mobility. This invokes an increased hydrogenation ability due to Co and presumes hydrogenation is the rate-limiting step in the reaction. As regards the first point, increased hydrogenation does occur, but the hydrogenation/hydrogenolysis ratio does not increase (18). The second point conflicts with many kinetic studies that show surface reactions to be rate limiting.

4. Due to $CoMoS_2$ intercalation. As has been discussed previously (see Section VII.B), this assigns the Co to a very specific site in the MoS_2

TABLE X
Role of Cobalt on Catalytic Activity

Effect	Ref.
Increases Mo area	56
Increase Mo reduction	9, 27
Increase H_2 mobility	126
Due to Co intercalation of MoS_2	7, 17
Due to MoS_2/Co_9S_8 synergism	9, 59, 61
Due to specific kinetic effects	27, 56
Prevents catalyst deactivation	27
Enhanced surface segregation	127
Prevents MoS_2 crystallization	74a

crystallite engendering formation of Mo^{3+} which is the active site for HDS. There is scant support for the presence of Mo^{3+} in these catalysts at present. Also, it cannot account for initial high activity of mildly reduced Mo/Al catalysts, where Mo^{3+} must be extremely low.

5. Due to MoS_2/Co_9S_8 synergism. This invokes an enhanced activity at the phase boundary between MoS_2 and Co_9S_8 for reasons as yet unknown. No quantiative measurements are available to test this hypothesis.

6. Specific kinetic effect. This would apply to the sulfided monolayer catalyst. Cobalt may affect adsorption-desorption properties or intrinsic activity of vacancies. No definitive data exist to support this proposal.

7. Prevent deactivation. The increased hydrogenating ability of Co prevents accumulation of coke deposits. There is some merit in this proposal as Mo/Al catalysts appear to deactivate faster than CoMo/Al catalysts (100). However, initial activity of the sulfided catalyst was higher for the CoMo/Al catalyst (100).

8. Enhanced surface segregation. The variation in catalytic activity with Co/Mo ratio is proposed to be due to a surface enhancement of Co species in a mixed CoMo-sulfide phase, the Co being considered a promoter (127). This model does not specify the promotion mechanism nor does it take account of any role of the support.

9. Prevents MoS_2 crystallization. The presence of Co hinders diffusion of Mo to form inactive (relatively) MoS_2, the active specie being a sulfided Mo monolayer (74a).

IX. Nickel–Molybdena Catalysts

Molybdena catalysts containing nickel have received relatively less attention as compared with cobalt. Nickel–molybdena catalysts show overall similar properties to those with cobalt, but there are some differences in details. The promotional effect on HDS reactions is well documented (*1*).

The minimum decomposition temperature for Ni(NO$_3$)$_2$ is reported to be 450° versus 350° for Co(NO$_3$)$_2$ (*110*). Thus, Ni-containing catalysts need to be calcined at higher temperatures to assure complete removal of nitrate. Nickel on a Ni/Al catalyst has been characterized in terms of two phases, a well-dispersed and a nondispersed phase (*111*). Lo Jacono and Schiavello (*54*) discuss interactions of Ni(and Co) ions with the Al$_2$O$_3$ support at 600° in terms of two competing reactions, viz., one leading to formation of a spinel and another to segregation to a bulk oxide phase. The relative amount of each depended on the transition metal level and calcination temperature. Lafitau *et al.* (*112*), however, report only the spinel form to be present for catalysts up to 15% Ni calcined between 550° and 650°. This phase was unreactive to sulfiding. Based on reflectance and infrared spectra (XRD lines were absent), Talipov *et al.* (*113*) claimed the following compounds were present on a 20% (NiO + MoO$_3$)/Al$_2$O$_3$ catalyst: NiAl$_2$O$_4$, Al(MoO$_4$)$_3$, and NiMoO$_4$.

Kiviat and Petrakis (*80*) found the infrared specta for adsorbed pyridine on NiMo/Al to be only slightly different than that on CoMo/Al. Both showed evidence of Lewis and Brønsted acid sites, but the Ni catalyst had a somewhat higher Brønsted-to-Lewis acidity ratio. Similar results were reported by Mone and Moscu (*15*), who also found that higher calcination temperatures produced more nickel aluminate spinel.

Ahuja *et al.* (*88*) report a similar promoting effect for Ni and Co molybdena catalysts for HDS and aromatic hydrogenation, whereas De Beer *et al.* (*114*) found a greater effect for Ni for thiophene desulfurization, and Ripperger and Saum (*91*) a greater effect for hydrogenation of aromatics and hydrodenitrification. Sultanov *et al.* (*115*) found the order of impregnation had a strong effect on catalytic activity for HDS of petroleum oil; Mo added first, followed by Ni was more active than the reverse impregnation procedure. Mone (*55*) reported that higher calcination temperatures produce less active catalysts.

X. Active Sites—Concluding Remarks

A. Reduced Catalysts

No one particular site or specie has been identified as the active site for the various reactions which occur on reduced molybdena catalysts. It may be that different sites are used for different types of reactions, e.g., hydrogenation and isomerization. Phase studies and measurement of Mo

CHARACTERIZATION OF MOLYBDENA CATALYSTS 305

valence states have not generally proved useful in correlating catalyst activity for these reactions. Perhaps the more fruitful correlations will come from measurements of vacancies and adsorbed hydrogen, as has already been started by Hall and co-workers. It would be desirable to have simpler techniques available to perform these kinds of measurements on a more continuous basis. Especially would a direct measurement of anion vacancies be helpful. The questions of strength (as well as number) of vacancies, their further characterization in terms of adsorbed complexes, and the specific role of poisons (and coke) all await further investigation. Finally, the promoting role of Co, found so important in hyrogenolysis reactions over sulfided catalysts, has hardly been investigated here.

B. Sulfided Catalysts

The picture here is even less clear than for reduced catalysts. In spite of the fine studies by Delmon and workers on the bulk mixed sulfides, and that of Schuit and De Beer and workers on various hybrid catalysts, it is not conclusively proved that bulk sulfides are the active ingredients for hydrodesulfurization reactions for the mildly sulfided catalysts employed in industry.

It is a fact that bulk MoCo-sulfides are active for HDS reactions, and therefore, are prime suspects in the search for the active sites. However, aside from arguments derived from characterization studies, there are two rather important differences in catalytic performance between bulk sulfides and sulfided molybdena. First, there is the fact that the reduced Mo/Al catalyst is initially very active (in fact more active than the sulfided state) for thiophene hydrogenolysis. Of course, this activity is not maintained due to catalyst sulfiding. Nevertheless, the basic features of the reaction occur even in this initial stage, and there is no evidence that reaction over the reduced catalyst (with or without Co) is fundamentally any different than that obtained in the steady-state sulfided condition. Second, the alumina-supported catalyst has appreciably higher hydrogenation activity than the bulk sulfides. This cannot be explained by a simple dual function, the Al_2O_3 supplying the hydrogenation, since Al_2O_3 itself has no intrinsic hydrogenation activity. This was further demonstrated by a mixture of MoS_2 Al_2O_3 which gave a low hydrogenation activity in thiophene desulfurization (100). The proposal of De Beer and Schuit (7), that enhanced hydrogenation is due to a small fraction of unreduced and unsulfided Mo left on the catalyst, is also difficult to accept in view of the fact that hydrogenation of ethylene occurred only on a reduced catalyst (104).

Undoubtedly, under certain industrial conditions (e.g., coal liquefaction) representing more severe sulfiding conditions, some bulk sulfides may be present. Even in those cases, it is not certain that all the Mo has been converted to MoS_2. The hydrogenation/hydrogenolysis activity may perhaps be taken as an indication of extent of conversion to MoS_2 (although account must be taken of the weak hydrogenation ability of MoS_2 itself). Furthermore, conversion to more bulk MoS_2 may occur gradually over a period of time during processing. Although this should cause a lowering in relative hydrogenation activity with time, it may be offset by increases in processing temperature as normally carried out in industrial practice.

With regard to identification of active sites, the search will depend on the models chosen. For the monolayer sulfide model, anion vacancies and adsorbed hydrogen sites, would seem to be of paramount importance. The remarks given above under reduced catalysts would apply here also. Especially important would be the role of strongly adsorbed and irreversibly adsorbed coke precursors on the number and activity of vacant sites and hydrogen chemisorption. Also, the promoting role of Co has not been systematically evaluated for this model vis-à-vis; its effect on number and strength of vacancies and H_2 chemisorption. For the bulk sulfides models, the role of Co is still not entirely clear, especially for the contact synergism model. Assessment of the nature and number of active sites needs attention, as does a simple means of measuring MoS_2 surface area (and perhaps $CoS-MoS_2$ interface area). The effect of coke and the nature of surface reaction intermediates has not been investigated for these systems.

ACKNOWLEDGMENT

The author thanks Dr. Charles Kibby for helpful comments on the manuscript.

REFERENCES

1. Weisser, O., and Landa, S., "Sulfide Catalysts, Their Properties and Applications." Pergamon, New York, 1973.
2. McKinley, J. B., in "Catalysis" (P. Emmett, ed.), Vol. 5, p. 405. Reinhold, New York, 1957.
3. Mitchell, P. C. H., "The Chemistry of Some Hydrodesulphurization Catalysts Containing Molybdenum." Climax Molybdenum Co., London, 1967.
4. Schuman, S. C., and Shalit, H., *Catal. Rev.* **4,** 245 (1970).
5. Schuit, G. C. A., and Gates, B. C., *AIChE J.* **19,** 417 (1973).
6. Amberg, C. H., *J. Less-Common Met.* **36,** 339 (1974).
7. De Beer, V. H. J., and Schuit, G. C. A., in "Preparation of Catalysts" (B. Delmon, P. A. Jacobs, and G. Poncelet, eds.), p. 343. Elsevier, Amsterdam, 1976; *Ann. N.Y. Acad. Sci.* **272,** 61 (1976).
8. Massoth, F. E., *J. Less-Common Met.* **54,** 343 (1977).
9. Delmon, B., *Am. Chem. Soc., Div. Pet. Chem., Prepr.* **22**(2), 503 (1977).

10. Ashley, J. H., and Mitchell, P. C. H., *J. Chem. Soc. A* p. 2821 (1968).
11. Ratnasamy, P., Mehrotra, R. P., and Ramawamy, A. V., *J. Catal.* **32**, 63 (1974).
12. Trunov, V. K., Lutsenko, V. V., and Kovba, L. M., *Izv. Vyssh. Uchebn. Zaved., Khim. Khim. Tekhnol.* **10**, 375 (1967); [*Chem. Abstr.* **68**, 73212j (1968)].
13. Stork, W. H. J., Coolegem, J. G. F., and Pott, G. T., *J. Catal.* **32**, 497 (1974).
14. Pott, G. T., and Stork, W. H. J., in "Preparation of Catalysts" (B. Delmon, P. A. Jacobs, and G. Poncelet, eds.), p. 537, Elsevier, Amsterdam, 1976.
15. Mone, R., and Moscou, L., *Am. Chem. Soc., Div. Pet. Chem., Prepr.* **20**(2), 564 (1975).
16. Kabe, T., Yamadaya, S., Oba, M., and Miki, Y., *Int. Chem. Eng.* **12**, 366 (1972).
17. De Beer, V. H. J., Van Der Aalst, M. J. M., Machiels, C. J., and Schuit, G. C. A., *J. Catal.* **43**, 78 (1976).
18. Gajardo, P., Declerck-Grimee, R. I., Delvaux, G., Olodo, P., Zabala, J. M., Canesson, P., Grange, P., and Delmon, B., *J. Less-Common Met.* **54**, 311 (1977).
19. Sontag, P., Kim, D. Q., and Marion, F., *C. R. Acad. Sci., Ser. C* **259**, 4704 (1964).
20. Hall, W. K., and Massoth, F. E., *J. Catal.* **34**, 41 (1974).
21. Massoth, F. E., *J. Catal.* **36**, 164 (1975).
22. von Destinon-Forstmann, J., *Can. Metall. Q.* **4**, 1 (1965).
23. Massoth, F. E., *J. Catal.* **30**, 204 (1973).
24. Dollimore, D., and Rickett, G., *Proc. Int. Congr. Therm. Analy., 3rd* **2**, i43 (1971).
25. Giordano, N., Castellan, A., Bart, J. C. J., Vaghi, A., and Campadelli, F., *J. Catal.* **37**, 204 (1975).
26. Kabe, T., Yamadaya, S., Oba, M., and Miki, Y., *Kogyo Kagaku Zasshi* **74**, 1566 (1971); [*Chem. Abstr.* **67**, 101646u (1971)].
27. Ratnasamy, P., Ramaswamy, A. V., Banerjee, K., Sharma, D. K., and Ray, N., *J. Catal.* **38**, 19 (1975).
28. Holm, V., and Clark, A., *J. Catal.* **11**, 305 (1968).
29. Hall, W. K., and Lo Jacono, M., *Int. Congr. Catal., 6th, London* Pap. A16 (1976).
30. John, G. S., Den Herder, M. J., Mikovsky, R. J., and Waters, R. F., *Adv. Catal.* **9**, 252 (1957).
31. Adamska, B., Haber, J., Janas, J., and Lombarska, D., *Bull. Acad. Pol. Sci., Ser. Sci. Chim.* **23**, 753 (1975).
32. Verhoeven, W., and Delmon, B., *Bull. Soc. Chim. Fr.* p. 3065 (1966).
33. Seshadri, K. S., Massoth, F. E., and Petrakis, L., *J. Catal.* **19**, 95 (1970).
34. Badger, E. H. M., Griffith, R. H., and Newling, W. B. S., *Proc. R. Soc., A* **197**, 184 (1949).
35. Gautherin, J., and Colson, J., *C. R. Acad. Sci., Ser. C* **278**, 815 (1974).
35a. Zabala, J. N., Grange, P., and Delmon, B., *C. R. Acad. Sci., Ser. C* **279**, 725 (1974).
36. Dufaux, M., Che, M., and Naccache, C., *J. Chim. Phys.* **67**, 527 (1970).
37. Lo Jacono, M., Verbeek, J. L., and Schuit, G. C. A., *J. Catal.* **29**, 463 (1973).
38. Masson, J., and Nechtschein, J., *Bull. Soc. Chim. Fr.* p. 3933 (1968).
39. Seshadri, K. S., and Petrakis, L., *J. Phys. Chem.* **74**, 4102 (1970).
40. Seshadri, K. S., and Petrakis, L., *J. Catal.* **30**, 195 (1973).
41. See discussion following paper in ref. 7.
42. Masson, J., Delmon, B., and Nechtschein, J., *C. R. Acad. Sci., Ser. C* **266**, 428 (1968).
43. Ramaswamy, A. V., Sivasanker, S., and Ratnasamy, P., *J. Catal.* **42**, 107 (1976).
44. Asmolov, G. N., and Krylov, O. V., *Kinet. Catal. (USSR)* **13**, 161 (1972).
45. Abdo, S., Lo Jacono, M., Clarkson, R. B., and Hall, W. K., *J. Catal.* **36**, 330 (1975).
46. Lo Jacono, M., Verbeek, J. L., and Schuit, G. C. A., *Proc. Int. Congr. Catal., 5th* p. 1409 (1973).

47. Ueda, H., and Todo, N., *J. Catal.* **27**, 281 (1972).
48. Ashley, J. H., and Mitchell, P. C. H., *J. Chem. Soc. A* p. 2730 (1969).
49. Asmolov, G. N., and Krylov, O. V., *Kinet. Catal. (USSR)* **12**, 403 (1971).
50. Lipsch, J. M. J. G., and Schuit, G. C. A., *J. Catal.* **15**, 174 (1969).
51. Giordano, N., Bart, J. C. J., Castellan, A., and Vaghi, A., *J. Less-Common Met.* **36**, 367 (1974).
52. Praliaud, H., *J. Less-Common Met.* **54**, 387 (1977).
53. Lo Jacono, M., Cimino, A., and Schuit, G. C. A., *Gazz. Chim. Ital.* **103**, 1281 (1973).
54. Lo Jacono, M., and Schiavello, M., *in* "Preparation of Catalysts" (B. Delmon, P. A. Jacobs, and G. Poncelet, eds.), p. 473. Elsevier, Amsterdam, 1976.
55. Mone, R., *in* "Preparation of Catalysts" (B. Delmon, P. A. Jacobs, and G. Ponelet, eds.), p. 381. Elsevier, Amsterdam, 1976.
56. Martinez, N. P., Mitchell, P. C. H., and Chiplunker, P., *J. Less-Common Met.* **54**, 333 (1977).
57. Mitchell, P. C. H., and Trifiro, F., *J. Catal.* **33**, 350 (1974).
58. Van der Aalst, M. J. M., and De Beer, V. H. J., *J. Catal.* **49**, 247 (1977).
59. Richardson, J. T., *Ind. Eng. Chem., Fundam.* **3**, 154 (1964).
60. Richardson, J. T., and Vernon, L. W., *J. Phys. Chem.* **62**, 1153 (1958).
61. Ben-Yaacov, R., Ph.D. Thesis, Univ. of Houston, Houston, Texas, 1975.
62. Cimino, A., and De Angelis, B. A., *J. Catal.* **36**, 11 (1975).
63. Miller, A. W., Atkinson, W., Barber, M., and Swift, P., *J. Catal.* **22**, 140 (1971).
64. Armour, A. W., Mitchell, P. C. H., Folkesson, B., and Larsson, R., *J. Less-Common Met.* **36**, 361 (1974).
65. Grimblot, J., Bonnelle, J. P., and Beaufils, J. P., *J. Electron. Spectrosc. Relat. Phenom.* **8**, 437 (1976).
66. Okamoto, Y., Nakano, H., Imanaka, T., and Teranish, S., *Bull. Chem. Soc. Jpn.* **48**, 1163 (1975).
67. Ratnasamy, P., *J. Catal.* **40**, 137 (1975).
68. Walton, R. A., *J. Catal.* **44**, 335 (1976).
69. Parekh, B. S., and Weller, S. W., *J. Catal.* **47**, 100 (1977).
70. Grimblot, J., and Bonnelle, J. P., *J. Electron. Spectrosc. Relat. Phenom.* **9**, 449 (1976).
71. Patterson, T. A., Carver, J. C., Leyden, D. E., Hercules, D. M., J. Phys. Chem. **80**, 1700 (1976).
72. Brinen, J. S., *J. Electron. Spectrosc. Relat. Phenom.* **5**, 377 (1974).
73. Stevens, G. C., and Edmonds, T., *J. Less-Common Met.* **54**, 321 (1977).
74. Friedman, R. M., Declerck-Grimee, R. I., and Fripiat, J. J., *J. Electron Spectrosc. Relat. Phenom.* **5**, 437 (1974).
74a. Okamoto, Y., Nakamo, H., Shimokawa, T., Imanaka, T., and Teranishi, S., *J. Catal.* **50**, 447 (1977).
75. Sonnemans, J., and Mars, P., *J. Catal.* **31**, 209 (1973).
76. Fransen, T., Van der Meer, O., and Mars, P., *J. Catal.* **42**, 79 (1976).
77. Brown, F. R., and Makovsky, L. E., *Appl. Spectrosc.* **31**, 44 (1977); Brown, F. R., Makovsky, L. E., and Rhee, K. H., *J. Catal.* **50**, 162, 385 (1977).
78. Nicholson, D. E., *Anal. Chem.* **34**, 370 (1962).
79. Lipsch, J. M. J. G., and Schuit, G. C. A., *J. Catal.* **15**, 179 (1969).
80. Kiviat, F. E., and Petrakis, L., *J. Phys. Chem.* **77**, 1232 (1973).
81. Ratnasamy, P., Sharma, D. K., and Sharma, L. D., *J. Phys. Chem.* **78**, 2069 (1974).
82. Dollimore, D., Galway, A., and Rickett, G., *J. Chim. Phys.* **72**, 1059 (1975).
83. Giordano, N., Bart, J. C. J., Vaghi, A., Castellan, A., and Martinotti, G., *J. Catal.* **36**, 81 (1975).
84. Lo Jacono, M., and Hall, W. K., *J. Colloid Interface Sci.* **58**, 76 (1977).

85. Fransen, T., Van Berge, P. C., and Mars, P., in "Preparation of Catalysts" (B. Delmon, P. A. Jacobs, and G. Poncelet, eds.), p. 405. Elsevier, Amsterdam, 1976.
85a. Samuel, P., and Yeddanapalli, L. M., *J. Appl. Chem. Biotechnol.* **24,** 777 (1974).
86. Massoth, F. E., *J. Catal.* **47,** 316 (1977).
87. Desikan, P., and Amberg, C. H., *Can. J. Chem.* **42,** 843 (1964).
87a. Cowley, S. W., and Massoth, F. E., *J. Catal.* **51,** 291 (1978).
88. Ahuja, S. P., Derrien, M. L., and LePage, J. F., *Ind. Eng. Chem., Prod. Res. Dev.* **9,** 272 (1970).
89. Armour, A. W., Ashley, J. H., and Mitchell, P. C. H., *Am. Chem. Soc., Div. Pet. Chem., Prepr.* **16**(1), A116 (1971).
90. Slager, T. L., and Amberg, C. H., *Can. J. Chem.* **50,** 3416 (1972).
91. Ripperger, W., and Saum, W., *J. Less-Common Met.* **54,** 353 (1977).
92. De Beer, V. H. J., Bevelander, C., Van Sint Fiet, T. H. M., Werter, P. G. A. J., and Amberg, C. H., *J. Catal.* **43,** 68 (1976).
93. Owens, P. J., and Amberg, C. H., *Adv. Chem. Ser.* **33,** 182 (1961).
94. Massoth, F. E., and Kibby, C. L., *J. Catal.* **47,** 300 (1977).
95. Todo, N., Muramatsu, K., Kurita, M., Ogawa, K., Sato, T., Ogawa, M., and Kotera, Y., *Bull. Jpn. Pet. Inst.* **14,** 89 (1972).
96. Massoth, F. E., *J. Catal.* **50,** 190 (1977).
97. Kotera, Y., Ogawa, K., Oba, N., Shimomura, K., Yonemura, M., Ueno, A., and Todo, N., in "Preparation of Catalysts" (B. Delmon, P. A. Jacobs, and G. Poncelet, eds.), p. 371. Elsevier, Amsterdam, 1976.
98. Yamagata, N., Owada, Y., Okazaki, S., and Tanabe, K., *J. Catal.* **47,** 358 (1977).
99. Burwell, R. L., Jr., Haller, G. L., Taylor, K. C., and Read, J. F., *Adv. Catal.* **20,** 1 (1969).
100. De Beer, V. H. J., Van sint Fiet, T. H. M., Van der Steen, G. H. A. M., Zwaga, A. C., and Schuit, G. C. A., *J. Catal.* **35,** 297 (1974).
101. Morris, R. V., Waywell, D. R., and Shepard, J. W., *J. Less-Common Met.* **36,** 395 (1974).
102. Giordano, N., Padovan, M., Vaghi, A., Bart, J. C. J., and Castellan, A., *J. Catal.* **38,** 1 (1975).
103. Lombardo, E. A., Lo Jacono, M., and Hall, W. K., *J. Catal.* **51,** 243 (1978).
104. Lombardo, E. A., Houalla, M., and Hall, W. K., *J. Catal.* **51,** 256 (1978).
105. Ueda, H., Todo, N., and Kurita, M., *J. Less-Common Met.* **36,** 387 (1974).
106. Kotera, Y., Todo, N., Muramatsu, K., Ogawa, K., Kurita, M., Sato, T., Ogawa, M., and Kabe, T. *Int. Chem. Eng.* **11,** 752 (1971).
107. De Beer, V. H. J., Dahlmans, J. G. J., and Smeets, J. G. M., *J. Catal.* **42,** 467 (1976).
108. Farragher, A. L., and Cossee, P., *Proc. Int. Congr. Catal., 5th* p. 1301 (1973).
109. Canesson, P., and Grange, P., *C. R. Acad. Sci., Ser. C* **281,** 757 (1975).
110. Keely, W., and Maynor, H., *J. Chem. Eng. Data* **8,** 297 (1963).
111. Tomlinson, J. R., Keeling, R. O., Jr., Rymer, G. T., and Bridges, J. M., *Actes Congr. Int. Catal., 2nd, 1960* **2,** 1831 (1961).
112. Lafitau, H., Neel, E., and Clement, J. C., in "Preparation of Catalysts" (B. Delmon, P. A. Jacobs, and G. Poncelet, eds.), p. 393. Elsevier, Amsterdam, 1976.
113. Talipov, G. S., Khakimov, U. B., Vorob'ev, V. N., Samigov, K. A., and Shchekochikhin, Y. M., *Kinet. Catal. (USSR)* **13,** 1384 (1972).
114. De Beer, V. H. J., Van Sint Fiet, T. H. M., Engelen, J. F., Van Haandel, A. C., Wolfs, M. W. J., Amberg, C. H., and Schuit, G. C. A., *J. Catal.* **27,** 357 (1972).
115. Sultanov, A. S., Abdurakhmanov, M., Talipov, G. S., Inoyatov, N. S., Samigov, K. A., and Kayumov, A. A., *Kinet. Catal. (USSR)* **13,** 1207 (1972).
116. Keeling, R., Jr., *J. Chem. Phys.* **31,** 279 (1959).

117. Maksimovskaya, R. I., Anufrienko, V. F., and Kolovertnov, G. D., *Kinet. Catal. (USSR)* **9,** 984 (1968).
118. Asmolov, G. N., and Krylov, O. V., *Kinet. Catal: (USSR)* **11,** 847 (1970).
119. Asmolov, G. N., and Krylov, O. V., *Bull. Acad. Sci. USSR, Div. Chem. Sci.* p. 2279 (1970).
120. Krylov, O. V., and Margolis, L. Y., *Kinet. Catal. (USSR)* **11,** 358 (1970).
121. Ueda, H., and Todo, N., *Bull. Chem. Soc. Jpn.* **43,** 3698 (1970).
122. Galliasso, R., and Menguy, P., *Bull. Soc. Chim. Fr.* p. 1331 (1972).
123. Naccache, C., Bandiera, J., and Dufaux, M., *J. Catal.* **25,** 334 (1972).
124. Grimblot, J., Pommery, J., and Beaufils, J., *J. Less-Common Met.* **36,** 381 (1974).
125. Salageanu, I., Trestianu, D., and Segal, E., *Rev. Roum. Chim.* **18,** 1537 (1973).
126. Smith, G. V., and Hinckley, C. C., *J. Catal.* **30,** 218 (1973).
127. Phillips, R. W., and Fote, A. A., *J. Catal.* **41,** 168 (1976).

Poisoning of Automotive Catalysts

M. SHELEF, K. OTTO, AND N. C. OTTO

Research Staff
Ford Motor Company
Dearborn, Michigan

I. Introduction	311
II. Catalysts	313
III. Poisons	314
A. Potential Poisons in Fuel	315
B. Potential Poisons in Oil	316
C. Other Potential Poisons	317
IV. Analysis and Examination of Poisons on Catalysts	317
V. Early Observations of Catalyst Poisoning	318
VI. Contaminant Retention	321
A. Lead Retention	321
B. Phosphorus Retention	324
C. Sulfur Retention	324
D. Retention of Other Contaminants	326
VII. Poison Distribution	327
A. Contaminant Distribution along the Catalyst Bed	327
B. Contaminant Distribution within the Porous System	330
VIII. Thermal Deactivation versus Poisoning	334
IX. Mass Transfer Effects in Poisoned Catalysts	337
X. Poisoning Effects of Individual Elements	341
A. Effect of Lead	341
B. Effect of Phosphorus	345
C. Effect of Sulfur	349
D. Effect of Lead Scavengers	350
E. Effect of Manganese	351
XI. Interaction between Poisons and Active Components	352
XII. Poison-Resistant Automotive Catalysts	357
XIII. Catalyst Rejuventaion	358
XIV. Future Developments	361
References	361

I. Introduction

The prime requirement for a useful catalyst is maintenance of activity for extended periods. Since activity is associated with the presence of certain sites on the catalyst, its maintenance is synonymous with the preservation of such sites and the access of reacting species to them. During the catalytic operation several chemical and physical processes take place which tend to obliterate active sites or to obstruct their accessibility. A reduction in the number of active sites can result from coalescence of the active phase particles, brought about by high temperatures. The surface structure could also rearrange so as to cancel certain active sites. Atoms of the active surface can be transported away by reacting with gas phase constituents, and larger particles can be dislodged by the gas flow. The active phase can interact with its support and literally sink into it. Finally, the active sites can adsorb, temporarily or permanently, gas phase species to the exclusion of further catalytic events.

Of these deterioration processes only the last one is defined as catalyst poisoning in the strictest sense. In its broader definition, poisoning includes also the deposition of extraneous matter directly onto the catalytically active surface or, perhaps more commonly, at the entrances of channels or pores of the support onto which the catalytic phase is dispersed. Such deposits impede access to a whole assembly of active sites rather than to individual ones. Since the outcome is the same, namely a decrease in activity, it is not easy to distinguish between these two deactivation processes. It is worth noting that both have anthropomorphic analogies. The true poisoning by site blocking has its counterpart in the biological poisons which bond to catalytic (enzymatic) sites. The obstruction of transport by pore clogging has its simile in clogging of the biological transport system, such as the arteries. A quarter century ago Maxted (1) summarized in this series the surface chemistry of poisoning, and Wheeler (2) in the same volume treated the influence of poisoning on mass transfer in porous catalysts.

This review deals with the more specific topic of poisoning of just one category of catalysts, namely those used for purification of automotive exhaust. A general review of automotive catalysts, which appeared recently in this series (3), provides background information on the subject as a whole. Poisoning is a specific but particularly important aspect with far reaching technological and even economic consequences. It has been the subject of much detailed and thorough investigation. Ample documentation of poisoning has, by government regulation, required the alteration of long accepted fuel formulations, beginning with 1975 automobile models, in order to permit installation of catalytic devices on vehicles. Automotive fuels marketed under the label of "unleaded" have, indeed, proved to be nearly devoid of lead. In turn, the availability of these fuels made it possible to design even more effective catalytic emission-control systems. One might argue that the question of poisoning of automotive catalysts has been largely resolved now, since the amount of contaminants has been decreased by orders of magnitude. It should be borne in mind, though, that this circumstance gives rise to more sophisticated and sensitive catalytic systems. Therefore, the catalyst–poison interaction is also being reinvestigated at new levels of sophistication, bringing into play the most modern analytical tools available.

This review, furthermore, should document and illustrate the considerations and importance of basic principles to a technology of universal importance.

II. Catalysts

Several classifications of automotive catalysts are necessary for the purpose of the subsequent discussion. One important classification is according to the physical form and divides them into pelleted or monolithic catalysts. Both embodiments are currently in use on automobiles. The *pelleted catalysts* are spheres or cylinders with characteristic dimensions between 3 and 8 mm. The *monolithic catalysts,* as their name implies, represent a single body inserted into the flow path of the exhaust stream. Presently, the pelleted catalysts are alumina or alumina–silica beads onto which a certain amount of active catalytic material has been deposited. The monolithic catalysts in present use are ceramic bodies (although metallic ones have been proposed) with a honeycomb array of channels oriented in the exhaust flow direction. The inner channel walls of the low porosity ceramic skeleton are coated with a high surface area material, usually some form of alumina, into which the active material has been incorporated. The thickness of this layer, referred to as "washcoat" in the trade jargon, is typically 10–50 μm; it constitutes 5 to 15% of the total weight of the catalyst. Other dimensional parameters of the monolithic catalyst, such as channel-wall thickness, number of channels per unit cross-sectional area, channel shape, total length and diameter, etc., may vary widely. Sometimes, these parameters are of consequence for the poisoning behavior. Wei (3), Yolles and Wise (4), and Dwyer (5) have additional detailed descriptions of automotive catalysts for the interested reader.

The amount of catalyst used per vehicle depends on engine size, catalyst location, desired efficiency, and several other considerations. Since it is necessary to relate the amount of catalyst to that of the poison which may come in contact with it, we indicate that on a typical U. S. eight-cylinder vehicle, made in 1977, two monolithic catalysts, each weighing about 1 kg, are employed. The weight of a pelleted catalyst on a similar vehicle is of the order of 3 kg. Furthermore, the pelleted catalysts are totally made of porous materials with high surface areas, while the monolithic catalysts contain only 15% by weight, or less, of such materials. Thus, there is more than an order of magnitude difference in the total surface area available for poison deposition and retention. Therefore, a different poisoning behavior can be anticipated for pelleted and monolithic catalysts.

Stabilizers can be introduced into the pellets or the washcoats with the intention of slowing down the thermally induced decrease in the surface area of the porous structure itself, or of the active component. Both, the active materials and the stabilizers, are put sometimes only on the outer layers of the pellets or monoliths, while, in other cases they penetrate the porous structures completely. Such preferential distributions have very specific aims, the utilization of the active materials and their protection from poisoning being the most important ones. There exists a vast body of patent literature on such designs.

The second classification of automotive catalysts is according to the nature of the active ingredient, the two broad classes being noble metal and base metal catalysts. The first class encompasses those catalysts where the active ingredient is one or more of the following metals: Pt, Pd, Rh, Ru, or Ir. Although these elements should be termed more precisely precious or platinum-group metals, we shall refer to them as *noble metals* in conformity with the usage in automotive catalysis. The second class contains all the other metals, which are used as deposits on the substrate. Numerous "mixed" formulations containing active ingredients of both classes have been tried out. Nevertheless, for the purpose of this discussion of poisoning behavior the distinction is important since there are sharp differences between these two general classifications.

Finally, the third classification is according to usage, i.e., CO and hydrocarbon oxidation, NO reduction, and combined oxidation and reduction on so-called *three-way* catalysts. The oxidation catalysts operate in an overall oxidizing atmosphere, while the reduction catalysts are subjected to reducing conditions. The three-way catalysts designed to remove all three pollutants simultaneously would, ideally, operate at a stoichiometric air/fuel ratio, but in practice are subjected to oxidizing and reducing oscillations with the gas composition moving rapidly back and forth across the stoichiometric point. The different operating conditions influence the chemical state of the poisons in contact with the catalyst.

III. Poisons

The chemical interaction of the vehicle system with the catalyst depends on the chemical makeup of the materials which are consumed in the system operation, such as fuel, oil, and various additives, and also on the chemical composition and inertness of the construction materials with which fuel, engine oil, and their combustion products come into contact upstream of the catalytic device. Each of these constituents can be retained to some extent by the catalyst, either permanently or in a transitory fashion, and have an effect on the catalytic activity, more often than not a detrimental one.

A. POTENTIAL POISONS IN FUEL

Catalyst-poisoning studies have concentrated on the potential poisons introduced with fuel, particularly on lead, which has been added intentionally to improve the combustion characteristics at the high compression ratios employed in modern internal combustion engines prior to the introduction of exhaust purification catalysts. The lead is usually introduced as "motor mix" which contains tetraethyllead in a mixture with organic halides, chlorides or bromides. These halides transport the lead in the form of volatile halides out of the engine and into the exhaust, and are hence termed *lead scavengers*. Thus the potential catalyst-poison elements associated with the additive mixture are Pb, Br, and Cl.

It is instructive to assess the potential amounts of these elements that may come into contact with the catalyst. The range, depending on the additive concentration, gas consumption, and vehicle usage is extremely wide. Nevertheless, it is quite simple to estimate the extremes of the range for a typical useful lifetime of the vehicle exhaust system, which is taken by federal regulation to be 50,000 miles.

A large U. S. passenger car consuming 1 gal of fuel per 10 miles of travel and using premium gasoline with 3.0 g Pb/gal (on a noncatalytic vehicle) will provide one extreme. A 1985 U. S. vehicle whose mandated average fuel economy will be 27.5 miles/gal using gasoline with a trace amount of lead (0.004 g Pb/gal) will provide the other. It should be mentioned that field surveys of 1976 unleaded gasolines have yielded an average value of 0.006 g Pb/gal. The values of the maximum lead deposits in these cases are 15,000 and 8 g, respectively.

Lest the value of the lower extreme conjures a notion of its complete insignificance, one should remember that the typical amount of the active noble metal per vehicle in present day oxidation catalysts is of the order of 3 g, of which usually 10% or less is exposed on the surface.

If the amount of lead in the fuel is known, it is simple to estimate the amounts of Br (as ethylene dibromide, EDB) and Cl (as ethylene dichloride, EDC) associated with Pb by assuming that the lead was introduced as "TEL motor mix", unless explicitly stated otherwise. The atomic ratio of Pb:Cl:Br in this additive package is 1:2:1. These scavenger-to-lead ratios are also sometimes called 1.0 and 0.5 "theories" of EDC and EDB, respectively.

The amount of sulfur in automotive fuels varies presently in the range from 0.01 to 0.10 wt%; a typical value is 0.03 wt%. Even in a vehicle with a fuel economy of 27.5 miles/gal, mentioned earlier, the amount of sulfur passing through the catalyst in 50,000 miles will be of the order of 1500 g. Since sulfur emissions from vehicles equipped with oxidation catalysts are a matter of environmental concern (6, 7), one might anticipate some reduction of the fuel sulfur in the future through more extensive desulfurization in refining. On the other hand, such trends may be offset by energy-utilization considerations. However, even a tenfold reduction in fuel sulfur will not completely nullify the effect of this potential poison with respect to some catalysts.

The phosphorus concentration in fuels used presently for the certification of engine and exhaust systems is 0.0008 g P/gal, according to EPA regulations; it represents an average level of current gasolines on the U. S. market. Thus the amount of P passed through the catalyst over 50,000 miles from gasoline totals only a few grams. The actual P level in some gasolines can be as low as 0.0002 g/gal.

The desire to improve the octane rating of fuels after the refining process without the use of lead has intensified a search for other "antiknock" compounds. For example, compounds of Mn (8) and Ce (9) have been explored for this purpose. Methylcyclopentadienyl manganese tricarbonyl, in particular, is already being marketed for octane improvement. At a recommended level of 0.125 g Mn/gal, 200–500 g of Mn can be expected to pass the exhaust system within 50,000 miles. The main criteria in accepting such fuel additives are their compatibility with catalytic systems, and, of course, health–safety considerations.

B. Potential Poisons in Oil

Most lubricating oils for engine use contain additives designed to improve such properties as lubricity, detergency, oxidation resistance, and viscosity. The additives contain elements that could be potentially harmful to catalysts. Table I lists these elements and their typical concentration in lubrication oils of 1973. The first three elements are combined usually in one compound, zinc dialkyldithiophosphate. Thus, before combustion, sulfur and phosphorus in oil are in a different chemical state than the same elements are in fuel. Little is known whether combustion nullifies these differences partially or fully. Some data, to be discussed subsequently, are available on the separate poisoning effects of these elements as derived either from the fuel or from the oil.

TABLE I
Cumulative Amounts of Potential Poisons in Engine Oil in 50,000 Miles of Operation[a]

Element	Wt% Range	Wt% typical	Grams
Zn	0.08–0.16	0.12	50
P	0.07–0.19	0.13	54
S	0.21–0.80	0.35	145
Ba	0–0.28	—	120
Ca	0–0.48	—	200
Mg	0–0.18	—	75
B	0–0.02	—	8
Ash	0.75–1.88	1.3	540

[a] Assuming oil consumption of 1 quart/1000 miles. From Shelef et al. (10) and McConnell and McDonnell (11).

C. Other Potential Poisons

The principal elements deriving from the construction materials of exhaust system and catalyst can are iron, nickel, chromium, and copper. Iron is the major component of the debris retained by the catalyst; nickel and chromium are usually components used to fabricate high-temperature materials for thermal reactors incorporated in some systems upstream of the catalyst. Copper may originate in engine bearings or in the copper lines used for air injection. As it is known that metals often cause deterioration of the high catalytic activity of platinum, all of them must be regarded as potential poisons.

It was noticed in some instances that alkali-metal impurities in alumina, or in the ceramic honeycombs, adversely influence the catalyst activity. For example, spodumene (lithium aluminum silicate) was found unsuitable for use as catalyst support, although it had good thermal properties.

IV. Analysis and Examination of Poisons on Catalysts

The characterization of fresh and used automotive catalysts, which includes the examination of poisons accumulated in the catalysts, uses a variety of modern analytical techniques. The two principal tools, besides conventional chemical methods, are atomic absorption and, most important, X-ray fluorescence (XRF). The latter technique has been refined and adapted for the analysis of automotive catalysts to permit rapid and accurate determination of all constituents, including the inadvertent contaminants. An example of a simultaneous XRF analysis of

TABLE II
X-Ray Fluorescence Analysis of 15 Components in a Catalyst Calibration Standard Sample[a]

Element	Nominal wt%	Calculated wt%
Pb	0.998	0.969
Zn	1.496	1.450
P	1.496	1.471
Pt	0.534	0.513
Ru	0.100	0.101
Ir	0.000	0.002
S	0.698	0.658
Pd	0.399	0.392
Cu	0.187	0.183
Ni	0.249	0.238
Cr	0.100	0.103
Rh	0.101	0.101
Fe	1.496	1.469
La	1.558	1.599
Ba	0.000	0.000

[a] From Mencik et al. (12). (Reprinted with permission of Plenum Press.)

15 elements found in an automotive catalyst is shown in Table II. The analysis was carried out by a multiple regression procedure with a minicomputer (13) to handle the large amount of data produced by the XRF spectrometer during the analysis.

Complete characterization of poisoned catalysts, of course, requires much more than chemical analysis. For example, the interaction of poisons with catalyst constituents and with each other has been studied by X-ray diffraction and by electron microscopy, the morphology of the poison deposits by optical methods, the distribution within the catalyst pellets and washcoats by the microprobe, and the distribution of poison on the surface of the active metals by Auger spectroscopy.

V. Early Observations of Catalyst Poisoning

It is advisable to consider only briefly the deactivation of automotive catalysts by lead, as observed in early attempts of the implementation of catalytic converters on automobiles. These observations are summarized adequately in the earlier reviews (3–5). In particular Table 5 of the review by Yolles and Wise (4) shows that researchers engaged in the development of automotive catalysts before the middle 1960s did not contemplate lead removal from gasoline, and most of the work was performed with fuels containing massive amounts of TEL, upwards of 2 g Pb/gal. Under this circumstance the designers of catalysts had to contend with excessive loading of very large beds of pelleted catalysts by lead compounds. For example, in a study of supported vanadia pellets for hydrocarbon oxidation (14), a fuel containing 2.9 g Pb/gal was used. The 820-in^3 (approximately 20 pounds) catalyst bed accumulated lead in excess of 10% of its own weight within 13,000 miles. The retained lead accounted for 54% of that emerging from the engine. Such massive loading with lead caused a rapid decrease in catalyst activity. Modern noble metal catalysts are still more vulnerable than base metals such as vanadia (14–16), under the conditions used, due to the comparatively small amounts of active material.

It should be stressed that the early catalyst development was motivated by the 1964 California standards which stipulated, by today's yardstick, a rather mild removal of hydrocarbons to be fulfilled for 12,000 miles of vehicle operation only. This task was met by the use of large beds of catalysts, which were roughly, by an order of magnitude, larger than present-day catalysts. The large beds provided a reserve of activity as the lead deposits successively deactivated zones of the catalyst beginning at the bed inlet.

Examination of the spent pellets from these early catalyst beds showed already then the deleterious effect of lead on the surface (17, 18), the covering of the pellet exterior by impervious lead containing layers (14), the interaction of lead with base metal active components (14), etc., as the causes of activity decrease.

Weaver (19) has examined in a systematic manner the effect of four levels of TEL in customer-type vehicle operation on a V_2O_5–CuO–Pd/alumina pelleted catalyst. Figure 1 shows the clear correlation between the amount of lead in the fuel and the efficiency of hydrocarbon oxidation measured on the EPA cold-start seven-cycle Federal Test Procedure, which was in force when the work was performed. A remarkable observation in Weaver's work is how the rate of activity decrease is related to the amount of lead passing through the engine. The decrease is much steeper when the system is exposed to smaller lead dosages, as shown in Fig. 2. The obvious implication from these results is that either the larger part of the lead emitted from engines operating on high lead concentrations is not retained on the catalyst or, perhaps, is less harmful to catalyst activity. Thus, these early results pointed out the important question of poisoning specificity.

A companion paper of Su and Weaver (20) presents the results of detailed postmortem examinations of the catalysts, including X-ray

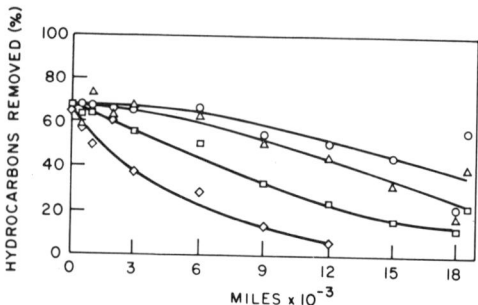

Fig. 1. Effect of Pb level on hydrocarbon removal (○:lead free; △:0.5, □:1.5, ◇:3 g Pb/gal). [From Weaver (19).] (Reprinted with permission of the Society of Automotive Engineers.)

diffraction analyses of the various crystallographic phases into which the lead was incorporated. Not surprisingly, in a catalyst containing substantial amounts of vanadia, a large fraction of the lead ended up as vanadate, $Pb_3(VO_4)_2$, and the rest in progressively decreasing amounts as oxide, sulfate, halide, and phosphate. Sulfur, halogens, and phosphorus originated in the sources discussed above. The most pernicious effect of the lead, for this particular catalyst, was ascribed to the formation of lead vanadate since the active component, vanadia, was rendered inert. Cannon and Welling who used a similar catalyst in a much earlier investigation (14) arrived at a similar conclusion.

Already these early results have shown the incompatibility of heavily leaded gasolines with catalysts designed to achieve a rather moderate goal of automotive emission control. The added stringent requirements of the

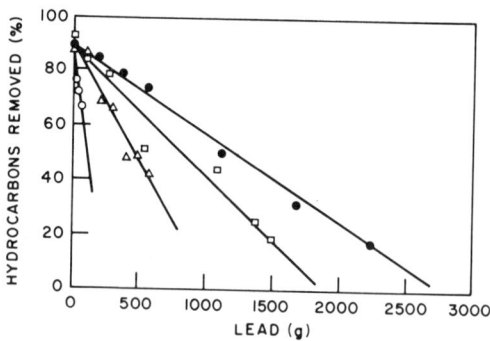

Fig. 2. Relation between lead consumption by the engine and hydrocarbon removal (○:lead free; △:0.5, □:1.5, ●:3 g Pb/gal). [From Weaver (19).] (Reprinted with permission of the Society of Automotive Engineers.)

1970 Clean Air Act redirected the interest toward the inherently more active noble metal catalysts and stimulated an interest in detailed examination of the poisoning phenomena.

VI. Contaminant Retention

An obvious question is: How much of a contaminant passing through the engine is retained on the catalyst? Certainly, not the whole of the nongaseous exhaust constituents reaches the catalyst, since a part is deposited in the bends and crevices of the exhaust system upstream of it, and some of it accumulates in the crankcase. Also, only a fraction of the contaminants reaching the catalyst is retained by it. Retention in this chapter means the fraction of a given contaminant found in a postservice examination in the catalyst as a whole, without any reference to its distribution, which will be discussed separately.

Several factors will conceivably influence the retention. Not all poisons will be retained to the same extent. Retention of any given element might depend on its amount in the fuel and oil; on the composition of fuel and oil; on the operation variables of the engine; on the design of the exhaust system; on temperature, shape, size, position of the catalyst, and the atmosphere to which it is exposed; the service time of the system; etc. It may, or may not, vary linearly with any of these parameters.

A. Lead Retention

A compilation of available lead retention data is given in Table III, indicating only the most important conditions of a given test, where the information was available. With the exception of the work with pelleted vanadia catalysts (14), all the data pertain to lead levels in the range of 0.01–0.5 g Pb/gal, which was the range of interest in the years from 1972 to 1975, when the limits for contaminant levels in gasoline compatible with catalyst operation were actively considered. The data in the table cover an extremely wide set of conditions and include laboratory simulation, dynamometer, and vehicle fleet tests. Nevertheless, with a few exceptions, the lead retention falls within a relatively narrow band between 13 and 30%.

As to the effect of the operating parameters on the retention of lead, there exists only a limited amount of information on some of these factors, such as TEL concentration in the fuel, catalyst temperature, and position.

In a relatively narrow range (0.01–0.06 g Pb/gal) of TEL concentration

TABLE III
Lead Retention on Automotive Catalysts

Type of test	Redox conditions	Temperature (°C)	Lead in fuel (g/gal)	Retention (%)	Equivalent mileage	Refs.
Burner	Oxidizing	600	0.03–0.07	13	—	21
Burner	Oxidizing	900	0.03–0.07	20–40	—	21
Dynamometer	Oxidizing	650–850	0.01–0.5	24–31	10,000	22
Dynamometer	Oxidizing[a]	540–620	0.007–0.075	3	12,000	23
Dynamometer	Oxidizing[a]	732–773	0.007–0.075	25	12,000	23
Dynamometer	Oxidizing	Varied	0.03–0.07	16	24,000	24, 25
Dynamometer	Reducing	565–600	0.03	13	40,000	26
Dynamometer	Reducing	649–704	0.03	13	40,000	27
Dynamometer	Reducing	565–730	0.03	10–14	40,000	26
Dynamometer	Cyclic	790–870	0.03	3	24,500	26
7 cars	Oxidizing[a]	550	0.012–0.063	17–25	24,500	28
15 cars	Oxidizing	Varied	Varied	15	Varied	10
10 cars	Oxidizing[a]	450	0.05–3.0	42–70	18,000	18
20 cars	Oxidizing	Varied	0.03	32	50,000	29
450 cars	Oxidizing	Varied	0.03	14–29	12–24,000	29

[a] Indicates pellets, all the other catalysts are monoliths.

in the fuel, concentration alone does not influence lead retention, either in monolithic (22) or pelleted (28) noble metal catalysts. The validity of this statement could possibly extend beyond this range (19), but more precise data are needed.

When varying the temperature of the catalyst, while keeping all other variables constant, it was noticed in laboratory devices (30), burning iso-octane containing TEL, that the retention of lead on monolithic catalysts does increase with temperature (10) in the 350°–760°C range. In burner experiments with monolithic base metal catalysts (21) lead retention doubled when the temperature was increased from 600° to 850°C. In dynamometer studies of pelleted catalysts, again, a temperature increase from 550° to 750°C caused a sevenfold increase in lead retention (23).

There are indications, however, that maximum retention is attained in the temperature range around 850°C. Further increase in temperature results in much lower values of lead retention. Thus, McArthur (26) has observed that subjecting a catalyst to an oxy-reduction treatment at temperatures up to 870°C drastically reduced lead retention.

As is often the case, there is contradictory evidence, as well. Cooper and co-workers (31, 32) registered a lower lead retention at 695–770 (13%) than at 590°–660°C (47.6%) with a monolithic catalyst placed close to a dynamometer-mounted engine. However, in order to attain the lower temperature, they had to lean out the combustion and thereby also changed the catalyst atmosphere. It is difficult to disentangle the effect of temperature from other factors involved in the lead deposition.

Since the retention of lead is primarily determined by the thermodynamic stability of the lead deposits, the influence of the temperature may vary with the composition of the lead deposits, which in turn, depends on the fuel and catalyst composition. We shall briefly discuss how such reasoning pertains to lead retention.

Due to the presence of sulfur, halogen, and phosphorus in the system, the lead deposits consist, as indicated before (14, 26), of a mixture of lead compounds with these elements and oxygen, and also with catalyst constituents. McArthur (26) has constructed tables of thermal stability of these compounds (see Table IV), following earlier work on engine lead deposits (33, 34). From stability considerations a conclusion can be reached that at temperatures above 870°C even lead sulfates or oxysulfates are not stable and decompose into free lead oxide. According to McArthur, these compounds can be again volatilized as halides, which explains the decreased retention in this temperature range. The last assumption of volatilization as a halide is probably incorrect since, as stated by the author himself, the halides are extremely unstable at these high temperatures. As the lead oxide also has an appreciable vapor pressure [0.1 torr at 850°C (21)], it can be transported away as such. Furthermore, the volatile oxide is highly reactive with most inorganic oxides, such as vanadia, chromia, silica, zirconia, and even alumina. These compounds have somewhat differing stability ranges, and thus the temperature at which the lead retention passes through a maximum depends on catalyst formulation. Here we add only in passing that the lead com-

TABLE IV

Reactants and Reaction Temperatures for Compounds Formed by Solid-State Reactions Involving Lead[a]

Reactants	Compound formed	Reaction temperature (°C)
$PbO + PbX_2$[b]	$PbO \cdot PbX_2$[b]	200–300
	$PbO \cdot PbX'X$[b]	
$2PbO + PbX_2$[b]	$2PbO \cdot PbX_2$[b]	350–400
$PbO + PbSO_4$	$PbO \cdot PbSO_4$	500
$4PbO + PbSO_4$	$4PbO \cdot PbSO_4$	600

[a] From McArthur (26). (Reprinted with permission of Plenum Press.)
[b] X = Br, Cl.

pounds formed with the active phases of base metal catalysts vanadia or chromia, for example, are more often than not inactive for the desired oxidation reactions.

In connection with the separate effects of lead and its scavengers, several investigators have observed that without the use of scavengers, lead retention on the catalyst drops off markedly. On reflection, this could well be anticipated, since the volatile compounds with scavengers are the main lead transport entities from the locus of combustion to the catalyst.

B. Phosphorus Retention

Information on phosphorus retention (Table V) is less abundant than that on lead. The presence of lead phosphates in used catalysts has been noted (26, 35). The retention and possibly its ability to poison a catalyst, as well, of phosphorus originating from fuel, will depend on the presence of lead. The work of McArthur (26) shows very low P retention from the fuel as compared with that from oil. The ad hoc explanation offered is that whereas P_2O_5 is the most likely form for the transport of fuel phosphorus, other forms may prevail for the oil phosphorus. This, indeed, may be so if one realizes that the oil contains species such as Zn and alkaline earth metals which form very stable phosphates. The harmful effects and the distribution of phosphorus might well be influenced by such differences, as will be discussed subsequently.

C. Sulfur Retention

The retention of sulfur is also influenced to a large degree by the lead level in the fuel that is related to the mentioned tendency to form lead sulfates and oxysulfates. Decomposition, especially that on catalysts

TABLE V
Phosphorus Retention on Monolithic Automotive Catalysts

Type of test	Source of P	Redox conditions	Catalyst temperature (°C)	Equivalent mileage	Retention (%)	Refs.
Dynamometer	Oil	Oxidizing	650	10–30,000	12–17	22
Dynamometer	Oil	Reducing	649–704	40,000	35	27
Dynamometer	Oil	Reducing	565–600	31,300	26	26
Dynamometer	Fuel	Reducing	565–600	11,600	4	26
15 cars	Fuel + Oil	Oxidizing	Varied	Varied	9	10

containing base metals, will influence the retention, since the sulfur can be combined as sulfates of the components that may or may not be stable under the conditions of operation. But in the case of noble-metal catalysts, supported on alumina, the retained sulfur is confined to the alumina surface only, and the upper limit of its retention is determined by the total available surface area of the catalyst. Therefore, during a 50,000-mile test, the retained sulfur is a small fraction of the total amount going through the system. This was observed in several instances, both in monolithic (*10*) and pelleted catalysts (*23, 28*). As the available BET area of the monolithic catalysts is substantially less, the corresponding limiting retention of sulfur is also smaller. Although the storage of sulfur is proportionately small, it is quite important in determining the extent of sulfur conversion to sulfate (*36, 37*) during cyclic vehicle operation.

If one assigns, after Yao (*38*), the area occupied by a surface sulfate group as 30 Å^2, and if the sulfur content and the BET area are known, it is possible to estimate whether, indeed, sulfur retention is kept below one monolayer. Table VI shows a compilation of sulfur retention, recalculated as number of monolayers. The buildup of sulfur at a given temperature takes place during relatively short exposures to the exhaust, typically of the order of a few thousand miles of vehicle operation. When the temperature is changed, the extent of retention will change. Thus, a certain pelleted catalyst (*42*) accumulated in 3000 miles about 0.5% of sulfur,

TABLE VI
Sulfur Retention

Catalyst	Type[a]	Surface Area (m^2/g)	Sulfur (%)	Number of monolayers	Refs.
Copper chromite	P	92	0.86	0.53	*39*
Base metal	P	100	0.78	0.44	*40*
Noble metal	P	250	0.8	0.18[b]	*41*
Noble metal	P	250	1.97	0.44[c]	*41*
Noble metal	M	10	0.02	0.12	*29*
Noble metal	M	3.0	0.06	0.11	*29*
Noble metal	M	8.8	0.05	0.32[d]	*50*
Noble metal	M	4.9	0.017	0.20[d]	*50*
Noble metal	M	5.2	0.005	0.05[d]	*50*

[a] M, monolith; P, pellets.
[b] 0.004 g Pb/gal.
[c] 0.5 g Pb/gal.
[d] 0.03 g Pb/gal.

which was not changed when the catalyst was kept for additional 9000 miles at a constant temperature of 540°–620°C. Raising the temperature by 150°–200° drastically depressed the sulfur retention. There are numerous references (43, 44) attesting to such behavior. Obviously, the retention is dependent on the stability of the surface sulfate groups, much in the same manner as the form of the retained lead, in the case of heavily leaded fuel, depends on the stability of the bulk sulfate and oxysulfate (see above). As a rule of thumb, base metal catalysts do retain more sulfur than do noble metal catalysts supported on insulator oxides (γ-Al_2O_3), as the surface sulfates of the divalent base metals are more stable than aluminum sulfate (36, 45, 46).

D. Retention of Other Contaminants

The retention of the scavenger elements Cl and Br on the catalyst is usually insignificant with the exception of catalysts used with heavily leaded fuels. Even then, the amount of retained halogen is relatively small because the catalysts operate at temperatures above the stability limits of the lead halogen compounds (Table IV). Nevertheless, the halogenated compounds can have a substantial role in the poisoning process, as will be discussed. One must mention, if only in passing, that in base metal catalysts the halogens can form volatile compounds with the active ingredients and carry them away. This has been well documented for Cu-containing catalysts (47, 48).

Retention of such oil components as Zn and the alkaline earth metals has been reported in relatively few instances. Deposits on the catalysts have not always been related to the amounts of these elements consumed in the operation. Analysis of deposits on catalysts originating from lubricants indicated that Ba, Ca, Zn, and P are present in the same proportion as given by the particular formulation of the oil employed for engine lubrication. The average retention of the amount of Zn passing over the catalysts of several test vehicles was reported as 3% (49).

Finally, it is practically impossible to assess the retention of metallic deposits from construction materials because the amounts in the feed gas entering the catalyst are totally unknown. Only in a few instances have analyses of used catalysts for elements deriving from construction materials been made available (10, 25). Significant differences in the content of Fe, Cr, or Ni of used catalysts have been noted when upstream of the examined catalyst there were situated metallic devices operating at high temperatures, such as thermal reactors, or reduction base metal catalysts (50).

VII. Poison Distribution

Two kinds of poison distributions must be distinguished. One distribution is that along the catalyst bed, the other one is within the porous system of the catalyst. It may be reasonably anticipated that under most conditions there will be a gradient of contaminant concentration which decreases in the direction from inlet to outlet; also that there will be a decreasing concentration of contaminants from the outer confines of each separate catalyst body inwards into the pore system. The contaminant distribution will, however, differ for different types of catalysts and contaminants.

A. Contaminant Distribution along the Catalyst Bed

Not surprisingly, all the data pertaining to axial distribution of contaminants in the bed were obtained for monolithic catalysts, where such determination is performed simply by successive sectioning (see Fig. 3) and analysis of each separate section. In pelleted catalysts there is considerable spatial mixing of the pellets during operation, and the sampling is also difficult.

From the numerous examples of the axial gradient of poisons we have

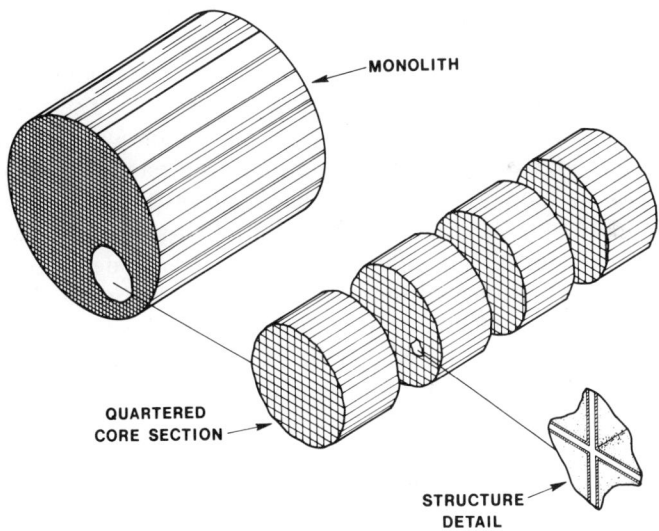

FIG. 3. Sectioning of monolith for deposit analysis. [From McArthur (27).] (Reprinted with permission from Advances in Chemistry Series. Copyright by the American Chemical Society.)

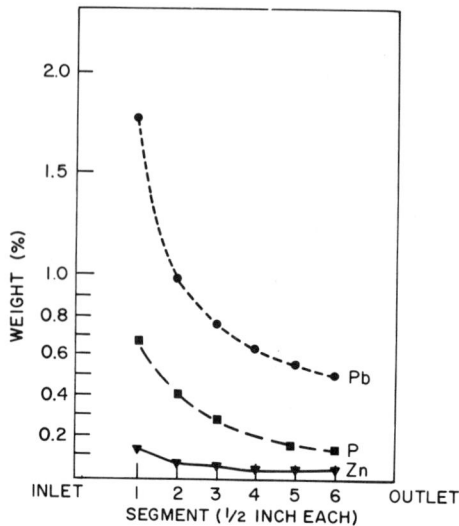

Fig. 4. Axial contaminant profile on oxidation catalyst after 24,000 miles. [From Weaver et al. (29).] (Reprinted with permission of the American Institute of Chemical Engineers.)

chosen a few representative cases for illustration. Figure 4 gives the distribution of Pb, P, and Zn in one of the catalysts removed from experimental fleets using oxidation catalysts. Figure 5 shows the axial distribution of the same poisons plus Ca and S in an NOx catalyst whose operation was predominantly in a reducing atmosphere. The main difference is in the gradient of the phosphorus distribution which is flatter in the case of the NOx catalyst. In addition, a higher retention is noted, as shown already in Table V. The steep gradient suggests that, at least in the monolithic catalysts, the locus of catalytic activity can shift with use along the bed axis under the influence of the poisoning deposits.

With increasing volatility of the poison compounds, i.e., with increasing temperature, the axial concentration gradient can be flattened or even slightly reversed, as shown for lead in Fig. 6. Temperatures of the order of 850°C or above are quite uncommon, and the gradients in automotive catalysts employed under realistic conditions will be usually like those depicted in Fig. 4.

An important factor that influences the lead distribution, especially in the channeled monolith bodies, is the character of flow. There is strong evidence that an induced change of the flow from laminar to turbulent augments the deposition of lead. An example is given in Fig. 7 by the deposition pattern of lead in a dual catalyst. Within each of the catalysts, the deposit pattern is in agreement with those shown in Figs. 4 and 5. However, the lead deposit on the inlet of the oxidation catalyst exceeds con-

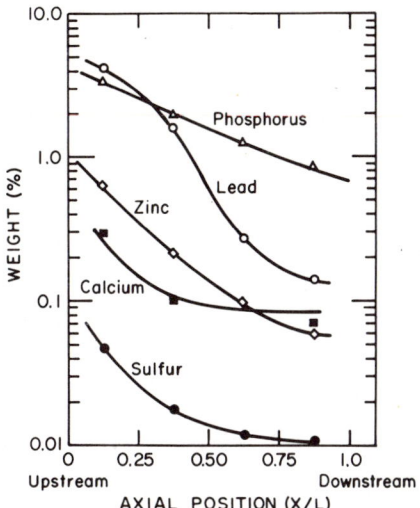

FIG. 5. Axial contaminant profile on an NO reduction catalyst after 40,000 miles. [From McArthur (27).] (Reprinted with permission from Advances in Chemistry Series. Copyright by the American Chemical Society.)

siderably that on the outlet of the upstream reduction catalyst. The same type of pattern has been observed in unpublished tests with one monolith unit cut into three equal cylindrical segments before installation. Insertion of an object, for example, a thermocouple, into the flow has a

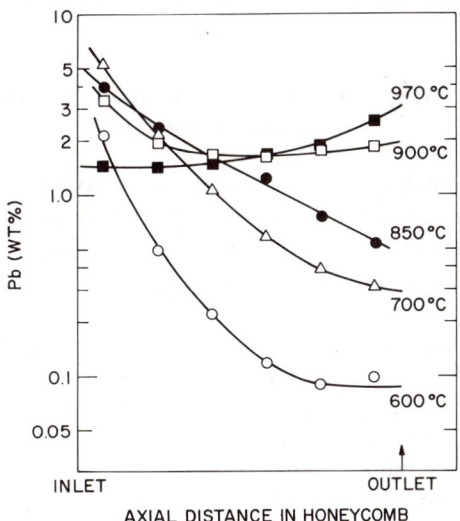

FIG. 6. Change of the axial lead profile with temperature on a base metal oxidation catalyst. [From Kummer et al. (21).] (Reprinted with permission of the Society of Automotive Engineers.)

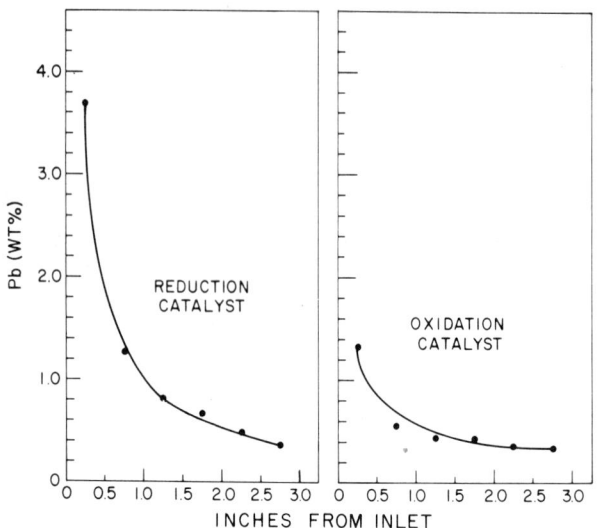

FIG. 7. Axial lead profiles on dual-bed catalysts. [From Gandhi et al. (50).] (Reprinted with permission of the Society of Automotive Engineers.)

similar effect (25). Effects of flow on lead deposition have also been described by Howitt and Sekella (51).

Uneven radial distribution of contaminants can be expected if the design does not ensure uniform flow distribution across the inlet face of the catalyst bed.

B. Contaminant Distribution within the Porous System

The distribution of contaminants within the porous layer again has to be considered separately for monolithic and pelleted catalysts. Gradients of the contaminant concentration in both cases can be very steep or relatively flat. Some inferences on the poison-carrying species can be deduced from such gradients.

The most useful tool for the study of penetration is the electron microprobe, which was first used for this purpose by Cannon and Welling (14). They have shown the buildup of relatively thick layers of lead compounds around pellets exposed to exhaust originating from heavily leaded fuels. More recently, penetration profiles in pellets have been published for exhaust produced by fuels with lower amounts of lead. Figure 8 shows such a profile in a pellet exposed to exhaust with 0.5 g Pb/gal for 400 hr. Note, that at the exterior of the pellet, lead concentration (by weight) has

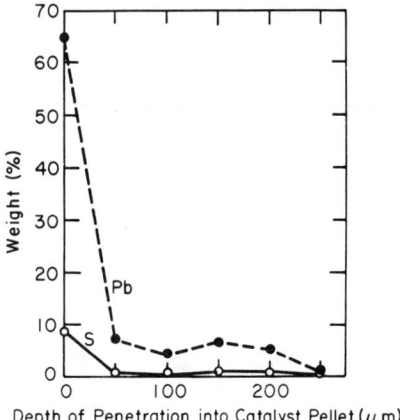

FIG. 8. Radial lead and sulfur distribution in poisoned catalyst pellet. [From Klimisch et al. (42).] (Reprinted with permission from Advances in Chemistry Series. Copyright by the American Chemical Society.)

reached 65%. A similar profile, with lead and sulfur penetration parallel to each other has been noted by McArthur (27), who used fuel with only 0.06 g Pb/gal and 350 ppm sulfur by weight. Penetrations of lead and sulfur into pellets to depths of several hundred micrometers have been noted even at exposures to fuels with low lead concentrations. It is easy to discern the potential harm of such accumulations in catalysts specifically designed to minimize mass transfer limitations, which contain the active material in the outermost layer (52).

The distribution of contaminants within the washcoat layer of monolithic catalysts has been studied by the electron-probe analyzer in considerable detail (35, 53). Figure 9 shows the washcoat structure on one type of ceramic skeleton. Cracking of the washcoat and its tendency to concentrate in the corners of the channel are clearly noted. When there is heavy contamination in the channels, e.g. from fuel with considerable TEL concentrations, dendritic lead deposits, mainly lead sulfate, are observed close to the inlet, as shown in Fig. 10. This indicates growth from the vapor phase. Elemental maps by the electron probe were taken from this heavily contaminated sample (Fig. 11). Since the washcoat does not contain Si it can be identified by comparing the outer edges of the Al and Si images. The outer surface of the washcoat coincides with the outer edge of the Al image, while the interface of washcoat and ceramic substrate coincides with the outer edge of the Si image.

Lead and sulfur uniformly penetrate the washcoat up to the substrate interface, as was also observed by Kummer and co-workers (21) with the

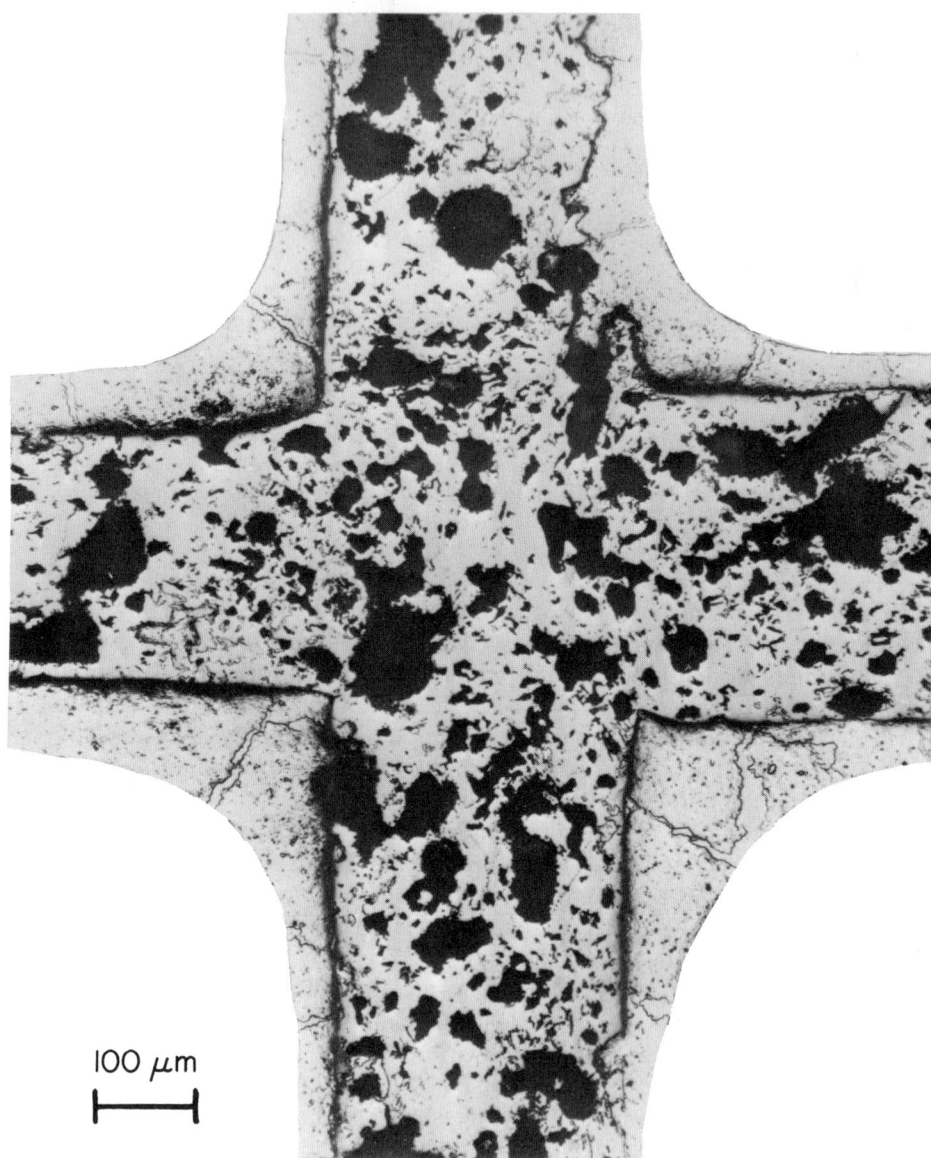

Fig. 9. Details of a monolithic catalyst showing washcoat deposited on a substrate.

same technique in the case of a completely different washcoat (zirconia). Phosphorus penetrates to a lesser extent, while zinc is contained in a continuous layer on the washcoat surface. The zinc-rich layer extends

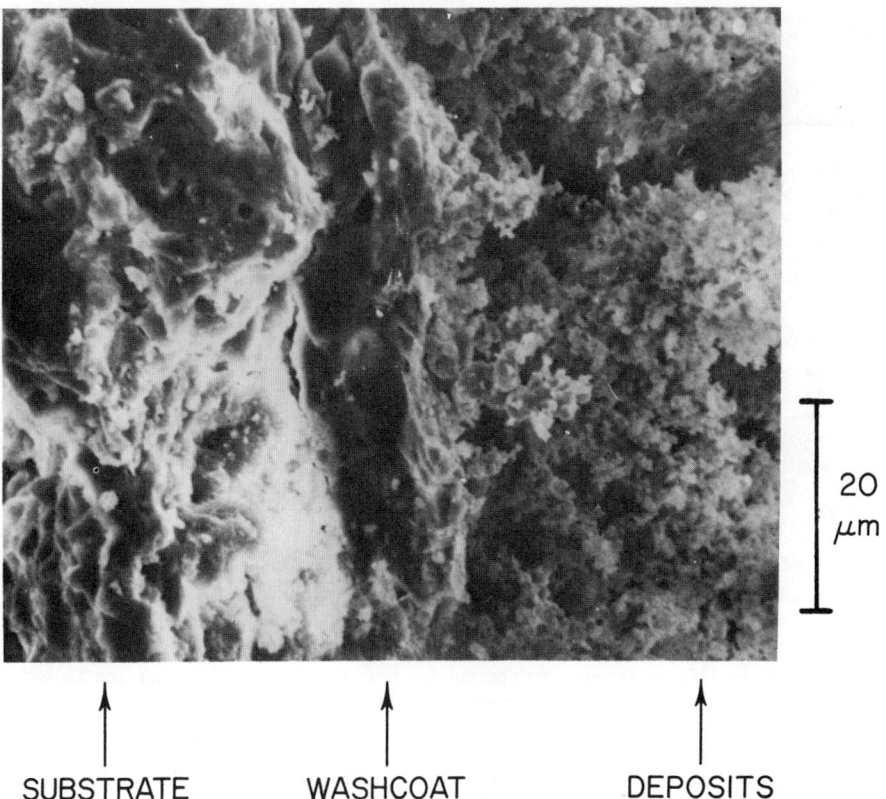

SUBSTRATE   WASHCOAT   DEPOSITS

FIG. 10. Fracture cross section normal to honeycomb channels at inlet; surface deposits grow outward from washcoat surface. [From Bomback *et al.* (*35*).] (Reprinted with permission from Environmental Science and Technology. Copyright by the American Chemical Society.)

throughout the entire inlet region of this converter. The penetration of lead, phosphorus, and sulfur into the washcoat suggests that these elements come from the gas phase, whereas zinc and iron are deposited as particulates.

The relative penetration of lead, zinc, phosphorus, and sulfur into the washcoat is seen in Fig. 12 at a corner where the washcoat is relatively thick. Lead and sulfur penetrate over 100 μm, phosphorus between 30 and 50 μm, and zinc less than 2 μm. All of the elements, except zinc, penetrate down into the deep crack and laterally into the washcoat. This supports again the supposition that zinc arrives at the surface in particulates, while lead, phosphorus, and sulfur arrive as gaseous entities. However, the penetration profiles do not prove the modes of deposition conclusively. Contaminant concentrations expressed in terms of lead

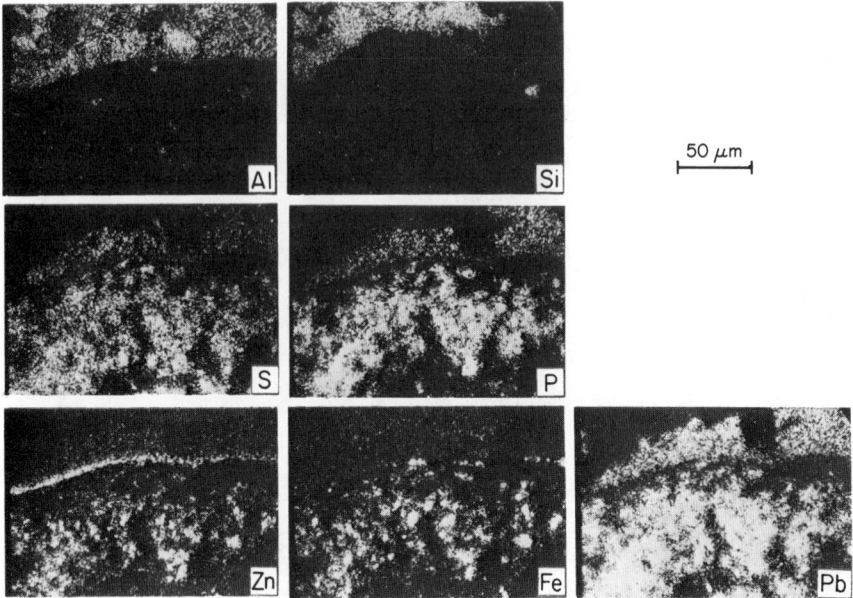

Fig. 11. Electron-probe elemental maps of polished cross section at catalyst inlet. [From Bomback et al. (35).] (Reprinted with permission from Environmental Science and Technology. Copyright by the American Chemical Society.)

salts in these heavily leaded samples can account for 16% of the lead as sulfate, and 14% as phosphate (35).

An X-ray intensity vs. energy spectrum shows that these deposits, in addition to the contaminants of Figs. 11 and 12, also contain Cr and Ni. This fact is understandable, as the catalyst was employed with an upstream thermal reactor made of stainless steel.

As will be discussed later, the particular form of contaminant distribution within the porous layer, as observed by the microprobe, does in some cases correlate with the contaminant effect on catalytic activity.

VIII. Thermal Deactivation versus Poisoning

It is simple to demonstrate that a catalyst operated in the absence of poisons (54, 55) still can show significant activity loss, albeit to a much smaller degree than in their presence. This deactivation process is induced by thermal effects. A separation of chemical and thermal deactivation requires considerable efforts.

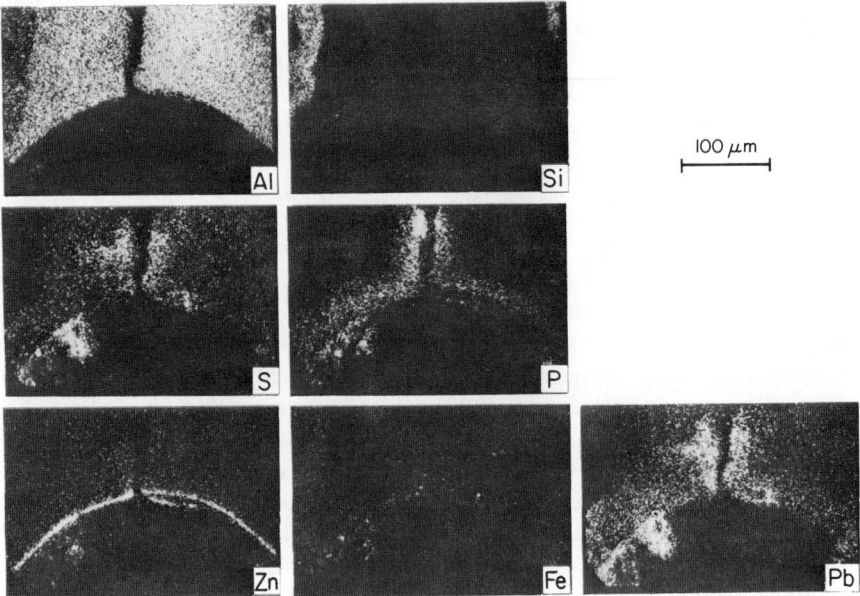

Fig. 12. Electron-probe elemental maps of polished cross section at catalyst inlet near monolith corner. [From Bomback et al. (35).] (Reprinted with permission from Environmental Science and Technology. Copyright by the American Chemical Society.)

If one could disregard the complicated influence of poisons on mass transfer processes, it would be possible to state in a first approximation that catalyst activity for a selected reaction is a monotonic function of the surface area occupied by the active component. The problem that arises is the measurement of the catalytic surface area in the presence of a support material. In the case of Pt such a measurement is relatively simple, done by hydrogen chemisorption (56, 57) or "titration" (58), although even in this case there are uncertainties associated with surface stoichiometry (59, 60). These problems become more complicated when Pd, or other noble metals are incorporated at the same time, and still more so, when the catalysts have been contaminated (61).

Examination of automotive catalysts by various chemisorption techniques has shown that a loss in noble metal surface area caused by higher temperatures correlates monotonically with various activity indices (62, 63). Moreover, Dalla Betta and co-workers (64) were able to separate the additional effect of poisons on the surface of the precious metal by painstaking attention to detail. They developed techniques for accurately measuring the crystallite-size distribution in used automotive catalysts by transmission-electron microscopy (65). A parallel measurement of the *accessible* surface area by CO chemisorption afforded the means for separating the chemical and thermal effects on the active area and hence on activity. During 50,000 miles of operation in a car fleet test (64) the noble metal particles (Pt, Pd, and Pt + Pd catalysts) of a total of 12 samples grew from a diameter of about 60 Å to 1000 Å, causing an associated 20-fold decrease in noble metal atoms on the surface. The accessible surface area, measured by chemisorption, was lower by an additional factor of 20.

Such a sharp drop in surface area of the noble metals does not result in a corresponding activity decrease. As measured by various empirical criteria, such as conversion at a certain temperature, it is found that activity loss is initially not nearly as steep as the indicated loss in site accessibility. The reason is that such measurements are usually carried out under conditions of mass transport control, when the vast majority of the active surface is not utilized in the catalytic process. However, once the active surface has dropped below a certain value, catalytic activity diminishes rapidly (66). These results emphasize that to begin with, a huge reserve of activity is required if the statutory service life of 50,000 miles is to be achieved. How large this reserve has to be is determined to a large extent by the poison levels.

In Fig. 13, the temperature that is required to convert 50% CO under

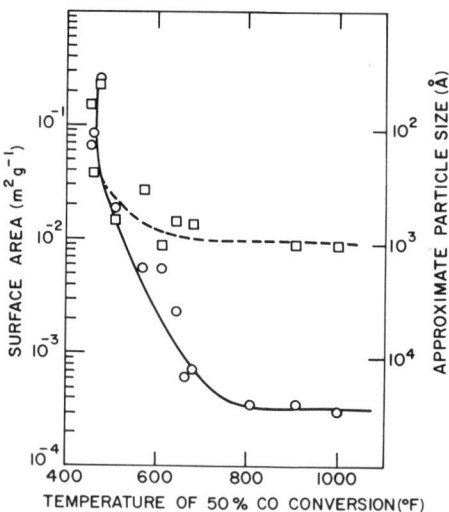

FIG. 13. Actual surface area (CO adsorption, ○) and expected surface area if chemical contaminants were not present (electron microscopy by tissue grinding, □) vs. observed catalyst activity. [From Dalla Betta *et al.* (64).] (Reprinted with permission of the American Chemical Society.)

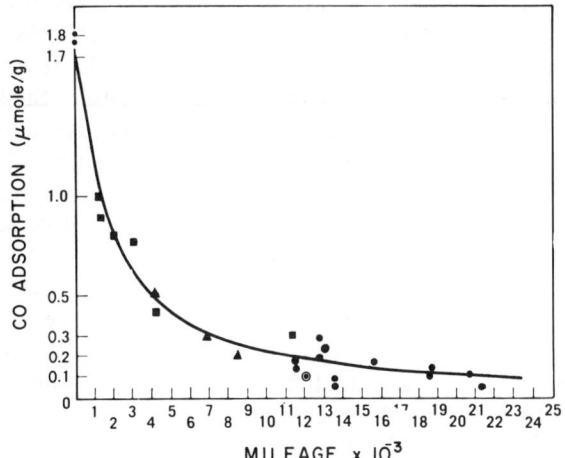

FIG. 14. Decrease of noble metal surface area with mileage. (●:450 car fleet. Simulated exhaust, ■: with pure iso-octane, ▲: with 0.5 g Pb/gal). [From Weaver *et al.* (*29*).] (Reprinted with permission of the American Institute of Chemical Engineers.)

certain arbitrary conditions is used as an activity criterion. The data show the decrease in activity as the number of accessible sites decreases, expressed by an increase in conversion temperature. While physical dimensions of the noble metal crystallites can be almost the same, a difference of 200°C may be required for a given degree of CO oxidation. This difference is, of course, attributable to poisoning.

Figure 14 gives the decrease of accessible area, measured by chemisorption, as a function of mileage. One μmole of CO adsorbed per gram is equivalent to a combined surface area of Pt plus Pd of 0.05 m^2/g. Data from car fleets and from laboratory experiments are incorporated. The laboratory data were obtained with both lead free and lead containing fuels. The steep drop below 4000 miles, obtained with lead-sterile iso-octane, is due to the initial loss of area by particle growth.

IX. Mass Transfer Effects in Poisoned Catalysts

In the previous section the effects of poisons on reaction rates were related to the active component surface, while the influence of mass transfer was disregarded. It has long been recognized, of course, that the overall rate and selectivity of chemical reactions in porous systems involves the coupling of chemical reactions with convective and diffusive mass transfer processes. Beginning with the pioneering work of Thiele (*67*), an entire discipline has evolved in which model systems are used to interpret and analyze mass transfer effects and their relation to the pore structure of the support system, reaction kinetics, etc. (*68*). Within this framework, models have also been developed which attempt to explain how poisoning affects the overall rate process. Such simplified approaches cannot be expected to apply exactly to the complex poisoning schemes present in automotive systems. It is instructive, however, to review these approaches and indicate where they are applicable to the poisoning of automotive catalysts, at least in a qualitative sense.

In these model systems the reaction process is viewed as being composed of three rate processes in series, namely the convective interphase mass transfer from the bulk fluid to the external surface of the catalyst, the intraphase mass transfer (usually by diffusion) through idealized pores, and the surface reaction at the active sites, distributed throughout this pore. At times, a fourth process, surface diffusion along the internal pore surface, is considered. The rate of reaction is, of course, controlled by the slowest of these steps, giving rise to three regimes of operation (*69*), complete kinetic control by the surface-reaction rate, pore-diffusion control, and interphase mass transfer control. The dependence of reaction rate on concentration and temperature will vary with the regime. Wei (*3*) outlines these effects on automotive catalysts. It is well to note that since automotive catalysts must operate over a wide range of conditions, any of the three regimes might be rate controlling at one time or another.

The work of Wheeler (*2*) forms the basis for including the effects of poisons into this model. Uniform (homogeneous) and pore-mouth poisoning are considered. In the first case, the sites rendered inactive by the poison are considered to be uniformly distributed around the pore. For complete kinetic control, the fractional activity of the poisoned catalyst is inversely proportional to the fraction of the sites poisoned. As the surface-reaction rates increase, the fractional activity drops less than linearly with the fraction poisoned because the poisoned portion in the inner pore is not used due to diffusion limitations. For complete external control, the rates would be the same for the unpoisoned catalyst. In pore-mouth poisoning, the poison is assumed to concentrate selectively in the region of the pore entrance. Thus a pore of length L with fractional poisoning α is pictured as being made up of a completely inactive region of length αL, followed by the active portion of length $(1 - \alpha)L$. As in the case of the uniform model, fractional activity decreases directly with the fraction poisoned for kinetic control. However, as pore-diffusion effects become important, the activity relative to the unpoisoned catalyst will drop markedly. The reason is that diffusion limitations have made the active inner portion of the pore less accessible, while the outer portion is completely inactive. Thus, the pore mouth presents no active surface for the reaction, but instead represents an additional mass transfer resistance. It should be noted in regard to automotive applications, as pointed out by Wei and Becker (70), that some oxidation reactions on noble metal catalysts exhibit negative order kinetics, in which case pore-diffusion limitations tend to increase rather than decrease the reaction rate. In such cases, the effect of pore-mouth poisoning might be diminished somewhat. Finally, for complete external mass transfer control, the reaction rates would be lower than for the unpoisoned case due to the resistance of the inactive layer. This type of poisoning is often discernible from the apparent activation energy of the reaction. As pointed out by Rajadhyaksha and Doraiswamy (71), catalysts with pore-mouth poisoning will show apparent activation energies characteristic of a diffusion process (essentially zero activation energy) in regions where the unpoisoned or uniformly poisoned catalyst would have activation energies somewhere between zero and the intrinsic value. Nonselective effects of poisons can also be envisioned, such as plugging of the pores, or even fouling of the external surface by particulates or dendritic growth.

The presence of nonselective pore plugging in automotive catalysts has been examined by Hegedus and Baron (72). They determined effective diffusivities of pelleted catalysts. Included were catalysts poisoned by both dynamometer aging, as well as by impregnation with lead solutions. No significant change in effective diffusivity was found with the addition of poisons, except for very high levels of lead. Dwyer and Morgan (73) found similar results for dynamometer-aged base metal catalyst pellets. Thus pore plugging does not seem to be important in normal catalyst usage. Dwyer and Morgan also looked at the effect of poisons on the kinetics of carbon monoxide and propylene oxidation for base metal pellets. These samples had been aged on a dynamometer cycle for the equivalent of 3000 and 6000 miles (at 3 g Pb/gal) and 46,000 miles (at 0.5 g Pb/gal). The data were analyzed in terms of both uniform and pore-mouth poisoning. Estimates of the fraction poisoned were obtained from the ratio of the intrinsic rate constant for the poisoned sample to that of another sample, aged similarly with lead-free fuel (equal extents of thermal deactivation are assumed). The pore-mouth model was found to correlate very well with both the CO and propylene reaction rates in the pore-diffusion region.

Acres et al. (22) have speculated on the modes of phosphorus and lead poisoning in monolithic catalysts, based on data obtained in simulated aging. Conversion vs. time-of-exposure curves for catalysts poisoned by either lead or phosphorus show quite different shapes, which the authors attribute to pore-mouth poisoning for phosphorus, and uniform poisoning for lead. This claim is substantiated by electron-microprobe analysis (35), mentioned above, which shows phosphorus concentrated at the surface, while lead is distributed throughout the washcoat. Uniform poisoning by lead is contrary to the results of Dwyer and Morgan (73), and other work described below. It should be noted that microprobe results are not conclusive in differentiating between the two types of poisoning. Lead deposits throughout the washcoat do not necessarily mean that the poisoning is uniform, as the lead in the interior might be in a nondeleterious form. The work of Klimisch et al. (42) provides perhaps the clearest and most important example of the effect of poisons on mass transfer. Pelleted catalysts containing noble metal were aged on an engine with fuels of different contaminant levels, and tested for activity on the engine, as well as in the laboratory. The results of the laboratory tests, where conversion was measured as a function of temperature, are most instructive. They show that as the temperature is increased, the maximum conversion attained by the poisoned samples is much lower than that found for fresh catalysts. In addition, the magnitude of the maximum conversion decreases as the lead exposure increases; it can be quite low for severe poisoning. As the authors point out, this effect is exactly what one would expect from pore-mouth poisoning. Similar behavior has also been observed in our laboratory on samples from several vehicle fleets (74), and a semi-empirical pore-mouth poisoning model has been developed which relates the poisoned fraction of the pore to the amount of lead on the catalyst (75).

In summary, there is evidence to suggest that the behavior of poisoned automotive catalysts displays characteristics of a model system in which the contaminants are assumed to deactivate the pore-mouth region selectively. In such a system, the effect of poisons is not simply the removal of active catalyst sites, but also the creation of an inactive region which decreases the accessibility to the unpoisoned region of the catalyst. The subsequently lower mass transfer becomes rate limiting, and in cases of severe poisoning may greatly reduce the overall rate. For example, it is found that automotive catalysts which are not poisoned, invariably give close to complete conversion of CO and of unsaturated hydrocarbons, even at high space velocities. With poisoning, however, the decreased mass transfer through the inactive region makes the system much more sensitive to space velocity (42), and limited conversions well below 100% are found even at moderate space velocities. The effect of poisons on mass transfer has obvious implications for the designing of catalyst systems to minimize the drop in performance during extented usage. This, in particular, is important for the choice of total catalyst volume needed for a desired performance.

X. Poisoning Effects of Individual Elements

Here we shall briefly summarize the effects of individual poisons on various catalytic reactions taking place on automotive catalysts. There are three main catalytic processes: oxidation of carbon monoxide and hydrocarbons and reduction of nitric oxide. Among secondary reactions there are undesirable ones which may produce small amounts of unregulated emissions, such as NH_3, SO_3 (6), HCN (76, 77), or H_2S under certain operating conditions. Among other secondary processes which are important for overall performance, in particular of three-way catalysts, there are water-gas shift, hydrocarbon-steam reforming, and oxygen transfer reactions. Specific information on the effect of poisons on these secondary processes is scarce.

Recent evaluations of SO_2 oxidation over noble metal catalysts (Pt, Pd, and Rh) have given some information on one particular secondary reaction. It was observed in car tests that SO_3 formation under the conditions of automobile exhaust is highly vulnerable to catalyst deactivation either by thermal sintering or by poisoning (78, 79). At the same time, the data indicated a lesser sensitivity of CO and hydrocarbon oxidation to catalyst aging. The results were confirmed in laboratory experiments (80). This is one example of preferential suppression of an undesirable side reaction. Obviously, the importance of a given poison on the different secondary reactions will vary widely with catalyst formulation and operating conditions.

A. Effect of Lead

All the surface processes on automotive catalysts which have been tested for the effects of lead poisoning are affected by the access of lead to the catalyst surface. The effect will differ, though, for different surface processes. Oxidation of hydrocarbons has been found repeatedly to be more vulnerable than oxidation of carbon monoxide to lead poisoning (10, 19, 25). The initial oxidation activity of noble metal catalysts, never exposed to poisons, is higher for CO than for hydrocarbons (54). Therefore, it is best to use the effect of lead on hydrocarbon oxidation for assessing the susceptibility of a given oxidation catalyst to this type of poisoning.

The individual hydrocarbons in the exhaust differ widely in the ease of both catalytic and noncatalytic oxidation. Thus, the poisoning effect may vary, and indeed does, with the hydrocarbon composition of the exhaust itself, and with the necessarily limited choice of particular hydrocarbons used in various laboratory tests. In many instances the hydrocarbons are subdivided, quite arbitrarily, into two broad classes: nonreactive, which encompass the lower alkanes, acetylene, and benzene; and reactive species, which include most of the other organic constituents of the exhaust, even partially oxygenated ones. It has been found that the oxidation of the least reactive hydrocarbon, methane, is vulnerable to relatively small amounts of lead. A fresh catalyst which oxidizes 45% of the methane at 500°C loses this ability completely after a short-time operation with an exhaust generated by fuel with a relatively low concentration of lead (Fig. 15). The oxidation activity for the other hydrocarbons is maintained much longer. The exhaust, produced by combustion of iso-octane, had a high methane content (20–25%) in its hydrocarbon composition.

Although catalyst deterioration for methane oxidation was clearly demonstrated in the laboratory test under steady-state conditions at 500°C, as shown in Fig. 15, it seems to be less discernible in car tests based on various driving cycles. Thus, data obtained on Ford vehicles, subjected to the 1977 Federal Certification Test over 50,000 miles, indicated that catalysts can be essentially inactive in removing methane even at zero miles, under these conditions.

The effect of lead on hydrocarbon oxidation over noble metal catalysts has been variously presented as a relation between catalyst activity and lead content in gasoline (*19, 22*), or of lead supplied to the engine (*31, 32*), etc. The most meaningful correlation is between hydrocarbon activity and lead deposit on the catalyst. Two examples of such correlations, one for laboratory-tested samples and the other for fleet-tested catalysts, will follow.

Figure 16 shows that hydrocarbon oxidation activity decreases logarithmically with the lead content of noble metal catalysts. Activity of these

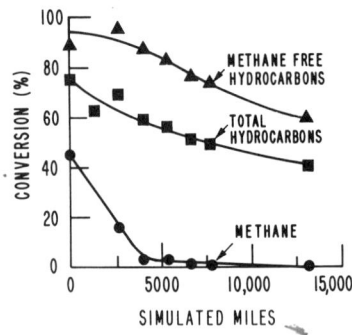

FIG. 15. Comparison of the deterioration of methane oxidation activity with total hydrocarbon oxidation activity at 500°C. Lead level in fuel is 0.05 g/gal. [From Shelef *et al.* (*10*).]

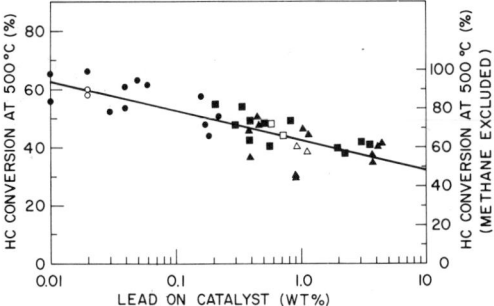

FIG. 16. Hydrocarbon activity as a function of lead concentration on catalyst, lead concentration in fuel 0.003–0.05 g/gal. [From Otto et al. (30).] (Reprinted with permission of the Air Pollution Control Association.)

catalysts was measured by the pulsator test (30). Figure 17 gives a similar correlation for a large number of catalysts used in field tests, with mileage accumulation up to 50,000 miles. It should be noted that the correlation in Fig. 17 includes another catalyst parameter, i.e., the total surface area as measured by the BET method. This parameter accounts mainly for the thermal deactivation which can vary widely in fleet tests from sample to sample. The points in Fig. 17 to the left of the line are associated with catalysts which have been exposed to severe, identifiable thermal damage. The catalytic activity decreases with a higher lead content and more

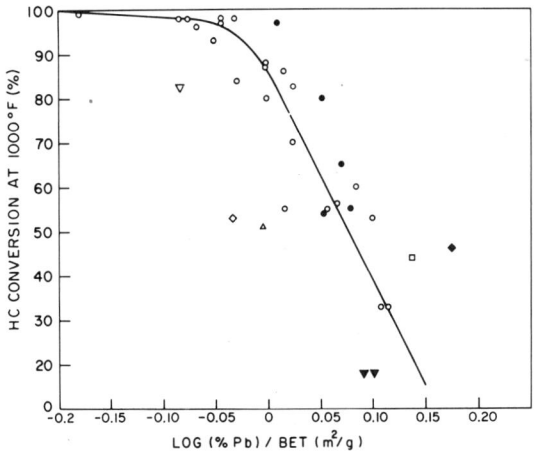

FIG. 17. Correlation of hydrocarbon conversion with lead concentration and surface area of catalyst. [From Weaver et al. (29).] (Reprinted with permission of the American Institute of Chemical Engineers.)

steeply with a lower BET area. In essence, the ratio plotted on the abscissa is the specific lead loading per unit of available catalyst surface area. For a surface reaction this is a more meaningful measure of lead contamination than the loading per unit weight. In Fig. 16, all the pulsator samples underwent an equal thermal treatment without overheating, and the BET areas are approximately the same for all the catalysts.

Hydrocarbon oxidation on base metal catalysts is also susceptible to lead poisoning, especially if the catalysts are exposed to relatively high temperatures, for at least part of their service time. It was noted above that lead retention, especially on base metal catalysts, also increases with temperature up to a certain point. This behavior is shown by the results of Yao and Kummer (81) in Fig. 18. One should note that the hydrocarbon used for testing catalyst activity, namely propylene, was quite reactive. With a less reactive test hydrocarbon one could expect a still sharper effect. The comparison with a reference production noble metal catalyst, given in Fig. 18, is quite instructive.

Work directed specifically toward the study of lead poisoning of NO reduction activity is relatively scant. It was observed (82) that the reaction is particularly sensitive to lead when the active component is a noble metal such as Ru or Rh. Commercially feasible catalysts can contain only small amounts of such active components, of the order of a few hundred ppm, which may explain the observed sensitivity.

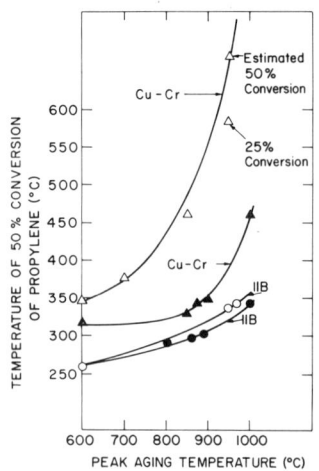

FIG. 18. Comparison of a copper–chromite catalyst (Cu–Cr) and a production noble metal catalyst (II B). Aging time 140 hr, open symbols for fuel containing 0.2 g Pb/gal, closed symbols for lead-sterile fuel. [From Yao and Kummer (81).] (Reprinted with permission of the American Chemical Society.)

TABLE VII
Sensitivity of Three-Way Catalysts to Poisoning[a]

	Fresh	After durability run[b]		Fuel B[d]
		Sterile fuel	Fuel A[c]	
80% Window width at 550°C (A/F units)	0.16	0.13	0.06	0
Temperature at which 80% simultaneous conversion is reached (°C)	285	300	350	Could not be reached

[a] From Gandhi et al. (83).
[b] 25,000 simulated miles on a pulse-flame apparatus (30).
[c] 0.007 g Pb/gal, 0.0008 g P/gal (76/77 Certification Fuel).
[d] 0.03 g Pb/gal, 0.003 g P/gal (74/75 Certification Fuel).

Three-way catalysts designed to remove simultaneously hydrocarbons, carbon monoxide, and nitric oxide in a narrow range of exhaust composition are an example of extreme sensitivity to poisoning. This sensitivity is best defined by the "window width" on the air/fuel-ratio scale in which the catalyst can remove simultaneously a given proportion of all three pollutants. For a graphic representation of the concept, the reader is referred to Fig. 8 in the review of Wei (3). Table VII shows, by means of the pulse-flame test, the narrowing of the "window" of a typical three-way catalyst in runs without and with two low levels of lead and phosphorus. These two levels, as indicated in the table, are associated with certification fuels. The levels of contaminants of the 1974/1975 fuels, which were quite compatible with activity requirements of oxidation catalysts at that time, are very detrimental to the activity of three-way catalysts. Even the extremely low concentrations of contaminants in current certification fuels have a pronounced effect. Indeed, the success of three-way catalysts is to a large degree predicated on the premise of availability of fuels essentially free of lead.

B. Effect of Phosphorus

The evidence that phosphorus in fuel is detrimental to the oxidation activity of noble metal catalysts is quite convincing. Data from engine dynamometer runs with pelleted catalysts (28) show that raising the level of phosphorus from 60 to 130 mg/gal (added as cresyl diphenylphosphate, CDP) at a low concentration of lead, markedly suppresses the oxidation of hydrocarbons and of carbon monoxide. The effect on the oxidation of

CO is more pronounced which is exactly opposite to the poisoning by lead. Also opposite to the lead poisoning behavior is the observation, by the same workers, that phosphorus poisoning affects Pt more strongly than Pd.

Using a monolithic noble metal catalyst, Gagliardi and co-workers (84) have shown the strong poisoning effect of fuel phosphorus. Also, in this case, the effect was prominent when the fuel was essentially free of lead (approximately 3 mg/gal). The poisoning is shown in Fig. 19. In 44 hr of durability testing a large difference in the hydrocarbon oxidation activity develops, as measured at 425°C, with fuels containing increasing amounts of phosphorus (added as CDP). The poisoning is partially reversible upon reverting to phosphorus-free fuel for longer periods of operation. Per unit weight of poison going through the catalyst, the effect of phosphorus poisoning was deemed greater than that of lead to Pt catalysts.

Acres and co-workers (22, 85) demonstrated by direct comparison that phosphorus, added to the fuel as tributylphosphate, is 1.5 times as toxic as lead on a weight basis, which makes the lead fourfold more toxic on an atomic basis.

The effect of fuel phosphorus on the activity of NO reduction catalysts has been found to be small. The phosphorus is found on the catalyst as P_2O_5 (26), notwithstanding the reducing conditions prevailing over a typical NO reduction catalyst for most of its operating time. There appears to be a need to establish with greater precision the effects of phosphorus on these catalysts and, in particular, on three-way catalysts.

There are indications in the literature that the simultaneous presence of lead and phosphorus in fuel is less detrimental since chemical action can form nonpoisonous lead phosphates. The data of Acres (22) support this point. The complex interaction between lead and phosphorus can pos-

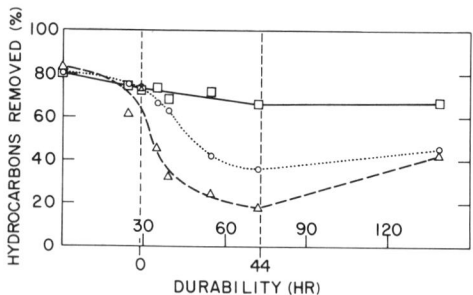

FIG. 19. Effect of phosphorus on hydrocarbon removal at 800°F (□: phosphorus free, ○: 0.10, △: 0.48 g P/gal). [From Gagliardi et al. (84).]

sibly explain why in one case investigators have reported a deactivating influence of fuel phosphorus and no effect in another instance (*62, 63*). The practical implication of this interaction is that in fuels with low levels of P and Pb, the presence of one might be more detrimental than the simultaneous presence of both.

Most commercial engine oils contain the antiwear agent zinc dialkyldithiophosphate (ZDP). The effect of phosphorus derived from oil is of particular concern when the poison concentration in the fuel is very low. One should keep in mind that the ZDP additive contains within its own chemical formula a potential scavenger for phosphorus, i.e., zinc. Other potential scavengers which form stable phosphates, such as alkaline earth metals, are also present in the commercial oil additive "package."

A large number of investigations have been devoted to establish the degrading influence of oil-derived poisons on automotive catalysts (*11, 41, 49, 86–88*). In several engine and fleet tests, the difference in catalyst performance was insignificant when either phosphorus-free or ZDP-containing oils were used, provided the fuel did contain certain amounts of Pb and P, even in low concentrations. When the fuel was essentially free of contaminants, some deterioration due to the use of ZDP-containing oil could be discerned (*41*), which was again overshadowed if fuel contaminants were added. In another case (*49*), the effect of ZDP in oil on catalysts was noticed at P levels in twofold excess of those used commercially (0.32%), when contaminant free fuel was used. Although not stated explicitly, it appears that the analysis of the oil high in P points to a P-containing additive free of Zn.

Indeed, in the work of Fitzgerald and Wilson (*88*), the introduction of tributyl phosphate into the oil gave a rather steep deterioration of hydrocarbon activity. A similar sharp deterioration was also noted when an unidentified ashless antioxidant additive by itself was employed. The conclusion reached by the authors is that the presence of phosphate-binding metals, mainly Zn and Ca, is responsible for the inhibition of the poisoning influence of phosphorus from engine oils.

Nevertheless, the following question still lingers: Can the fully formulated phosphorus containing oil (SE) be considered harmless with the potentially more sensitive catalyst systems required for meeting the most stringent emission standards? Caracciolo and Spearot (*89*) attempted to answer this question by a detailed dynamometer study with oil containing ZDP, or tricresylphosphate (TCP), or no phosphorus at all. The runs with oil containing ZDP were obtained with or without a sulfonate detergent. The authors have endeavored to establish a correlation not only with the amount of phosphorus fed into the system, but also with that actually retained on the catalyst, as noted above in the case of lead. The fuel was essentially phosphorus-free with a small amount of lead (0.014 g/gal).

The important poisoning results of the work, expressed by hydrocarbon efficiency of a pelleted Pt–Pd catalyst after 200 hr of dynamometer testing, are summarized in the composite graph of Fig. 20. The effect of phosphorus on the deactivation is plotted in a threefold manner: (a) vs. the amount supplied to the engine, (b) vs. the amount retained, and (c) vs. the amount that actually reached the catalyst. Only 10–15% of the P is retained on the catalyst, which is in fair agreement with the data in Table V. However, a much larger proportion reaches the catalyst without being retained. When plotted against the amount of retained P, there appears to be a difference in toxicity: phosphorus derived from ZDP being more toxic. When plotted against phosphorus which has reached the catalyst, all phosphorus additives have a similar poisoning effect. The differences in the retention, observed by Caracciolo and Spearot were discussed above. The important implication of this work, quite in contrast to the vehicle and fleet results cited, is that binding by Zn or Ca does not neutralize

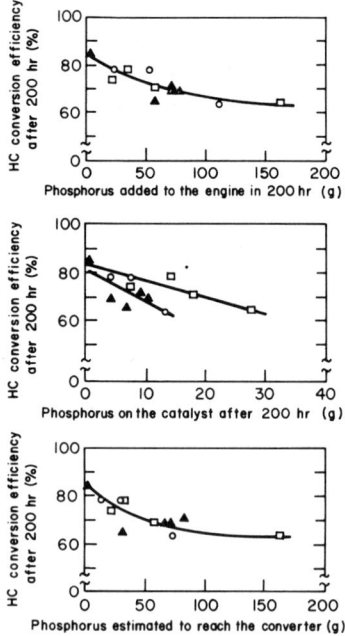

FIG. 20. Effect of fuel phosphorus on hydrocarbon efficiency. (□:TCP, ashless; ○:ZDP, ashless; ▲:ZDP, Ca Sulfonate). [From Caracciolo and Spearot (89).] (Reprinted with permission of the Society of Automotive Engineers.)

completely the poisoning tendency of phosphorus. Although the paper correlates mainly hydrocarbon oxidation activity with phosphorus poisoning, it indicates that the CO oxidation activity is affected even more, as was mentioned earlier. A point is made that the extent of the P-engendered deterioration of oxidation catalysts is not enough to justify the removal of ZDP from oil, since it provides much needed wear protection. Whether this will remain equally true for the next generation of catalysts, designed to meet much more stringent standards, remains to be determined in the future.

C. Effect of Sulfur

There is no evidence in vehicle operation that the oxidation activity of noble metal catalysts suffers from poisoning by SO_2 (24, 28, 84), although Hunter claims (43) that Pt can be poisoned below 900°F. In contrast, severe deactivation of base metal catalysts has been observed in many instances.

Yao (38, 90) has studied a series of unsupported base metals concerning their poisoning by SO_2 at levels of up to a few hundred ppm. Severe deactivation was observed for all the tested catalysts, which contained Co, Cu, Cr, Mn, Fe, or Ni. At a constant SO_2 feed rate, oxidation of CO and hydrocarbons decreased monotonically to a constant value, until the SO_2 concentration was changed. Sometimes sulfur poisoning is more severe for CO than for hydrocarbon oxidation, while in other cases the order is reversed (38, 44). In extreme cases, a few ppm of SO_2 in the gas phase caused practically complete poisoning of base metals, even before a sufficient amount of SO_2 for the formation of one monolayer on the active surface had reached the catalyst. Rate and degree of recovery that could be achieved in a sulfur-free environment varied strongly with catalyst composition, temperature, and other experimental conditions. For example, Co_3O_4, poisoned at 400°C could not be completely regenerated at 600°C, yet SO_2 poisoning of Co_3O_4 at 600°C was completely reversible at the same temperature.

Farrauto and Wedding (44) have identified by X-ray diffraction sulfate formation on base metals, such as Co and Cu, as one mode of poisoning. At the same time, they have demonstrated by infrared spectroscopy that sometimes poisoning can be caused solely by adsorption of SO_2 or SO_3 without sulfate formation, which was the case for a $CuCr_2O_4$ catalyst at 500°C. An important point to be considered here is the hydrothermal stability of a sulfate, since different sulfates vary considerably in this respect.

Fishel et al. (39) concluded in the case of catalysts containing Cu and Cr on alumina that poisoning by S from 1975 fuels far outweighs the detrimental effects of Pb and P.

Finally, there is evidence (48, 90, 91) that the presence of small amounts of noble metal on the surface of base metal catalysts can suppress catalyst deactivation by SO_2 to some extent.

Different modes of poisoning can be expected in catalysts operating under reducing conditions. The chemical species transporting the poison is likely to be, at least in part, hydrogen sulfide, resulting in the formation of sulfides rather than sulfates. Metals which exhibit this poisoning pattern in the temperature range from 550°–800°C are Cu, Fe, Zn, Ru, and others (27). McArthur concluded from a comparison of total sulfur deposit and active metal surface area that sulfide formation of the bulk metal does not take place but is confined to the metal surface.

On some catalysts simultaneous presence of S and Pb causes deterioration to a higher degree than the separate effects of the two poisons predict, as was reported by Jackson et al. (92) for Cu–Ni and Pt–Ni reduction catalysts. In contrast, laboratory tests of a Ru reduction catalyst (82) indicated that the combined effects of Pb and S was less detrimental than the sum of their separate effects.

D. Effect of Lead Scavengers

Permanent retention of the scavenger elements Cl and Br, as such, on noble metal oxidation catalysts is usually insignificant because of their volatility. In spite of this fact, it has been demonstrated [(66) and references therein] that scavengers by themselves can suppress the oxidation activity of Pt and Pd.

The very strong influence of EDB (ethylene dibromide, 1,2-dibromoethane) is illustrated by the data of Table VIII. A concentration of EDB, which would have been associated with 1.5 g Pb/gal in TEL motor mix, was used in this experiment. A drastic decrease in hydrocarbon and CO conversion was observed immediately when EDB

TABLE VIII
Comparison of EDB Effect on Pt and Pd[a]

| | Baseline | | Conversion percentages at 500°C | | | |
| | | | 10,000 miles | | Recovery | |
	CO	HC	CO	HC	CO	HC
Pt	94	73	90	64	94	71
Pt–Pd	94	73	82	58	95	74
Pd	94	67	75	49	96	68

[a] From Otto and Montreuil (66). (Reprinted with permission from Environmental Science and Technology. Copyright by the American Chemical Society.)

was added to the fuel. The deactivation persisted at the same level, and is shown in Table VIII at the endpoint of the EDB exposure, corresponding to 10,000 simulated miles. The table shows also that the poisoning action is stronger on Pd than on Pt. The catalysts recover their activity completely after EDB is eliminated from the fuel.

Deactivation by EDC (ethylene dichloride, 1,2-dichloroethane) was found to be less important. Although there is some evidence of reversible poisoning by this scavenger in engine and laboratory tests (25, 66, 93), data on poisoning by EDC are not consistent from one investigation to another. The indication is that compared to EDB, poisoning by EDC is not significant at the usual temperatures of catalyst operation.

The transitory poisoning by scavengers is explained by competitive adsorption of a halogen-containing species on catalyst sites that are needed for the oxidation of CO and hydrocarbons. In the case of EDB it is thermodynamically probable that HBr (33), or Br_2 is the actual adsorbed species (66). The possible interactions of EDB and EDC with TEL and the resulting loss in noble metal surface area on the one hand, and catalyst activity on the other, are very complex (66).

E. Effect of Manganese

Methylcyclopentadienyl manganese tricarbonyl (MMT) is an additive in many commercial unleaded gasolines, and thus its potential effect on catalyst performance is an important consideration. It was reported by Faggan et al. (8) that MMT at a recommended level of 0.125 g Mn/gal shows no adverse effect on emissions, when compared to unleaded gasoline, in tests on cars operated on the 50,000-mile EPA certification schedule. These findings have been confirmed by a number of unpublished test reports from several industrial laboratories. In fact, it is indicated that some of the manganese deposits can aid catalytically in the removal of CO, and possibly to some extent also of hydrocarbons. The consequence of such a catalytic effect of Mn has still to be explored.

While the absence of Mn poisoning in the case of Pt and Pd has been established, the potential poisoning of reduction and three-way catalysts has yet to be investigated in detail. In our laboratory studies, it was observed that any detrimental effect of Mn on a reduction catalyst was reversed by a brief catalyst treatment under oxidizing conditions. Two formulations of three-way catalysts were tested recently for their resistance against Mn poisoning. Again, no deterioration was noticed beyond that caused by the same fuel, containing Pb, P, and S at the level used for engine certification, but without Mn. A slight decrease in NO activity attributable to Mn, cannot be excluded at this time. Further details of this investigation have been reported by Otto and Sulak (94).

These considerations disregard a possible plugging of catalyst pores, or monolith channels by Mn-containing particulates, which has been observed occasionally, when cars were operated on relatively high levels of MMT, or when the catalyst was subjected to severe thermal stress (8).

XI. Interaction between Poisons and Active Components

While discussing retention, distribution, and effects of the various poisons, nothing was said about the underlying causes of deactivation. The immediate questions that arise are: What are the species that carry the poisons to the catalyst surface? Do the poisons tend to associate with active components with a higher specificity than with the inert surface of the support? How does a poison bind to an active site, and what surface species is formed thereby? The answers to these questions are, indeed, very fragmentary for two reasons. First, they are difficult to establish without ambiguities and, second, in this applied field, efforts to provide practical solutions take precedence over those directed to gain fundamental understanding.

In certain instances of poisoning, especially in the case of base metal catalysts, the deactivation can be simply explained by the formation of new bulk solid phases between the base metal and the poison. Examples are the formation of lead vanadates (14) in vanadia catalysts, or of sulfates in copper–chromite and other base metal catalysts (81). These catalytically inactive phases are identifiable by X-ray diffraction. Often, the conditions under which deactivation occurs coincide with the conditions of stability of these inert phases. Thus, a base metal catalyst, deactivated as a sulfate, can be reactivated by bringing it to conditions where the sulfate becomes thermodynamically unstable (45). In noble metal catalysts the interaction is assumed to be, in general, confined to the surface, although bulk interactions have also been postulated.

Let us consider first whether there is a preferential affinity of the poison for the active metal surface. This has been investigated in some detail only for lead interacting with noble metals. Williams and Baron (95) inserted very pure Pt and Pd foils into beds of catalyst pellets and examined them by Auger electron spectroscopy after exposure to exhaust gas from an engine burning fuel with lead, scavengers, phosphorus, and sulfur. Short exposures to fuels with small amounts of lead and phosphorus produced measurable Auger peaks of Pb and P from the surface of the exposed foils. Examination of foils placed in two different positions in the bed confirmed the gradients observed by postmortem analysis mentioned before. Thus, upstream foils had more heavily con-

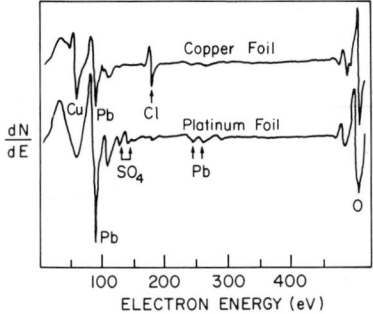

FIG. 21. Comparison of Pt- and Cu-foil probes simultaneously exposed to exhaust. [From Williams and Baron (95).]

taminated surfaces than those placed downstream in a catalyst bed. A copper foil exposed simultaneously with a Pt foil produces considerably weaker Pb signals than the Pt sample, as shown in Fig. 21. This in itself hints at a specificity for the lead–platinum interaction, at least as compared to lead on copper.

Figure 22 shows that, after simultaneous treatment, Pt differs from Pd in that lead is absent from the surface of Pd. The authors explain this behavior by the diffusion of lead ions into the oxidized surface layer of palladium, forming a solid solution of lead and palladium oxides (96). The presence of PO_4^{3-} is evident on the surface of both foils. The following

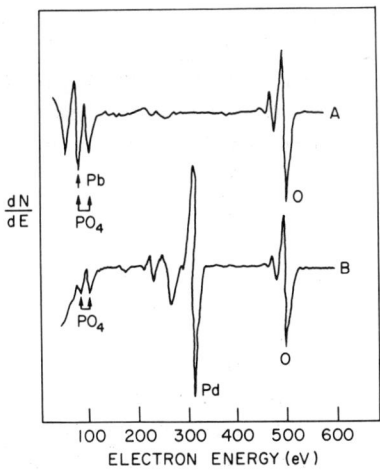

FIG. 22. Lead and phosphorus accumulation on Pt (A) and Pd (B). [From Williams and Baron (95).]

general mechanism of interaction was postulated: the lead halides, transported to the catalyst surface, decompose on Pt with lead remaining in the surface and the halogen desorbing. As a result the halides are unlikely to contribute to the irreversible poisoning, as was indicated above. The noble metal surface becomes saturated with the lead species. Sulfur trioxide (and phosphorus pentoxide probably, as well) can interact with excess lead and grow overlayers outward from the surface. Although not explicitly stated, an inference exists that the support surface is not capable of decomposing the lead carrying species, and thus there exists a special affinity for the preferential association of lead with the active metal.

An attempt was made in our laboratory (97) to examine the specificity of the association of noble metal and lead, using the carefully controlled exhaust produced by the pulse-flame generator (30). A microscope slide was covered by a 100-Å-layer of alumina. Two separate strips of Pt (or Pd), 50 or 300 Å thick, were sputtered on top of the alumina layer. A constant temperature in the 350°–700°C range was kept during the experiment.

There are four Al_2O_3/Pt (or Pd) interfaces. By burning fuel containing certain amounts of lead with and without scavengers and passing the exhaust over the slide, we expected to find out whether there is a greater affinity for lead exhibited by the metal surface than by the Al_2O_3 surface. As will be seen below, the results from these experiments are far from being unambiguous. They serve as a good illustration of the difficulties encountered in designing such experiments.

The surface of the exposed specimens was examined by the electron microprobe. As the electron-microprobe traces of a few of the experiments indicate (Fig. 23), an enhanced Pb affinity for platinum seems to be, indeed, the case. The lower trace (A) shows that there is much more lead retained on the areas covered by Pd, while the middle trace (B) indicates the same for the areas covered by Pt. This phenomenon recurs at each of the four interfaces between the metal and Al_2O_3 segments of the slide. In the experiment, the slides were exposed to approximately 72 hr of treatment (which is equivalent to 2000–2500 miles of simulated vehicle operation), and the fuel contained 0.5 g Pb/gal. In case of the traces (A) and (B), there was no scavenger in the fuel.

Prima facie, the evidence in Fig. 23 seems to be convincing, but the difficulty in interpretation comes in, when it is realized that ample Pb is found on the glass slide on regions not covered by Al_2O_3 due to a strong lead–silica interaction. Furthermore, it was found that the difference in lead deposition between an Al_2O_3 surface and an adjoining Pt strip is much less pronounced on solid Al_2O_3 (sapphire) slides, although it does exist.

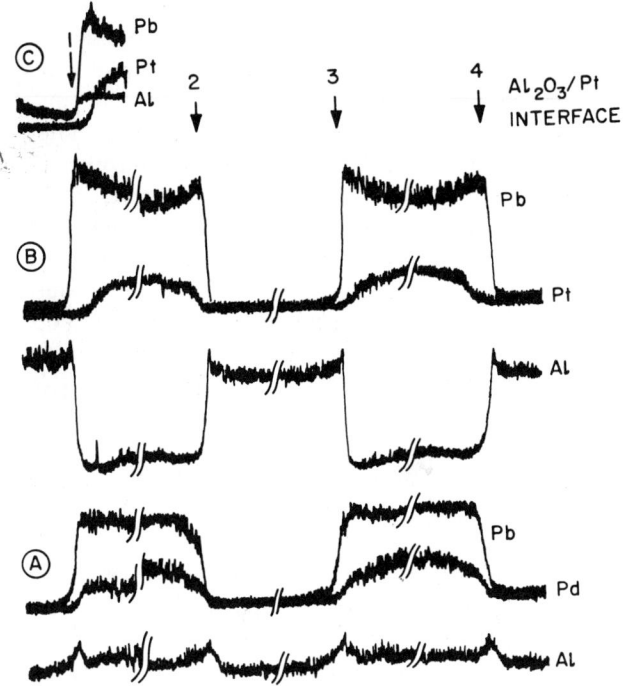

FIG. 23. Microprobe traces of Pt (Pd), Pb, and Al on slides treated in pulse-flame exhaust. [From Tabock et al. (97).]

The most plausible interpretation of the traces in Fig. 23 is as follows: the volatile species transporting lead decomposes on the noble metal, and the lead diffuses copiously through cracks in the Al_2O_3 film into the glass slide underneath.

Additional evidence was obtained from Auger surface analysis of the Pt-sapphire specimens. Peak-to-peak heights of the lead line at 94 eV were taken from six spots on the Pt-covered area and from six spots on the adjoining sapphire area after a 24-hr exposure to pulsator exhaust. There was, after correction of the 94-eV line for backscattered electrons, more lead on the Pt-covered area in five out of six comparisons. On the average there was a 50% higher lead content on the Pt-covered areas. At lower lead exposures the differences are difficult to discern, since the accumulation of the ubiquitous particulate lead is not expected to vary from one kind of surface to another. Such behavior is essentially in accord with the findings of Williams and Baron (95). At present, however, we still must consider the preferential association of lead with Pt or Pd only as eminently plausible but in need of further, still more direct confirmation.

Neither is the nature of the volatile species, which transports the lead, and possibly decomposes on the noble metal, completely clear, even after several experiments. On the one hand, addition of motor mix, EDC, or EDB to the TEL-containing fuel generally lowers the amount of lead found on the noble metal-covered strips; on the other hand, however, introduction of $PbBr_2$ (in aqueous solution) into the burnt gases still shows preferential association of the lead with the noble metal strips. Again, as seen in trace (C) of Fig. 23, introduction of TEL into the postcombustion exhaust also shows preferential Pb association with the noble metal regions. Without trying to rationalize the contradictions, one could plausibly assume that all volatile Pb species may preferentially decompose on noble metals. Acres *et al.* (*22*) have suggested that small amounts of TEL, which escape the combustion process, do preferentially poison the active sites of noble metals.

Johnson and co-workers (*62*) have come to the conclusion that interaction of lead with Pt crystallites results in the formation of an inactive phase in which the Pt atoms are ionized and soluble in HCl. These data were derived from engine tests, in which the catalysts were exposed to fuels with 0.03–0.1 g Pb/gal. The amount of crystalline Pt in these catalysts was smaller than in catalysts run on lead-free fuels. The authors indicate that noncrystalline forms of Pt are present on Al_2O_3 supports under certain conditions, and that lead stabilizes such forms. The question whether the noncrystalline, ionic Pt is a surface or a bulk phase remains unanswered. Bulk mixed Pt–Pb oxides have been described (*98, 99*), but, again, the dispersed forms of noble metals supported on Al_2O_3, which lead (and other elements) may stabilize, are known to be associated with the surface only. Palladium can be expected to form such noncrystalline dispersed phases to a still greater extent since it is more easily oxidized than Pt.

In contrast to lead, the possible poisoning by metallic elements, derived from the vehicle system, is not easily documented. Many formulations of automotive catalysts contain both base and noble metals, but the detailed effect of such combinations on the particular reactions is rarely known with precision. One study was concerned with the effect of Cu on noble metal oxidation catalysts, since these, placed downstream from Monel NOx catalysts, could accumulate up to 0.15% Cu (*100*). The introduction of this amount of Cu into a practical catalyst containing 0.35% Pt and Pd in an equiatomic ratio has, after calcination in air, depressed the CO oxidation activity, but enhanced the ethylene oxidation. Formation of a mixed Pt–Cu-oxide phase is thought to underlie this behavior. This particular instance shows an example, when an element introduced into a given catalyst serves as a poison for one reaction, and as a promoter for

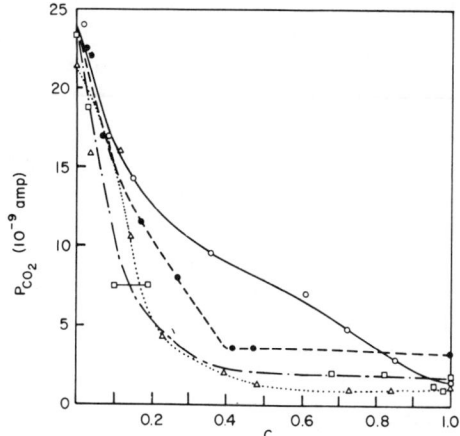

FIG. 24. Maximum catalytic activity for CO oxidation as a function of alloy composition for Pt–Cu (△), Pt–Ni (○, ●), and Pt–Fe (□) systems. [From Bonzel and Wynblatt (101).] (Reprinted with permission of North-Holland Publishing Company.)

another. Such instances are not that rare, but their discussion is outside of the scope of the present review.

Poisoning of noble metal catalysts by base metals is probably best studied on model surfaces accessible to methods able to measure the surface composition. A clear example was provided by Bonzel and Wynblatt (101) in the study of CO oxidation on films of Pt containing increasing proportions of the surface covered by Cu, Fe, or Ni. Figure 24 shows that the effect of replacement of Pt on the surface is very marked and highly nonlinear; it is similar for all three base metals. It is fair to state that the extent of the work concerned with the understanding of poisoning of automotive and other catalysts on the microscopic level is meager. It can be anticipated that the advent and broader use of modern tools of surface examination will remedy this situation.

XII. Poison-Resistant Automotive Catalysts

The use of TEL has afforded production of automotive fuels with higher octane rating at lower refining expenditures, both in cost and in energy. The removal of sulfur, especially of the residual amounts, in the refining process requires considerable outlays. These considerations provide a powerful economic incentive to design catalysts compatible with lead and sulfur. Similar reasoning explains the incentive to design catalyst systems compatible with phosphorus, especially that from oil.

Efforts of this nature only rarely find their way into the technical literature, and therefore the survey here is necessarily brief. A well-publicized effort is based on the use of mixed base metal oxides that crystallize in a perovskite-type lattice. Only recently Katzman and co-workers (*102*) have claimed that perovskite compounds, such as $LaCoO_3$, were immune to lead or bromine poisoning, when the poisons were applied by laboratory techniques and the catalysts tested also in the laboratory. On the other hand, a Pt-on-Al_2O_3 catalyst is poisoned after similar treatment. These results are, however, hardly relevant to automotive catalysts, since unpoisoned Pt has a specific activity for CO or hydrocarbon oxidation higher by several orders of magnitude than any base metal perovskite, and it takes extreme poisoning to equalize the activity of these two vastly different types of catalysts. It was established that unusual activity of base metal perovskites has been associated with small admixtures of unintentionally introduced noble metal (*91, 103*). Base metal perovskites are made still less acceptable as poison-resistant catalysts by the fact that their susceptibility to sulfur poisoning is as high as that of simple base metal oxides. Again, the sometimes quoted claim of sulfur resistance of the perovskites (*104*) was explained by noble metal impurity (*90, 91, 103, 105*).

Obviously, the above facts have led to attempts to introduce noble metals intentionally into base metal perovskites to make them poison resistant in this way. Several patents (*106*) described the formulation of such perovskites containing varying amounts of noble metals. DuPont has claimed in press releases (*107*) that these catalysts can be used with fuels containing high amounts of Pb, P, and S. Since most of the data are not in the public domain, one must wait whether this claim will be proven true by the introduction of such catalysts on vehicles. Limited work in our laboratory on perovskite-type catalysts, which contain noble metals, indicates that the susceptibility of a noble metal atom on a perovskite surface to poisoning by lead does not differ greatly from that on an alumina surface. This was studied in detail for Ru-containing perovskites designed for NO reduction (*50*).

XIII. Catalyst Rejuvenation

If the avoidance of deactivation by poisoning is difficult, could a partially deactivated catalyst be regenerated by chemical or physical means? In the oil and chemical industry, catalyst reactivation is a standard procedure, done sometimes *in situ* or after removal from the reactor. Logistically, it does not seem to be feasible to remove an automotive catalyst for external regeneration. Also, *in situ* treatment, for instance in a service garage, will require elaborate arrangements. Nevertheless, the technological possibilities have been explored to some degree.

There are two main routes for the removal of poisons: one is based on chemical, the other on thermal treatment. One of the first patents for the regeneration of automotive catalysts was filed by Houdry and Calvert (*108*) in 1955; it covers two wet chemical methods, presently contemplated for removing lead. The lead is leached out, either by nitric acid, or by an acetate solution. In the early 1960s a patent was granted (*109*) covering a gaseous process of lead removal by halide volatilization at relatively low temperatures. Both patents were aimed at the heavily contaminated catalysts of the early era of catalyst application, as discussed previously. Consequently, large amounts of lead were removed, as explained in the examples cited in the respective patents. There was a substantial regeneration as measured by various activity indices. Still another patent (*110*) indicates the possibility of a gaseous lead removal process based on an elaborate oxidation–reduction cycle. It appears that this cycle is effective when applied at elevated temperatures (above 850°C), and therefore it impinges upon the thermal means of lead removal. It is quite difficult to understand the underlying chemistry from the description in the patent specification. One experimental fact, however, is quite significant. The lowering of the lead content of a used catalyst from 19% to 14–15% caused substantial activity regeneration. The *prima facie* implication is that the removable lead is more poisonous than that which is left on the catalyst. None of the patents, cited above, indicates how long the regenerated activity persists after reexposure to leaded fuel.

A recent paper (*111*) has examined the acetate leching of lead from noble metal catalysts exposed to exhaust from leaded fuels containing less than 0.1 g Pb/gal. Again, it was established that relatively small amounts of removed lead sharply increase the oxidation activity, especially for hydrocarbons. The duration of the rejuvenation process has obviously an effect on the depth of the catalyst thickness from which the lead is leached out. Therefore, a small amount removed from the surface can have a large transient regenerative effect, which disappears relatively quickly, when during reuse the operating temperature engenders diffusion of the remaining lead into the outer regions accessible to the reacting gas molecules. Therefore, the assumption that the removable lead is more poisonous can be refuted by the argument that it is removed from the regions where it is most harmful. Nevertheless, the fact that the removal of only a small amount of lead (in some cases, only a few percent of the amount present on the catalyst) gave a substantial activity improvement,

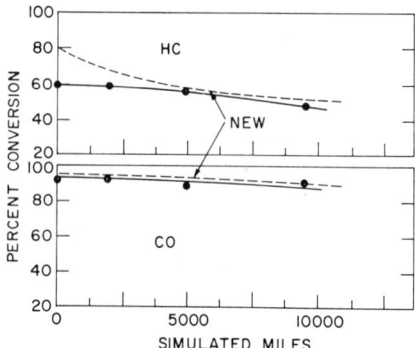

FIG. 25. Deactivation comparison of rejuvenated and fresh catalysts. [From Rothschild (*111*).] (Reprinted with permission of the Society of Automotive Engineers.)

suggests possible preferential removal from the surface associated with the noble metal, which would be in a sense a mirror effect to the suspected preferential deposition on these surfaces, mentioned above. A pulse-flame test of a thoroughly rejuvenated fleet catalyst is shown in Fig. 25 and is compared with the aging behavior of a fresh catalyst. With the exception of the initial high activity of the fresh catalyst, the subsequent activity decline is almost the same for the fresh and the rejuvenated samples, in this particular laboratory test. The thorough rejuvenation required two consecutive treatments with a 10% acetate-ion solution. The interested reader will find in Rothschild (*111*) a substantial amount of detail on the rejuvenation procedures.

Thermal rejuvenation is always carried out at high temperatures to bring the vapor pressure of the contaminant to a level where it can be swept out by the passing gas. This is relatively easily done for sulfur by bringing the sulfate in question, be it that of Pb, Al, Ni, Co, Cr, etc., to appreciable decomposition. As mentioned, the "sulfur storage" of catalysts is associated with surface aluminum sulfate which can be decomposed above 600°C. The removal of lead requires much higher temperatures. McArthur (*26*) indicates that in an oxy-reduction treatment in the 790°–870°C range a thermally "rejuvenated" monolithic NOx catalyst shows a very substantial removal of lead. The treatment is somewhat similar to that described in the patent (*110*), cited above. It is probably better to remove Pb in a reducing than in an oxidizing atmosphere. Although Pb and PbO have approximately the same volatility, metallic Pb is less reactive with inorganic oxides, such as alumina. It is to be expected that catalysts of widely varying compositions will respond differently to thermal rejuvenation. To regenerate a $Pd-Al_2O_3$ catalyst, the oxy-reduction thermal treatment was performed at 1000°C (*112*) which removed most of the Pb, S, and P. It must be cautioned that experience with vehicles indicates that the regenerated activity, acquired by drastic thermal means, is usually shortlived since the surface-area loss suffered in the process is prohibitively severe.

XIV. Future Developments

The future work on the poisoning of automotive catalysts will have to deal, primarily, with the specific interaction of particular poisons with the active components. The present trend toward more complex catalytic systems, containing several active components, will make the task still more difficult. One could foresee the use of modern, more sophisticated methods of surface analysis for studying the interactions between poison and active components.

The intentional design of model systems can be envisioned, as for instance binary or multiple assemblies (clusters) of active components and poisons, for the examination of their activity in chemisorption, or specific reactions. The results can then be compared with respective clusters containing the active species only. Perhaps, such model systems will be amenable to computational methods capable of predicting their chemisorptive behavior and their surface reactivity. Such approaches are now employed for the design of improved multicomponent catalysts and can, obviously, be used to study the reverse effect, i.e., the mutual deactivation of the cluster components.

On the other end of the spectrum, the testing of existing, and potentially new fuel components for their compatibility with newly developed catalysts will continue, as well. The field experience to be accumulated in the near future, may well present us with new challenges in the study of the long-term activity of automotive catalysts, and the role played by the contaminants.

Acknowledgments

As the perusal of the reference list will quickly indicate, a major part of the results reviewed here was the work of our numerous colleagues at Ford Motor Co. Special thanks for the critical reading of the manuscript are due S. Gratch, J. T. Kummer, E. E. Weaver, B. Weinstock, and P. Wynblatt. Of course, any errors are the responsibility of the authors.

References

1. Maxted, E. B., *Adv. Catal.* **3,** 129 (1951).
2. Wheeler, A., *Adv. Catal.* **3,** 249 (1951).
3. Wei, J., *Adv. Catal.* **24,** 57 (1975).

4. Yolles, R. S., and Wise, H., *Crit. Rev. Environ. Control* **2**, 125 (1971).
5. Dwyer, F. G., *Catal. Rev.* **6**, 261 (1972).
6. Pierson, W. R., "Sulfuric Acid Generation by Automotive Catalysts," *Chem. Tech.* **6**, 332 (1976).
7. Beltzer, M., Campion, R. J., Harlan, J., and Hochhauser, A. M., *SAE (Soc. Automot. Eng.), Pap.* No. 750095 (1975).
8. Faggan, J. E., Bailie, J. D., Desmond, E. A., and Lenane, D. L., *SAE (Soc. Automot. Eng.), Pap.* No. 750925 (1975).
9. Tischer, R. L., and Eisentraut, K. J., *Am. Chem. Soc., Div. Pet. Chem., Prepr.* **21** (4), 904 (1976).
10. Shelef, M., Dalla Betta, R. A., Larson, J. A., Otto, K., and Yao, H. C., "Poisoning of Monolithic Noble Metal Oxidation Catalysts in Automotive Exhaust Environment," *Am. Inst. Chem. Eng., New Orleans Meet.* (1973).
11. McConnell, R. J., and McDonnell, T. F., *SAE (Soc. Automot. Eng.), Pap.* No. 730597 (1973).
12. Mencik, Z., Berneburg, P. L., and Short, M. A., *Adv. X-Ray Anal.* **18**, 396 (1975).
13. Artz, B. E., Kelly, C. J., and Short, M. A., *Adv. X-Ray Anal.* **18**, 309 (1975).
14. Cannon, W. A., and Welling, C. E., *SAE (Soc. Automot. Eng.), Pap.* No. 59-29T (1959).
15. Vanderveer, R. T., and Chandler, J. M., *SAE (Soc. Automot. Eng.), Pap.* No. 59-29S (1959).
16. Cannon, W. A., and Welling, C. E., U. S. Patent 2,912,300 (1959).
17. Davis, D. L., and Onishi, G. E., *SAE (Soc. Automot. Eng.), Pap.* No. 62-486F (1962).
18. Yarrington, R. M., and Bambrick, W. E., *J. Air Pollut. Control Assoc.* **20**, 398 (1970).
19. Weaver, E. E., *SAE (Soc. Automot. Eng.), Pap.* No. 690016 (1969).
20. Su, E. C., and Weaver, E. E., *SAE (Soc. Automot. Eng.), Pap.* No. 730594 (1973).
21. Kummer, J. T., Yao, Y., and McKee, D., *SAE (Soc. Automot. Eng.), Pap.* No. 760143 (1976).
22. Acres, G. J. K., Cooper, B. J., Shutt, E., and Malerbi, B. W., *Adv. Chem. Ser.* No. 143, p. 54 (1975).
23. Hetrick, S. S., and Hills, F. J., *SAE (Soc. Automot. Eng.), Pap.* No. 730596 (1973).
24. Neal, A. H., Wigg, E. E., and Holt, E. L., *SAE (Soc. Automot. Eng.), Pap.* No. 730593 (1973).
25. Holt, E. L., Wigg, E. E., and Neal, A. H., *SAE (Soc. Automot. Eng.), Pap.* No. 740248 (1974).
26. McArthur, D. P., in "The Catalytic Chemistry of Nitrogen Oxides" (R. L. Klimisch and J. G. Larson, eds.), p. 263. Plenum, New York, 1975.
27. McArthur, D. P., *Adv. Chem. Ser.* No. 143, p. 85 (1975).
28. Giacomazzi, R. A., and Homfeld, M. F., *SAE (Soc. Automot. Eng.), Pap.* No. 730595 (1973).
29. Weaver, E. E., Shiller, J. W., and Piken, A. G., *AIChE Symp. Ser.* No. 72 (156), p. 369 (1976).
30. Otto, K., Dalla Betta, R. A., and Yao, H. C., *J. Air Pollut. Control Assoc.* **24**, 596 (1974).
31. Cooper, B. J., *Platinum Met. Rev.* **19**, 141 (1975).
32. Cooper, B. J., Renny, L. V., and White, R. J., "Lead Poisoning of Automobile Emission Control Catalysts—Influence of Emission System and Catalyst Design Characteristics on the Poisoning Mechanism," *Am. Chem. Soc., Symp. Automot. Catal., Chicago Meet., 1975.*
33. Newby, W. E., and Dumont, L. F., *Ind. Eng. Chem.* **45**, 1336 (1953).

34. Street, J. C., *SAE Trans.* **61,** 443 (1953).
35. Bomback, J. L., Wheeler, M. A., Tabock, J., and Janowski, J. D., *Environ. Sci. Technol.* **9,** 139 (1975).
36. Hammerle, R. H., and Truex, T. J., *Am. Chem. Soc., Div. Pet. Chem., Prepr.* **21** (4), 769 (1976).
37. Taylor, K. C., *Am. Chem. Soc., Div. Pet. Chem., Prepr.* **21** (4) 760 (1976).
38. Yao, Y.-F. Y., *J. Catal.* **39,** 104 (1975).
39. Fishel, N. A., Lee, R. K., and Wilhelm, F. C., *Environ. Sci. Technol.* **8,** 260 (1974).
40. Casassa, J. P., and Beyerlein, D. G., *SAE (Soc. Automot. Eng.), Pap.* No. 730558 (1973).
41. Gallopoulos, N. E., Summers, J. C., and Klimisch, R. L., *SAE (Soc. Automot. Eng.), Pap.* No. 730598 (1973).
42. Klimisch, R. L., Summers, J. C., and Schlatter, J. C., *Adv. Chem. Ser.* No. 143, p. 103 (1975).
43. Hunter, J. E., *SAE (Soc. Automot. Eng.), Pap.* No. 720122 (1972).
44. Farrauto, R. J., and Wedding, B., *J. Catal.* **33,** 249 (1973).
45. Stern, K. H., and Weise, E. L., "High Temperature Properties and Decomposition of Inorganic Salts. Part 1: Sulfates," Natl. Bur. Stand. NSRDS-NBS7, Category 5, Thermodyn. Transp. Properties. U. S. Gov. Print. Off., Washington D. C., 1966.
46. Kummer, J. T., *Adv. Chem. Ser.* No. 143, p. 178 (1975).
47. Roth, J. F., *Ind. Eng. Chem., Prod. Res. Dev.* **10,** 381 (1971).
48. Roth, J. F., and Gambell, J. W., *SAE (Soc. Automot. Eng.), Pap.* No. 730277 (1973).
49. Bouffard, R. A., and Waddey, W. E., *SAE (Soc. Automot. Eng.), Pap.* No. 740135 (1974).
50. Gandhi, H. S., Stepien, H. K., and Shelef, M., *SAE (Soc. Automot. Eng.), Pap.* No. 750177 (1975).
51. Howitt, J. S., and Sekella, T. C., *SAE (Soc. Automot. Eng.), Pap.* No. 740244 (1974).
52. Ohara, T., Ono, T., Ichihara, S., and Saito, K., U. S. Patent 3,953,369 (1976).
53. Liederman, D., Voltz, S. E., and Snyder, P. W., *Ind. Eng. Chem., Prod. Res. Dev.* **13,** 168 (1974).
54. Barnes, G. J., and Klimisch, R. L., *SAE (Soc. Automot. Eng.), Pap.* No. 730570 (1973).
55. Aykan, K., Mannion, W. A., Mooney, J. J., and Hoyer, R. D., *SAE (Soc. Automot. Eng.), Pap.* No. 730592 (1973).
56. Weidenbach, G., and Furst, H., *Chem. Tech.* (Berlin) **15,** 589 (1963).
57. Gruber, H. L., *Anal. Chem.* **34,** 1828 (1962).
58. Benson, J. E., and Boudart, M., *J. Catal.* **4,** 704 (1965).
59. Wilson, G. R., and Hall, W. K., *J. Catal.* **17,** 190 (1970).
60. Wilson, G. R., and Hall, W. K., *J. Catal.* **24,** 306 (1972).
61. Dalla Betta, R. A., *J. Catal.* **31,** 143 (1973).
62. Johnson, M. F. L., Mooi, J., Erickson, H., Kreger, W. E., and Breder, E. W., *SAE (Soc. Automot. Eng.), Pap.* No. 741079 (1974).
63. Mooi, J., Kuebrich, J. P., Johnson, M. F. L., and Chloupek, F. J., "Modes of Deactivation of Exhaust Purification Catalysts," *Am. Pet. Inst., Div. Refin., Prepr.* 01-73 (1973).
64. Dalla Betta, R. A., McCune, R. C., and Sprys, J. W., *Ind. Eng. Chem., Prod. Res. Dev.* **15,** 169 (1976).
65. Sprys, J. W., Bartosiewicz, L., McCune, R., and Plummer, H. K., *J. Catal.* **39,** 91 (1975).
66. Otto, K., and Montreuil, C. N., *Environ. Sci. Technol.* **10,** 154 (1976).

67. Thiele, E. W., *Ind. Eng. Chem.* **31,** 916 (1939).
68. Satterfield, C. N., "Mass Transfer in Heterogeneous Catalysis." MIT Press, Cambridge, Massachusetts, 1970.
69. Frank-Kamenetskii, D. A., "Diffusion and Heat Exchange in Chemical Kinetics." Princeton University Press, Princeton, New Jersey, 1955.
70. Wei, J., and Becker, E. R., *Adv. Chem. Ser.* No. 143, p. 116 (1975).
71. Rajadhyaksha, R. A., and Doraiswamy, L. K., *Catal. Rev.* **13,** 209 (1976).
72. Hegedus, L. L., and Baron, K., *J. Catal.* **37,** 127 (1975).
73. Dwyer, F. G., and Morgan, C. R., *Am. Chem. Soc., Div. Pet. Chem., Prepr.* **17** (4), H9 (1972).
74. Piken, A. G., unpublished observations (1973).
75. Su, E. C., unpublished observations (1974).
76. Voorhoeve, R. J. H., Patel, C. K. N., Trimble, L. E., and Kerl, R. J., *Science* **190,** 149 (1975).
77. Dunlevey, F. M., and Lee, C. H., *Science* **192,** 809 (1976).
78. Irish, D. C., and Stefan, R. J., *SAE (Soc. Automot. Eng.), Pap.* No. 760037 (1976).
79. Krause, B., Bouffard, R. A., Karmilovich, T., and Kayle, E. L., *SAE (Soc. Automot. Eng.), Pap. No.* 760091 (1976).
80. Gandhi, H. S., Otto, K., Piken, A. G., and Shelef, M., *Am. Chem. Soc., Div. Pet. Chem., Prepr.* **21** (4), 800 (1976); Environ. Sci. Tech. **11,** 170 (1977).
81. Yao, Y.-F. Y., and Kummer, J. T., *Am. Chem. Soc., Div. Pet. Chem., Prepr.* **21** (4), 807 (1976).
82. Shelef, M., *Catal. Rev.-Sci. Eng.* **11,** 1 (1975).
83. Gandhi, H. S., Piken, A. G., Stepien, H. K., Shelef, M., DeLosh R. G., and Heyde, M. E., *SAE (Soc. Autom. Eng.)* Pap. No. 770196 (1977).
84. Gagliardi, J. C., Smith, C. S., and Weaver, E. E. "Effects of Fuel and Oil Additives on Catalytic Converters," *Am. Pet. Inst., Div. Refin., Prepr.* (1972).
85. Acres, G. J. K., and Cooper, B. J., *Proc. Int. Clean Air Congr. 3rd* p. F14 (1973).
86. Meyers, A. T., *SAE (Soc. Automot. Eng.), Pap.* No. 750448 (1975).
87. Pless, L. G., *SAE (Soc. Automot. Eng.), Pap.* No. 750900 (1975).
88. Fitzgerald, W., and Wilson, J. V. D., *SAE (Soc. Automot. Eng.), Pap.* No. 750447 (1975).
89. Caracciolo, F., and Spearot, J. A., *SAE (Soc. Automot. Eng.), Pap.* No. 760562 (1976).
90. Yao, Y.-F. Y., *J. Catal.* **36,** 266 (1975).
91. Katz, S., Croat, J. J., and Laukonis, J. V., *Ind. Eng. Chem., Prod. Res. Dev.* **14,** 274 (1975).
92. Jackson, H. R., McArthur, D. P., and Simpson, H. D., *SAE (Soc. Automot. Eng.), Pap.* No. 730568 (1973).
93. Barnes, G. J., Baron, K., and Summers, J. C., *SAE (Soc. Automot. Eng.), Pap.* No. 741062 (1974).
94. Otto, K., and Sulak, R. J., *Environ. Sci. Techn.* **12,** 181 (1978).
95. Williams, F. L., and Baron, K., *J. Catal.* **40,** 108 (1975).
96. Muller, O., and Roy, R., *Adv. Chem. Ser.* No. 98, p. 28 (1971).
97. Tabock, J., Bomback, J. L., and Shelef, M., unpublished observations (1975).
98. Ostorero, J., and Makram, H., *J. Cryst. Growth* **24/25,** 677 (1974).
99. Sleight, A. W., *Matter. Res. Bull.* **6,** 775 (1971).
100. Kummer, J. T., *J. Catal.* **38,** 166 (1974).
101. Bonzel, H. P., and Wynblatt, P., *Surface Sci.* **36,** 822 (1973).
102. Katzman, H., Pandolfi, L., Pedersen, L. A., and Libby, W. F., "Lead Tolerant Auto Exhaust Catalysts," *Chem. Tech.* **6,** 369 (1976).
103. Croat, J. J., Tibbetts, G. G., and Katz, S., *Science* **194,** 318 (1976).

104. Trimble, L. E., *Matter. Res. Bull.* **9,** 1405 (1974).
105. Gallagher, P. K., Johnson, D. W., Jr., Vogel, E. M., and Schrey, F., *Mater. Res. Bull.* **10,** 623 (1975).
106. Lauder, A., U. S. Patent 3,897,367 (1975); Ger. Patent 2,446, 251 (1975); Ger. Patent 2,446,331 (1975).
107. *Chem. Eng. News Aug. 4,* p. 8 (1975).
108. Houdry, E. J., and Calvert, W. R., U. S. Patent 2,867,497 (1959).
109. Lester, G. R., U. S. Patent 3,090,760 (1963).
110. Br. Patent 981,152 (1965).
111. Rothschild, W. G., *SAE* (*Soc. Automot. Eng.*), *Pap.* No. 760321 (1976).
112. Hori, R., Kinoshita, H., and Uchida, K., Jpn. Patent 76 20,090 (1976); [*Chem. Abstr.* **85,** 67525y (1976)].

Author Index

Numbers in parentheses are reference numbers and indicate that an author's work is referred to although his name is not cited in the text. Numbers in italics show the page on which the complete reference is listed.

A

Abdo, S., 278(45), *307*
Abdurakhmanov, M., 304(115), *309*
Abeles, R. H., 258(83), *263*
Acres, G. J. K., 26, *56*, 322(22), 324(22), 339, 342(22), 346, 356, *362, 364*
Adams, C. R., 184(10), 185, 185(14, 15), 187, *222*
Adams, M. L., 184, *222*
Adamska, B., 270(31), 275(31), *307*
Adzhamov, K. Yu., 206(99, 100, 101), 207, 218(137), *224, 225*
Ahlrichs, J. L., 172(240), *182*
Ahuja, S. P., 123(84), *178*, 286(88), 298(88), 304, *309*
Akimoto, M., 195, 196, 214, 220(58), 221(58), *223*
Akiyama, M., 196(61), *223*
Alkhazov, T. G., 206, 207, 208(105), 218(137), *224, 225*
Allen, A. D., 261(99), *263*
Amberg, C. H., 267, 270(90, 114), 286(87, 98), 287(92, 93), 288(92, 93), 304(114), *306, 309*
Amundson, N. R., 74, *96*
Andre, J. M., 152(172), *180*
Andreu, P., 120(57), 134, *177*
Angell, C. L., 139, 140, 143(147), 159(147), 160, 161(214), 163(214), 165, *179, 181*
Annenkova, B., 208, *225*
Antipina, T. V., 125(110, 112), 125, *178*
Anufrienko, V. F., 269(117), *310*
Aomura, K., 121(65), 133, *177, 179*
Aoshima, A., 216(132), *225*
Arai, Y., 145(162), 146(162), *180*
Arens, H., 80(52), *96*
Arias, J. A., 31, 35(16), 37(16), 37(16), 41(24), 46, 46(27), 48(27), 50(16), 53(16), *57*
Aris, R., 60, 61, 64(14), 70, 74, 95, *96*
Arlman, E. J., 237(30, 31), *262*
Ariyartne, J. K. P., 234(27), 241(40), 244(40), *262*
Ariyoshi, J., 231(15), 238(34), 239(34), 250(65), *261, 262, 263*
Armour, A. W., 270(64, 89), 281(64), 286(89), *308, 309*
Armstrong, W. E., 207(102), *224*
Artz, B. E., 318(13), *362*
Aryoshi, J., 237(33), 249(33), *262*
Ashley, J. H., 267, 269(10, 48), 270(89), 278(48), 279(48), 280(10, 48), 286(89), 290(48), *307, 308, 309*
Ashmead, D. R., 25, 37, *56, 57*
Ashmore, P. G., 192(37, 42), 193(42), 194(42, 49), 215(125), 216(131), *223, 225*
Asmolov, G. N., 270(44, 49, 118, 119), 277, 278(49), 277, *307, 308, 310*
Asselin, G. F., 123(88), 129, *178*
Atkinson, W., 270(63), 281(63), *308*
Avakian, P., 56, *57*
Aykan, K., 191, 201, 203, 205, 206(96, 97, 98), 206, *223, 224*, 334(55), *363*

B

Babior, B. M., 258(86), *263*
Baddour, R. F., 93, *96*
Badger, E. H. M., 276(34), *307*
Bailie, J. D., 316(8), 351(8), 352(8), *362*
Baily, G. C., 246(56), *262*
Bakshi, K. R., 121(71), *178*
Ballivet, D., 132, *179*
Bambrick, W. E., 319(18), 322(18), *362*
Bandiera, J., 221(142), *225*, 270(123), *310*

367

AUTHOR INDEX

Banerjee, K., 271(27), 275(27), 283(27), 286(27), 291(27), 296(27), 299(27), 303(27), *307*
Banks, R. L., 246(56), *262*
Bánsági, F., 12(20), 18(20), *22*
Barber, M., 270(63), 281(63), *308*
Barelko, V. V., 70, 71, *96*
Barnes, G. J., 334(54), 341(54), 351(93), *363, 364*
Baron, K., 27(10), 29, 31, 56, 339, 351(93), 352, 353, 355, *364*
Bart, J. C. J., 270(25, 51, 83, 102), 274(25), 278(51), 279(25), 283(51), 285(83), 288(51, 83), 289(83), 295(102), *307, 309, 309*
Barthomeuf, D., 132(129), 152(178, 179, 180), 153(178, 179, 180), 153, 156(178, 179, 180, 201), 156, 157(178, 179, 180), 158(178, 179, 180), 158, *179, 180, 181*
Bartosiewicz, L., 336(65), *363*
Basila, M. R., 109(36), 112(40), *177*
Bass, J. L., 155(198), 155, *181*
Batist, Ph. A., 191, 208, 209, 209(109), 210, 213, 213(106), *223, 225*
Batta, I., 12, 18(20), *22*
Beaufils, J., 270(124), *310*
Beaufils, J. P., 271(65), 281(65), *308*
Beaumont, R., 152, 152(179, 180), 153, 156(179, 180, 201), 156, 157(179, 180), 158(179, 180), 158, *180, 181*
Becker, E. R., 339, *364*
Beer, A., 4(13), 7(13), *22*
Belen'kii, M. S., 208(105), *225*
Beljajev, V. D., 65, 67, *95*
Belov, N. V., 200(70), *224*
Beltzer, M., 316(7), *362*
Benesi, H. A., 101, 102, 104, 105, 106(26), 118(18, 26, 54), 120(18, 54), 143, 144(157), 144, 149(54), 166, 167(157), 169, 170, 171, *176, 177, 180*
Benson, J. E., 335(58), *363*
Ben-Yaacov, R., 270(61), 280(61), 285(61), 286(61), 302, 303(61), *308*
Berald, A. L., 68(27), *96*
Berkheimer, H. E., 105(30), *177*
Berneburg, P. L., 318(12), *362*
Bestian, H., 232(22), *262*
Bertsch, L., 159, *181*
Bethall, T., 184(5), *222*
Beusch, H., 65, *95*

Bevelander, C., 271(92), 286(92), 287(92), 288(92), *309*
Beyerlein, D. G., 325(40), *363*
Bielanski, A., 153, *180*
Bissot, T. C., 141(153), *179*
Black, E. R., 175, *182*
Blackburn, D. M., 108(33), 109(33), *177*
Blasse, G., 202(83), 202, *224*
Blazek, J. J., 131(127), 137(127), *179*
Bleijenberg, A. C. A. M., 200, *224*
Bloch, H. S., 123(88), 129(88), *178*
Block, F., 141(152), *179*
Block, J., 6(11), *22*
Boedeker, E. R., 107(32), 113(32), 114(32), 115(32), *177*
Bogdanovic, P., 250(62), 251(62), *263*
Bolton, A. P., 161, 162, 163(218), 164(218), 167(229), *181*
Bomback, J. L., 324(35), 331(35), 333, 334, 334(35), 335, 340(35), 354(97), 355(97), *363, 364*
Bonnelle, J. P., 271(65, 70), 281(65), 282, *308*
Bonzel, H. P., 357, *364*
Boreskov, G. K., 195(53), 195, *223*
Boreskova, E. G., 115(51), 148, 163, *177, 181*
Bosman, J., 123(90), *178*
Bottomley, F., 261(99), *263*
Boudart, M., 335(58), *363*
Bouffard, R. A., 326(49), 341(79), 347(49), *363, 364*
Bourne, K. H., 131, *179*
Bouwens, J. F. H., 213(119), *225*
Bozik, J. E., 211(114), *225*
Bozso, F., 211, *225*
Breck, D., 139(145), *179*
Breder, E. W., 335(62), 347(62), 356(62), *363*
Bremer, H., 132, *179*
Breslow, D. S., 231(21), *262*
Bridges, J. M., 304(111), *309*
Brieger, G., 261(97), *263*
Briend-Faure, M., 137(138), *179*
Brinen, J. S., 282, 286(72), *308*
Brintzinger, H., 232(18, 19, 20), 233(19, 20, 24), *261, 262*
Brown, D. G., 257(81), 258(81), *263*
Brown, D. J., 254(69), *263*
Brown, F. R., 271(77), 284, *308*
Brown, G., 168(233), *181*

Brown, H. C., 101(15, 16), *176*
Brown, H. T., 125(104), *178*
Brown, K. L., 258(88), 260(88), *263*
Broussard, L., 151(171), 152, *180*
Brouwer, D. M., 124(96), 127(117), *178, 179*
Bruncke, F., 20(30), *22*
Buben, N., 69, 70, *96*
Bulgakov, O. V., 125(110), *178*
Burbridge, B. W., 167(230), *181*
Burlamacchi, L., 215, 215(128, 129), *225*
Burwell, R. L., 122(79), 292, *178, 309*
Butterfield, R. O., 254(73), *263*
Buyanov, R. A., 51, *57*

C

Calderon, N., 246(54), 247(57, 58), 248(54), *262*
Callahan, J. L., 184, 184(3, 4), 204(3, 4, 87, 88), 205, *222, 224*
Calvert, W. R., 359, *365*
Campadelli, F., 270(25), 279(25), *307*
Campbell, D. H., 141(153), *179*
Campion, R. J., 316(77), *362*
Canesson, P., 271(18), 273(18), 281(18), 297(18), 300, 302(18), *307, 309*
Cannings, F. R., 131(128), *179*
Cannon, W. A., 319(14, 16), 320(14), 321(14), 323(14), 352(14), *362*
Cant, N. W., 141(154), 153, *180,* 187(22, 23), 188, 188(22, 23), 196, *222*
Capozzi, V. F., 29, *57*
Caracciolo, F., 347, 348, *364*
Cardin, D. J., 249(61), *263*
Cardoso, M. A. A., 70, *96*
Carter, J. L., 139(149), 159(149), 160, 161(149), *179*
Carty, T. J., 258(83), *263*
Carver, J. C., 271(71), 282(71), 286(71), 291(71), *308*
Casassa, J. P., 325(40), *363*
Castellan, A., 270(25, 51, 83, 102), 274(25), 278(51), 279(25), 283(51), 285(83), 288(51, 83), 289(83), 295(102), *307, 308, 309*
Causs, K., 232(22), *262*
Centola, P., 195(54, 55), 213(54, 55), *223*
Cesari, M., 200(75), 205, *224*

Cetinkaya, B., 249(61), *263*
Chandler, J. M., 319(15), *362*
Chang, G., 227(11), *261*
Chaussidon, J., 172(238), *182*
Che, M., 270(36), 277(36), 283(36), *307*
Chen, H. V., 247(57), *262*
Chen, N. Y., 138(144), *179*
Chen, T., 200(74), 201(74), *224*
Chernyshev, K. S., 200(71), 201(71), 202(71), *224*
Chinsoli, G. P., 261(96), *263*
Chiplunker, P., 271(56), 279(56), 288(56), 289(56), 297(56), 303(56), *308*
Chloupek, F. J., 335(63), 347(63), *363*
Christie, J. R., 192(45), 193, 193(45), *223*
Christner, L. A., 161, 165, *181*
Christner, L. G., 139(146), 140(146), 141(146), 142(146), 143(146), 152(146), *179*
Chu, P., 143, 143(159), *180*
Chuong, T., 132(130), *179*
Ciapetta, F. G., 123(83), 126(83), *178*
Cimino, A., 270(53, 62), 279(53), 281(62), *308*
Clark, A., 108, 108(34), 109(33, 34), 124(100), 125(108), *177, 178,* 269(28), 275(28), 284(28), *307*
Clarkson, R. B., 278(45), *307*
Clement, J. C., 304(112), *309*
Collman, J. P., 229(13), *261*
Colson, J., 276(35), *307*
Condon, F. E., 98(5), 99(5), *176*
Coolegem, J. G. F., 270(13), 272(13), *307*
Cooper, B. J., 322, 322(22), 324(22), 339(22), 342(22, 31, 32), 346(22, 85), 356(22), *362, 364*
Cody, I. A., 122, *178*
Copelowicz, I., 64(14), *95*
Cossee, P., 326(29), 237(31), *262,* 300(108, 109), *309*
Cotton, F. A., 242(47), 243(47), 244(49), 245(49), *262*
Cousins, M., 235(28), *262*
Covini, R., 125(109), *178*
Cowley, S. W., 296(87a), *309*
Cramer, R., 261(89), *263*
Croat, J. J., 350(91), 358(91, 103), *364*
Crowfoot-Hodgkin, D., 256(78), *263*
Csicsery, S. M., 143, 144(158), 145(158), *180*
Cutler, A., 241(41), 243(41), *262*

D

Dahler, J. S., 51, *57*
Dahlmans, J. G. J., 298(107), *309*
Dalin, M. A., 203, *224*
Dalla-Betta, R. A., 317(10), 322(10, 30), 324(10), 325(10), 326(10), 335, 335(61), 336(64), 336, 341(10), 342(10), 343(30), 354(30), *362, 363*
Danforth, J. D., 114, 115(48), *177*
Daniel, C., 189, 190(28, 32), 207, *223, 225*
Datka, J., 153, *180*
Davidtz, J. C., 173, *182*
Davies, W., 69, 70, *96*
Davis, C. F., Jr., 43(26), *57*
Davis, D. L., 319(17), *362*
De Angelis, B. A., 270(62), 281(62), *308*
De Beer, V. H. J., 267, 270(100, 144), 271(17, 58, 92), 273(17), 279(58), 286, 286(92), 287, 287(92), 288, 291(17), 294, 297, 297(17), 298(107), 299, 300, 301, 303(7, 17, 100), 304, 305, 305(100), *306, 307, 308, 309*
De Boer, N. K., 187, 195(19), 195, *222*
De Clerek, F. D., 132(131), *179*
De Clerck-Grimee, R. I., 270(74), 271(18), 273(18), 281(18), 282(74), 297(18), 302(18), *307, 308*
Delafosse, D., 137(138), *179*
Delmon, B., 267, 269(42), 271(18), 273(18), 275(32), 276(35a), 277(42), 281(18), 294, 297, 297(18), 301, 302(18), 303(9), *306, 307*
De Losh, R. G., 345(83), *364*
Delvaux, G., 271(18), 273(18), 281(18), 297(18), 302(18), *307*
Dempsey, E., 152, 153, 156, 156(202), 163(202), *180, 181*
Den Herder, M. J., 269(30), 275(30), *307*
Deno, N. C., 104(29, 30), 105, *177*
Derbentsev, Yu. I., 189(30), 190(30), *223*
Derleth, H., 15(26), *22*
Derrien, M. L., 123(84), *178,* 286(88), 298(88), 304(88), *309*
Desikan, P., 286(87), 297, *309*
Desmond, E. A., 316(8), 351(8), 352(8), *362*
Dewig, J., 220, *225*
Deyrup, A. J., 100(13), *176*
Diamond, H., 26, *56*
Ditmarch, R., 123(90), *178*

Dobrzhanskii, G. F., 200(71), 202(71), *224*
Dolejsek, Z., 194(51), *223*
Dolgov, V. Ya., 206(100), 218(137), *224, 225*
Dollimore, D., 270(24, 82), 274(24), 275(24), 285, 285(82), *307, 308*
Dolphin, D., 258(83, 84, 87), 260(87), *263*
Donaldson, G. R., 123(88), 129(88), *178*
Doraiswamy, L. K., 339, *364*
Doyle, M. J., 249(61), *263*
Dozono, T., 214(124), *225*
Drushel, H. V., 102(21), 103(21), 104(21), 105, 106(21), 118(21), 171, *176*
Ducros, P., 172(245), *182*
Du Faux, M., 221(142), *225,* 270(36, 123), 277(36), 283, *307, 310*
Dumont, L. F., 323(33), 351(33), *362*
Dunlevey, F. M., 341(77), *364*
Dupont, M., 172(245), *182*
Dwyer, F. G., 313, 318(5), 339, 340, *362, 364*
Dzisko, V. A., 102(20), *176*

E

Eberly, P. E. Jr., 141, 143(151), 145(151), 151, 153(170), 159, 159(151), 160, 160(151), 161, 169, *179, 180, 181*
Echigoya, E., 195, 214, 196(61), 220, 221(58), *223*
Eckert, E., 75, 76, *96*
Edmonds, T., 271(73), 282(73), 285(73), 286(73), 288, *308*
Eischens, R. P., 41(25), *57,* 109(35), 110, 110(37), *177*
Eisentraut, K. J., 316(9), *362*
Eley, D. D., 25(3, 4), 25, 26(8), 35, 36, 37, 37(19), 49, 50, 53, *56, 57*
Ellis, M., 43(26), *57*
Emmett, P. H., 115(49), *177*
Engelen, J. F., 270(114), 304(114), *309*
Erickson, H., 335(62), 347(62), 356(62), *363*
Erman, L. Ya., 200, 200(72), 201, 202, *224*
Evans, W. L., 105(30), *177*
Eyles, M. K., 167(230), *181*
Eyring, H., 113(45), *177*

F

Faggan, J. E., 316(8), 351, 352(8), *362*
Farkas, L., 25, *56*
Farragher, A. L., 300(108), *309*

AUTHOR INDEX

Farrauto, R. J., 326(44), 349(44), 349, *363*
Fattore, V., 125(109), *178*, 192(38), 202, 204, *223*
Ferroni, E., 215(128, 129), *225*
Fieguth, P., 65(15, 16), 81(53), *95, 96*
Figoli, N. S., 121(75), 133(132), *178, 179*
Finch, J. N., 124(100), *178*
Fink, P., 123(94), *178*
Firsova, A. A., 217, 217(134, 136), 218, *225*
Fischer, E. O., 227(3, 4), 231(3, 4), *261*
Fishel, N. A., 325(39), 349, *363*
Fitzgerald, W., 347(88), 347, *364*
Flockhart, B. D., 99, 121(72), *176, 178*
Foitzik, J., 4(13), 7(13), *22*
Folkesson, B., 270(64), 281(64), *308*
Foreman, R. W., 184(3), *222*
Forni, F., 98(3), 99(3), 100(3), 107(3), 113(3), *176*
Forrest, H., 25(4), 35(4), 36(4), 37(4), 49(4), 50(4, 31), 53(31), *56, 57*
Forzatti, P., 198(65), *224*
Fote, A. A., 303(127), *310*
Francis, J., 242(44, 46), 243(44, 46), 244(46, 48), *262*
Francis, J. N., 242(47), 243(47), *262*
Frank-Kamenetskii, D. A., 70, 71, *96*, 338(69), *364*
Frankel, E. N., 254(73, 74), *263*
Fransen, T., 271(76), 283, 286(85), 288, 292, *308, 309*
Frenkel, M., 171, 173, *181*
Frenz, B. A., 242(47), 243(47), *262*
Freude, D., 154, 161(216), 162(216), *180, 181*
Friedman, B. S., 164(223), *181*
Friedman, R. M., 270(74), 282, *308*
Fries, R. W., 240(39), *262*
Fripiat, J. J., 132(131), 152(172), 154(187, 188, 189), 154, 172(238), *179, 180, 182*, 270(74), 282(74), *308*
Fuhrman, Z. A., 192(38), 202(38), 204(38), *223*
Furst, H., 335(56), *363*
Furusawa, T., 65, *96*

G

Gagliardi, J. C., 346, 349(84), *364*
Gajardo, P., 271(18), 273(18), 281(18), 297, 297(18), 302(18), *307*
Gallagher, P. K., 358(105), *365*

Gallais, E., 51, *57*
Galliasso, R., 270(122), *310*
Gallopoulos, N. E., 325(41), 347(41), *363*
Gal'Perin, E. L., 200(71, 72), 201(71, 72, 77, 78), 202(71, 72, 77, 84), *224*
Galway, A., 270(82), 285(82), *308*
Gambell, J. W., 326(48), 350(48), *363*
Gamid-Zade, E. G., 211(116), *225*
Gamidov, R. S., 200(70), *224*
Gandhi, H. S., 325(50), 326(50), 330, 341(80), 345, 358(50), *363, 364*
Gates, B. C., 267, 298, *306*
Gati, G., 124, *178*
Gautherin, J., 276(35), *307*
Gavalas, G. R., 121(71), *178*
Gay, I. D., 121(75b), 136, *178*
Gerberich, H. R., 125(111), *178*
Germain, A. H., 93, *96*
Gertisser, B., 184(4), 204(4, 87), 205(91), *222, 224*
Ghorbel, A., 124, *178*
Giacomazzi, R. A., 322(28), 325(28), 345(28), 349(28), *362*
Giannetti, J. P., 123(87), 127(87), *178*
Giordano, N., 125(109), *178*, 270(25, 51, 83, 102), 274(25), 278(51), 279(25), 283, 285(83), 288, 288(51), 289(83), 295, *307, 308, 309*
Glass, C. A., 254(74), *263*
Glasstone, S., 113(45), *177*
Goble, A. G., 123(86), 127(86), 128, 128(86), 129(86), *178*
Godin, G. W., 186(17), 187(21), 189(17), *222*
Gogarin, S. G., 189(31), 190(31), *223*
Gold, V., 104(28), *176*
Gol'danskii, V. I., 217(135), 218(135), *225*
Goldstein, M. S., 98(1), 99(1), 100(1), 107(1), 113(1), 116, 117(53), 120(53), 148, 152, *176, 177*
Golebiewski, A., 221(140), *225*
Golodets, G. I., 195(52), *223*
Golovin, A. V., 51, *57*
Good, G. M., 120(56), *177*
Gorokhovatskii, Ya. G., 199, *224*
Gorshkov, A. P., 189(29, 30), 190(29, 30), *223*
Gossner, K., 14(25), *22*
Gough, G., 186(17), 187(17), *222*
Grange, P., 271(18), 273(18), 276(35a), 281(18), 297(18), 300(109), 302(18), *307, 309*

Granquist, W. T., 174(247), *182*
Grasselli, R. K., 184(7), 187(20), 192(20), 204(20, 88, 89), *222, 224*
Gray, P. F., 67(17), *95*
Green, M. L. H., 233(25), 234(27, 28), 235(27, 28), 241(40), 244(40), *262*
Greensfelder, B. S., 120(56), *177*
Gribov, I. M., 189(29), 190(29), *223*
Griffith, R. H., 276(34), *307*
Grimblot, J., 270(124), 271(65, 70), 281(65), 282, *308, 310*
Gruber, H. L., 335(57), *363*
Gryder, J. W., 187(26), 197(26), *223*
Grzybowska, B., 210(113), 211, 214(123), 218, 221(140), *225*
Gupta, S. K., 254(69), *263*

H

Haag, W. O., 125, 125(105, 106), *178, 262*
Haber, J., 196(59), 210, 211, 214(123), 218(138), 221(59, 140), *223, 225*, 270(31), 275(31), *307*
Habersbernia, K., 194(51), *223*
Habgood, H. W., 159, 159(207), *181*
Hadley, D. T., 184(5), *222*
Haensel, V., 123(88), 129(88), *178*
Hafner, W., 227(3, 4), 231(3, 4), *261*
Haines, R. J., 246(55), 248(55), 249(55), *262*
Hair, M. L., 110(39), 120(63), 121(63), 122(59), *177*
Hájková, M., 65(23), 68(23), *96*
Halgeri, A. B., 130, 138, *179*
Hall, W., 25, 26(5), 30, 37(5), *56*
Hall, W. K., 29, 30, 54, 54(13), 57, 125(111), 139(146), 140(146), 141, 141(146), 142(146), 143(146), 152, 152(146), 153, 161(215, 217), 165(217), *178, 179, 180, 181*, 187(22, 23), 188, 188(22, 23), 196, 222, 270(20), 271(29, 84), 274(20), 275(20, 29), 278(45), 280, 286, 292, 295, 295(104), 305(104), *307, 308, 309*, 335(59, 60), *363*
Haller, G. L., 122(79), *178*, 292(99), *309*
Halpern, J., 261(94), *265*
Halvorson, D., 205(94), *224*
Hamilton, L. A., 143(160), 144(160), 161(219), 162(219), 163(219), *180, 181*
Hammerle, R. H., 325(36), 326(36), *363*
Hammett, L. P., 100(13, 14), *176*

Hancock, M., 228(12), 229(12), 230(12, 14), 231(15), 234(14), 235(14), 236(14), 239(14), 244(14), 245(14), *261*
Hanika, J., 93, *96*
Hansford, R. C., 111(43), 113, 114(43), *177*
Hantzsh, A., 104(27), *176*
Hardman, A. F., 184(7), *222*
Hardt, P., 250(62), 251(62), *263*
Harlan, J., 316(7), *362*
Harmon, R. E., 254(69), *263*
Hart, D. W., 254(75), 255(75), *263*
Harrison, L. G., 51, *57*
Harter, R. D., 172(240), *182*
Hattori, H., 138(143), 165, 175, *179, 181, 182*
Hauffe, K., 4(6), 5, *22*
Haulicek, M., 65(22), 68(22), *95*
Hawes, B. W. V., 104(28), *176*
Hearne, G. W., 184, *222*
Heck, R. F., 239(37), 240(37), 261(93, 95), *262, 263*
Hegedus, L. L., 339, *364*
Heimbach, W., 250(62), 251(62), *263*
Heinemann, H., 98(8), 99(8), *176*
Hendra, P. J., 121(60), *177*
Henry, P. M., 261(90), *263*
Hercules, D. M., 271(71), 282(71), 286(71), 291(71), *308*
Hertl, W., 120(63), 121(63), *177*
Hetrick, S. S., 322(23), 325(23), *362*
Heyde, M. E., 345(83), *364*
Heylen, C. P., 111(44), 118, 120(44), 149, *177*
Hicks, G. S., 60, 64, *95*
Hickson, D. A., 143, 144(158), 145(158), *180*
Hightower, J. W., 124(103), 175(248), *178, 182*
Hill, H. A. O., 256(80), *263*
Hillar, S. A., 133(132), *179*
Hillery, H. F., 167(229), *181*
Hillis, J., 242(45, 46), 243(45, 46), 244(45, 46), *262*
Hills, F. J., 322(23), 325(23), *362*
Hilsch, P., 12, *22*
Hinckley, C. C., 303(126), *310*
Hines, L. F., 240(39), *262*
Hirschler, A. E., 104(24), 104, 105, 106(24), 126(24), *176*
Hlaváček, V., 60(6, 7), 61(6, 7, 9, 10, 11, 12, 13), 62(9, 10, 11, 12, 13), 64, 64(9, 10, 11, 12), 68(9, 10, 11, 13), 74(40), 75(43), 76(43), 77(47, 48, 49, 50), 80, 80(47, 48,

49, 50), 81, 81(47, 48, 49, 50), 85(47, 48, 49, 50, 54, 55, 56), 89, 89(57), *95, 96*
Hoang-Van, C., 124(102), *178*
Hobson, M. C., Jr., 102(19), 102, 103, *176*
Hochhauser, A. M., 316(7), *362*
Hockey, J. A., 192(37, 42), 193(42), 194(42, 49), 215(125), 215(125), 216(131), *223, 225*
Hofmann, H., 77(47, 50), 80(47, 50), 81(47, 50), *96*
Hoffmann, R., 248(59), *262*
Holden, J., 254(75), 255(75), *263*
Holm, V., 269(28), 275(28), 284(28), *307*
Holm, V. C. F., 108(33, 34), 109(33, 34), 125(108), *177, 178*
Holt, E. L., 322(24, 25), 326(24, 25), 330(24, 25), 341(24, 25), 349(24, 25), 351(24, 25), *362*
Homfeld, M. F., 322(28), 325(28), 345(28), 349(28), *362*
Honmaru, S., 184(8), *222*
Horák, J., 65, 65(21, 22, 23), 68, 68(23), 75(45), 76, *95, 96*
Hori, R., 360(112), *365*
Hopkins, P. D., 144(161), 164, 166, *180*
Hoser, H., 213(121), 214(121), *225*
Hoshi Thong, 116(52), 117(52), 148(52), *177*
Houalla, M., 295(104), 305(104), *309*
Houdry, E. J., 359, *365*
Houtman, J. P. W., 124(99), *178*
Howitt, J. S., 330, *363*
Hoyer, R. D., 334(55), *363*
Hucknall, D. J., 184(13), *222*
Hudman, C. E., 231(16, 17), 232(16, 17), *261*
Hughes, T. R., 111(41), 111, 112(41), 126, 141, 141(41), 143, 143(41), 145(41), 153, 153(41), 175(41), *177, 178*
Hugo, P., 72, 76, 77, *96*
Hunter, J. E., 326(43), 349, *363*
Hunter, F. D., 155(195), 158(195), *180*
Huo-Shih T'Huoang, 158, *181*
Huo-Shih Thoalt, 147(167), *180*

I

Ichihara, S., 331(52), *363*
Ichihashi, H., 184(8), *222*
Idol, J. D., Jr., 184, *222*
Ignat'eva, L. A., 159, 160, *181*
Ikeda, M., 133(134), *179*
Ikemoto, M., 145, 146(163), *180*

Ilisca, E., 51, *57*
Imai, J., 120(59), 121(59), 122(59), *177*
Imanaka, T., 271(74a), 281(66), 282(74a), 297(74a), 303(74a), *308*
Imokawa, T., 133(134), *179*
Imshennik, V. K., 217(136), 218(136), *225*
Ingraham, L. L., 258(88), 260(88), *263*
Inoyatov, N. S., 304(115), *309*
Irish, D. C., 341(78), *364*
Isaev, O. K., 199, *224*
Isgulyants, G. V., 189(30), 190(30), *223*
Ito, M., 138(142), *179*
Itoh, M., 131(125), 138(143), *179*
Ivanova, T. M., 147(166), *180*
Iwasawa, Y., 121(73, 73a), 135(73, 73a), *178*

J

Jackson, H. R., 350, *364*
Jacobs, P., 152(174), 153(174), *180*
Jacobs, P. A., 111(44), 118, 120(44), 149, 149(169), 150(169), 155(199, 200), 155, 158, *177, 180, 181*
Jaczewska, A., 122(81), *178*
Jakubith, M., 72(38), 76, *96*
James, D. E., 240(38, 39), *262*
Janas, J., 270(31), 275(31), *307*
Janowski, J. D., 324(35), 331(35), 333(35), 334(35), 335(35), 340(35), *363*
Jardine, F. H., 245(50, 51), *262*
Jeitschko, W., 209, *225*
Jelinek, J., 74(40), *96*
Jelli, A. N., 152(172), *180*
Jennings, T. J., 185, 185(14, 15), 187, *222*
Jiráček, F., 65, 68(21, 22, 23), 75(45), 76, *95, 96*
Jiru, P., 194, 195(55), 213(55), 215(127), *223, 225*
Johanneson, R. B., 101(15), *176*
John, G. S., 269(30), 275(30), *307*
Johnson, D. W. Jr., 358(105), *365*
Johnson, M. F. L., 335(62, 63), 347(62, 63), 356, *363*
Johnson, O., 104, *176*
Judy, W. A., 247(58), *262*
Justi, E., 47, *57*

K

Kabe, T., 270(16, 106), 273(16), 274(16), 275(16, 26), 286(16), 296(106), *307, 309*
Kaduskin, A. A., 197(64), *224*

AUTHOR INDEX

Kaerlein, C., 125(107), *178*
Kaesz, H. D., 261(98), *263*
Kagan, J. B., 93, *96*
Kageyama, Y., 121(65), *177*
Kaiser, A., 47, *57*
Kalckar, F., 51, *57*
Kamtner, T. R., 109(36), 112(40), *177*
Kanner, B., 101(16), *176*
Karakchiev, L. G., 102(20), *176*
Karmilovich, T., 341(79), *364*
Kasai, P. H., 161(214), 163(214), *181*
Kato, F., 145(164), 146(164), 147(164), *180*
Katz, S., 350(91), 358(91, 103), *364*
Katzman, H., 358, *364*
Kayle, E. L., 341(79), *364*
Kayumov, A. A., 304(115), *309*
Kazanskii, V. B., 196, 197, 197(60), *223*
Kealy, T. J., 227(2), *261*
Kearby, K. K., 130(120), *179*
Keeling, R., Jr., 269(116), *309*
Keeling, R. C., Jr., 304(111), *309*
Keely, W., 304(110), *309*
Keen, I. M., 167(230), *181*
Keim, M., 250(62), 251(62), *263*
Kelly, C. J., 318(13), *362*
Kennedy, J., 174(247), *182*
Kennedy, R. M., 98(6), 99(6), *176*
Kerl, R. J., 341(76), *364*
Kermarec, M., 137, *179*
Kerr, G. T., 154(192), 155(192, 194, 196), 157, *180*
Keulks, G. W., 189, 190, 190(28), 192(41, 47), 192, 193, 193(41), 194(41), 194, 207, 209(108), *223, 225*
Khakimov, V. B., 304(113), *309*
Khovanskaya, N. N., 217(134), *225*
Kibby, C. L., 271(94), 287, 296, 299(94), *309*
Kiguchi, S., 207(103), *224*
Kim, D. Q., 269(19), 274(19), *307*
Kimberlin, C. N., Jr., 166(228), *181*
Kimura, T., 145(164), 146(164), 147(164), *180*
Kinoshita, H., 360(112), *365*
Kirillov, V. A., 69(28), *96*
Kishiwada, S., 184(8), *222*
Kiso, Y., 251(66), 252(66), 253(66), 254(68), *263*
Kitagawa, J., 136(136), 137(137), *179*
Kiviat, F. E., 127(116), 130, *179,* 270(80), 284, 304, *308*
Kiyoura, T., 136(136), 137, 137(137), *179*

Kladnig, W., 163(221), *181*
Klimisch, R. L., 325(41, 42), 331, 334(54), 340, 340(41, 42), 341(54), 347(41, 42), *363*
Knözinger, H., 123(91), 124, 124(91), 125(107), *178*
Knox, K., 204(89), *224*
Knox, S. D., 255(76, 77), *263*
Kobayashi, A., 5(10), 18(10), *22*
Kobayashi, H., 121(61), *177*
Kokes, R. J., 115(49), *177,* 187(24, 25), 197(25), *223*
Kolchin, I. K., 189(29, 30, 31), 190(29, 30, 31), 200(71), 201(71), 202(71), *223, 224*
Koller, K., 17(27), *22*
Kolovertnov, G. D., 269(117), *310*
Komatsu, W., 5(10), 18(10), *22*
Komorek, J., 214(123), *225*
Kotera, Y., 270(95, 106), 271(97), 288(95, 97), 296, *309*
Kotsarenko, N. S., 102(20), *176*
Koubek, J., 121(68, 74), *177, 178*
Kouwenhoven, H. W., 130(119), 167, *179*
Kovba, L. M., 272(12), *307*
Koyano, T., 237(33), 238(35, 36), 249(33, 35, 36), *262*
Krause, B., 341(79), *364*
Kreger, W. E., 335(62), 347(62), 356(62), *363*
Krenzke, L. D., 192(47), 193, *223*
Kritikos, A., 20(32, 33), *22*
Kröner, W., 250(62), 251(62), *263*
Krudel, E. K., 258(83), *263*
Krylov, O. V., 197, 197(62, 63, 64), 206(100), 217(135), 218(135, 137), 221(141), *223, 224, 225,* 270(44, 49, 118, 119, 120), 277, 278 (49), *307, 308, 310*
Kubelkova, L., 195(57), 213(57), 214(57), *223*
Kubiček, M., 62(10, 11, 12, 13), 64(10, 11, 12, 13), 66(10, 11, 12, 13, 14), 68(10, 11, 12, 13), 74(40), *95, 96*
Kuebrich, J. P., 335(63), 347(63), *363*
Kugler, B. L., 187(25, 26), 197(25, 26), 197, *223*
Kuliyev, A. R., 211(116), *225*
Kumada, M., 251(66, 67), 252(66, 67), 253(66, 67), 254(68), *263*
Kummer, J. T., 322(21), 326(46), 329, 331(21), 344, 352(81), *362, 363, 364*
Kuniï, D., 65, *96*
Kuratek, J., 234(26), *262*

AUTHOR INDEX

Kurita, M., 270(95, 106), 288(95), 295(105), 296(106), *309*
Kushnarev, M. Ya., 199(66), *224*

L

Lafitau, H., 304, *309*
Lago, R. M., 113, *177*
Laidler, K. J., 113(45), *177*
Landa, S., 266, 267, 304(1), *306*
Landis, P. S., 143(160), 144(160), 161(219), 162(219), 163(219), *180, 181*
Lappert, M. F., 249(61), *263*
Larson, J. A., 317(10), 322(10), 324(10), 325(10), 326(10), 341(10), 342(10), *362*
Larsson, R., 270(64), 281(64), *308*
Lau, K., 240(39), *262*
Lauder, A., 358(106), *365*
Laukonis, J. V., 350(91), 358(91), *364*
Lawrence, P. A., 123(86), 127(86), 128, 128(86), 129(86), *178*
Lee, C. H., 341(77), *364*
Lee, R. K., 325(39), 349(39), *363*
Leeman, H. E., 149(169), 150(169), 158(205), *180, 181*
Lefebvre, A. G., 93, *96*
Leffler, A. J., 51, *57*
LeFrancois, M., 166, 167, *181*
Leftin, H. P., 102(19), 102, 103, *176*
La Page, J. F., 123(84), *178,* 286(88), 298(88), 304(88), *309*
Leigh, G. J., 246(55), 248(55), 249(55), *262*
Lemcoe, M. M., 172(244), 173, *182*
Lenane, D. L., 316(8), 351(8), 352(8), *362*
Leonard, A. J., 132, *179*
Lester, G. R., 359(109), *365*
Levy, M. N., 227(9), 230(14), 231(15), 234(14), 235(14), 236(14), 237(33), 239(14), 244(14), 245(14), 249(33), *261, 262*
Leyden, D. E., 271(71), 282(71), 286(71), 291(71), *308*
L'Homme, G. A., 93, *96*
Liang, S., 121(75b), 136, *178*
Libby, W. F., 358(102), *364*
Liederman, D., 331(53), *363*
Liengme, B. V., 141, 152(150), 161(217), 165(217), *179, 181*
Liew, K. Y., 121(72), *178*
Linett, J. W., 68(25), *96*
Lippens, B. C., 123, *178,* 200(73), *224*

Lipsch, J. M. J. G., 269(50, 79), 278(50), 283, 284, 285(79), 286(79), 298, *308*
Little, L. H., 110(38), 123(38), 141(154), *177, 180*
Loader, E., 121(60), *177*
Lo Jacono, M., 209, *225,* 270(37, 53, 54), 271(29, 53, 54, 84), 275(29), 277(37), 278(37, 45, 46), 279(53, 54), 280(53, 54), 280, 286, 287(45, 46), 295, 295(107), *304, 307, 308, 309*
Lombardo, E. A., 295(103, 104), 305(104), *309*
Lombarska, D., 270(31), 275(31), *307*
Long, W. P., 232(21), *262*
Low, M. J. D., 122, *178*
Lucchesi, P. J., 139(149), 159(149), 160(149), 161(149), *179*
Lugovskoi, B. I., 69, *96*
Lunsford, J. H., 121(67), 124, 167, *177, 181,* 191, 197(33), *223*
Luss, D., 61, 70, 93, *95, 96*
Lutinski, F. E., 125(111), *178*
Lutsenko, V. V., 272(12), *307*
Lygin, V. I., 163(222), *181*

M

Maatman, R. W., 113, 120(47), *177*
Maccache, C., 270(36), 277(36), 283(36), *307*
Machiels, C. J., 271(17), 273(17), 291(17), 297(17), 303(17), *307*
Madhusudhan, C. P., 27, 28(11), *56*
Magee, J. S., 131(127), 137(127), *179*
Maher, P. K., 154(191), 155(195), 158, *180*
Majewski, W., 122, *178*
Makay, K., 152(174), 153(174), *180*
Makovsky, L. E., 271(77), 284, *308*
Makram, H., 356(98), *364*
Maksimov, Yu. V., 217(135, 136), 218, 218(136), *225*
Maksimouskaya, R. I., 269(117), *310*
Malbois, G., 166, 167, *181*
Malebi, B. W., 322(22), 324(22), 339(22), 342(22), 346(22), 356(22), *362*
Malinowski, S., 122(81), *178*
Mamedov, E. A., 211(116), *225*
Mamedov, Kh. S., 200(70), *224*
Manara, A., 205(92), *224*
Manara, G., 192(38), 200(75), 202(38), 204(38), 205(75), *223, 224*
Mangasaryan, N. A., 203(85), *224*

Mannion, W. A., 334(55), *363*
Marak, E. J., 192(40), *223*
Marczewski, W., 218(138), *225*
Marek, M., 60(6), 62(10), 75(43), 76(43), 80(48), *95, 96*
Margolis, L. Ya., 184(12), 187, 187(27), 189(29, 30, 31), 190(29, 30, 31), 196, 199(66), 206(100), 217(134, 135, 136), 218(135, 136, 137), 221(141), *222, 223, 224, 225*, 270(120), *310*
Marion, F., 269(19), 274(19), *307*
Mapes, J. E., 109(35), 110, *177*
Mars, P., 191, *223*, 270(75), 271(76), 283(75, 76), 286(85), 288, 288(85), 289(75), 292(85), *308, 309*
Marshall, K., 120(64), 121(64), *177*
Martini, G., 120(57), 134(57), *177*, 215(126, 128, 129), *225*
Martinez, N. P., 271(56), 279, 279(56), 288, 289(56), 297, 303(56), *308*
Martinotti, G., 270(83), 285(83), 288(83), 289(83), *308*
Masson, J., 269(38, 42), 277(38, 42), *307*
Massoth, F. E., 127(116), *179*, 267, 270(20, 21, 23), 271(94), 274(20, 21, 23), 275(20, 21, 23), 276, 276(33), 278(33), 282(23), 283, 286(20, 21, 33, 86), 287, 287(33), 289(20, 21), 291(23), 292(20, 21), 293(20, 21), 296, 296(8, 87a), 298, 299(33, 86, 94, 96), 300(96), *307, 309*
Matros, J. S., 69(28), *96*
Matsuura, I., 208(106), 209(106, 110), 211, 212, 213(106, 119, 120), 216, 218, *225*
Matsuura, K., 131, *179*
Matthes, B., 14(24), *22*
Maxted, E. B., 312, *361*
Maynor, H., 304(110), *309*
McArthur, D. P., 322(26, 27), 322, 323(26, 27), 324(26, 27), 324, 327, 329, 331, 350(26, 27, 92), 360, *362, 364*
McCain, C. C., 186, 187(21), 189, 192(45), 193(45), *222, 223*
McClellan, W. R., 205, 208, *224*
McConnell, R. J., 317, 347(11), *362*
McCune, R., 336(65), *363*
McCune, R. C., 335(64), 336(64), *363*
McDaniel, C. V., 154(191), *180*
McDonnell, T. F., 317, 347(11), *362*
McDowell, C. A., 51, *57*
McKee, D., 322(21), 323, 329(21), 331(21), *362*

McKinley, J. B., 267, *306*
McLaren, A. D., 172(241), *182*
Medellin, P., 93, *96*
Medema, J., 124(99), *178*
Mehrotra, R. P., 270(11), 272(11), *307*
Mekhtiev, K. M., 200, 203(85), *224*
Mekhtieva, V. L., 203(85), *224*
Mencik, Z., 318, *362*
Menguy, P., 270(122), *310*
Messenger, J. V., 125(104), *178*
Mestdagh, M., 154(187), *180*
Mestdagh, M. M., 154, *180*
Meyers, A. J., 347(86), *364*
Miale, J. N., 138(144), *179*
Miki, Y., 270(16), 273(16), 274(16), 275(16, 26), 286(16), *307*
Mikovsky, R. J., 269(30), 275(30), *307*
Miller, A. F., 184(7), *222*
Miller, A. W., 270(63), 281(63), *308*
Millikan, T. H., 168(232), 170(232), *181*
Milliron, D. L., 175(248), *182*
Mills, G. A., 98(8), 99(8), 107, 113, 114, 115(32), 168(232), 170(232), *176, 177, 181*
Misano, M., 23(1), 39(22), 41(22), *56, 57*, 118, *177*
Mitchell, P. C. H., 213(122), 214, *225*, 267, 269(10, 48), 270(57, 64, 89), 271(56), 278(48), 279(48, 56, 57), 280(10, 48), 281(64), 283, 283(57), 286(89), 288(56), 289(56, 57), 290(48), 299(56), 303(56), *306, 307, 308, 309*
Miura, H., 192(44), 193(44), *223*
Miyake, N., 254(68), *263*
Mizuno, K., 133, *179*
Moffat, A. J., 192(40), *223*
Moffat, J. B., 137, *179*
Mone, R., 164(224), *181*, 270(15), 271(55), 273(15), 279(55), 284, 285, 294, 300, 304, *307, 308*
Montagna, A. A., 31, *57*
Montreuil, C. H., 336(66), 350(66), 350, 351(66), *363*
Mooi, J., 335(62, 63), 347(62, 63), 356(62), *363*
Mooney, J. J., 334(55), *363*
Morgan, C. R., 339, 340, *364*
Morgan, T. R., 116, 117(53), 120(53), 148, *177*
Morikawa, Y., 192(44), 193(44), *223*
Morimoto, T., 120, 121(59), 122(59), *177*

Morita, Y., 145, 146(164), 147(164). *180*, 207, *224*
Moro-oka, Y., 194, *223*
Morris, R. V., 294, *309*
Morrison, S. R., 217, *225*
Morrow, B. A., 122, *178*
Morterra, C., 122, *178*
Mortland, M. M., 172(238, 242, 243), 172, *182*
Moscou, L., 164(224), *181*, 270(15), 273(15), 284, 304, *307*
Moskovskaya, I. F., 159(208), 160, *181*
Mostecký, J., 89(58), *96*
Muetterties, E. L., 249(60), *262*
Muller, O., 353(96), *364*
Müller, W., 6(11), *22*
Murahashi, S., 241(42), 243(42), *262*
Murakami, Y., 192(39), *223*
Muramatsu, K., 270(95, 106), 288(95), 296(106), *309*
Murphy, A. J., 170(235), 171(235), *181*
Murray, H. H., 170(236), 171(236), *181*
Myers, J. W., 127(115), *179*

N

Naccache, C., 221(142), *225*, 270(123), *310*
Nagao, M., 120(59), 121(59), 122(59), *177*
Nagaska, Y., 184(8), *222*
Nagy, P. L. I., 233(25), *262*
Naka, I., 5(10), 18(10), *22*
Nakamo, H., 271(74a), 281(66), 282(74a), 297(74a), 303(74a), *308*
Nakamura, A., 254(70, 71, 72), *263*
Nakatomi, S., 131, *179*
Naugle, D. G., 12(22), *22*
Neal, A. H., 322(24, 25), 326(24, 25), 330(24, 25), 341(24, 25), 349(24, 25), 351(24, 25), *362*
Nechitailo, A. E., 217(135), 218(135), *225*
Nechtschein, J., 269(38, 42), 277(38, 42), *307*
Neel, E., 304(112), *309*
Neeleman, J. F., 137, *179*
Nestrick, T. J., 261(97), *263*
Neville Jones, D., 255(76, 77), *263*
Newby, W. E., 323(33), 351(33), *362*
Newling, W. B. S., 276(34), *307*
Newman, M. S., 104(29), *177*
Ng, C. F., 39, 45(23), *57*
Nicholson, D. E., 284, *308*
Nielsen, S. N., 51, *57*

Nishikawa, E., 207(103), *224*
Niwa, M., 192(39), *223*
Noller, H., 120(57), 134(57), *177*
Nomizo, Y., 107(31), 119(31), *177*
Notari, B., 192(38), 200(75), 202(38), 204(38), 205(75, 92), *223*, *224*
Notermann, T., 209(108), *225*
Novakova, J., 194, 215, *223*, *225*
Nozaki, F., 115(50), 116(50), 148(50), 163(50), 167, *177*
Nozakura, S., 241(42), 243(42), *262*

O

Oba, M., 270(16), 273(16), 274(16), 275(16, 26), 286(16), *307*
Oba, N., 271(97), 288(97), *309*
Oberkirch, K., 250(62), 251(62), *263*
Oblad, A. G., 98(8), 99(8), 107(32), 113(32), 114(32), 115(32), 125(104), 168(232), 170(232), *176*, *177*, *178*, *181*
Oehme, W., 154(190), 161(216), 162, *180*, *181*
Oelderik, J. M., 127(117), *179*
Ofstead, E. A., 247(58), *262*
Ogawa, K., 270(95, 106), 271(97), 288(95, 97), 296(106), *309*
Ogawa, M., 270(95, 106), 288(95), 296(106), *309*
Ogasawara, S., 121(73, 73a), 126, 135(73, 73a), *178*, *179*
Ohara, T., 331(52), *363*
Okamoto, Y., 271(74a), 281(66), 282, 297, 303(74a), *308*
Okazaki, S., 271(98), 288(98), *309*
Olah, G. A., 98(7), 99(7), 127(118), *176*, *179*
Olodo, P., 271(18), 273(18), 281(18), 297(18), 302(18), *307*
Olson, D. H., 152, 153, *180*
Ondrey, J. A., 211(114), *225*
Onishi, G. E., 319(17), *362*
Ono, T., 331(52), *363*
Ono, Y., 148, *180*
Ooki, I., 5(10), 18(10), *22*
Oppengeim, V. D., 159(208), 160, *181*
Ori, M., 242(44, 46), 243(44, 46), 244(46), *262*
Osborn, J. A., 245(50, 51), *262*
Ostorero, J., 356(98), *364*
Otouma, H., 145, 146(162), *180*
Otsubo, T., 192(44), 193, 193(44), *223*

Otsuka, S., 254(70, 71, 72), *263*
Otto, K., 317(10), 322(10, 30), 324(10), 325(10), 326(10), 336(66), 341(10, 80), 342(10), 343, 343(30), 350(66), 350, 351(66), 351, 354(30), *362, 363, 364*
Owada, Y., 271(98), 288(98), *309*
Owens, P. J., 287, 287(93), *309*
Ozaki, A., 194(50), *223*

P

Padberg, G., 80(52), 80, 81(51), *96*
Padovan, M., 270(102), 295(102), *309*
Pandolfi, L., 358(102), *364*
Parekh, B. S., 271(69), 282(69), 286(69), *308*
Parera, J. M., 121(75), 133, *178, 179*
Pariiskii, G. B., 197(63), *223*
Parker, A. J., 192(37), 215(125), 216(131), *223, 225*
Parkyns, N. D., 123(93), *178*
Parry, E. P., 104(23), 110, 123(23), 124(23), 125(23), *176*
Parry, R. W., 141(153), *179*
Parshall, G. W., 245(53), 246(53), 247(53), *262*
Pasek, J., 121(68, 74), *177, 178*
Pasquon, I., 195(54, 55, 56, 57), 213(54, 55, 56, 57), 214(54, 55, 56, 57), *223*
Patel, C. K. N., 341(76), *364*
Patinken, S. H., 164(223), *181*
Patterson, T. A., 271(71), 282, 286(71), 291(71), *308*
Pauling, L., 136(135), *179*
Pauson, P. L., 227(2), *261*
Peacock, J. M., 192, 215, 215(125), 216, *223, 225*
Pearce, D. R., 25(4), 35(4), 36(4), 37(4), 49(4), 50(4), *56*
Pedersen, L. A., 358(102), *364*
Pendleton, P., 192(46), 193, 193(46), *223*
Perego, G., 200(75), 205(75, 92), *224*
Peri, J. B., 123(92), 124(95, 98), 124, 130, 141(155), 155(197), *178, 179, 180*
Peterson, H. J., 105(30), *177*
Petrakis, L., 130, *179*, 270(33, 39, 40, 80), 276(33), 277(39, 40), 278(33), 284, 286(33), 287(33), 294(39, 40), 299(33), 304, *307, 308*
Petzinger, K. G., 51, *57*
Pfahler, G., 11(19), *22*
Pfeil, W., 12, *22*

Phillips, R. W., 303(127), *310*
Pichat, P., 121(58), 132(129), *177, 179*
Pickert, P. E., 156, 163(202), *181*
Pickett, A. G., 172(244), 173, *182*
Pierron, E. D., 203(86), *224*
Pierson, W. R., 316(6), *362*
Piguzova, L. I., 115(51), 148(51), *177*
Piken, A. G., 322(29), 325(29), 328(29), 337(29), 340(74), 341(80), 343(29), 345(83), *362, 364*
Pines, H., 125(105, 106), 125, *178*
Pink, R. C., 99, 121(72), *176, 178*
Pitkethyly, R. C., 131(128), *179*
Plank, C. J., 161(220), 162, 163, *181*
Pless, L. G., 347(87), *364*
Pliskin, W. A., 110(37), *177*
Plummer, H. K., 336(65), *363*
Pollitzer, E. L., 123(88), 129(88), *178*
Pommery, J., 270(124), *310*
Poore, A. B., 70(34), 74(41, 42), *96*
Popovskii, V. V., 195(53), *223*
Porter, E. A., 187(21), *222*
Portyanskii, A. E., 203(85), *224*
Pott, G. T., 270(13, 14), 272(13, 14), *307*
Praliaud, H., 271(52), 279(52), *308*
Prater, C. C., 113, *177*
Prette, H. J., 191(36), *223*
Prince, R. H., 258(85), 259(85), *263*
Putzar, R., 10(17, 18), *22*
Pyatnitzkiĭ, Yu. I., 195(52), *223*

Q

Quarterman, L. A., 141(152), *179*

R

Rabo, J. A., 156(202), 161(214), 161, 163, *181*
Rader, C. G., 70, *96*
Raghu, S., 241(41), 243(41), *262*
Rajadhyaksha, R. A., 339, *364*
Raman, K. V., 172(243), 172, *182*
Ramaswamy, A. V., 270(11), 271(27, 43), 272(11), 275(27), 277(43), 280(43), 283(27), 286(27), 291(27), 296(27), 299(27), 303(27), *307*
Rampino, L., 102(22), 122(22), 126(22), 127(22), *176*
Rashkin, J. A., 203(86), *224*
Ratnasamy, P., 130, 132(131), *179*, 270(11, 81), 271(27, 43, 67), 272, 275(27),

277(43), 280(43), 282(67), 283, 285, 286(27), 291, 296, 299(27), 303(27), *307, 308*
Ray, N., 271(27), 275(27), 283(27), 286(27), 291(27), 296(27), 299(27), 303(27), *307*
Ray, W. H., 60, 70, 74(42), *95, 96*
Read, J. F., 292(99), *309*
Renny, L. V., 322(32), 342(22), *362*
Reuben, B. G., 68(25), *96*
Richardson, J. T., 269(59, 60), 270(59, 60), 280, 280(59, 60), 303(59, 60), *308*
Rickett, G., 270(24, 82), 274(24), 275(24), 285(82), *307, 308*
Ridgewill, G. L., 120(64), 121(64), *177*
Rieck, G. D., 200(76), 201, *224*
Ripperger, W., 123(85), *178*, 271(91), 286(91), 304, *309*
Rizayev, R. G., 211(116), *225*
Rochester, C. H., 120(64), 121(64), *177*
Rogers, D. B., 205(93, 94, 95, 96, 97, 98), 206(95, 96, 97, 98), *224*
Roiter, V. A., 195(52), 195, *223*
Romanovskii, B. V., 147(166), *180*
Romanovskii, V. B., 116, 117(52), 148, *177*
Rosenblum, M., 227(5, 8), 241(41), 243(41), 261(100), *261, 262, 263*
Rossington, D. R., 29, *57*
Rosynek, M. P., 121(67), 124(67), 124, *177, 178*, 190(32), *223*
Roth, J. F., 326(47, 48), 350(47, 48), *363*
Rothschild, W. G., 359(111), 360, *365*
Rouxhet, P. G., 121(62), 135, 154(187), *177, 180*
Roy, R., 353(96), *364*
Rubaxlik, M. Ya., 199(68), *224*
Rudham, R., 25(3, 4), 35(4), 36(4), 37(4, 19), 49(4), 50(4, 31), 53(31), *56, 57*
Russell, J. D., 172(239), *182*
Ryland, L. B., 100(11), 102(22), 122(22), 126(22), 127(22), 131, *176*, 207(102), *224*
Rymer, G. T., 304(111), *309*

S

Sachsse, H., 25, *56*
Sachtler, W. M. H., 184(11), 187, 191, 195(18, 19), 195, *222*
Saillant, R. B., 261(98), *263*
Saito, K., 331(52), *363*
Saito, Y., 118(55), *177*
Salageanu, I., 270(125), *310*

Samigov, K. A., 304(113, 115), *309*
Sampson, R. J., 184(9), *222*
Samuel, P., 271(85a), 286(85a), *309*
Sancier, K. M., 192(43), 193(43), 193, 196, 214, 216, *223, 225*
Sandler, Y. L., 31, 54(14), *57*
Sato, M., 138(142), *179*
Sato, T., 121(69), 135(69), *177*, 270(95, 106), 288(95), 296(106), *309*
Satterfield, C. N., 338(68), *364*
Saum, W., 123(85), *178*, 271(91), 286(91), 304, *309*
Sazonov, V. A., 195(53), *223*
Scalapino, D. J., 51, *57*
Scarle, R. D., 213(121), 214(121), *225*
Schaffer, P. C., 139, 140, 143(147), 159(147), 160, 165, *179*
Scherzer, J., 155, 155(195, 198), 158(195), *180, 181*
Schiavello, M., 271(54), 279(54), 280(54), 304, *308*
Schieber, M. M., 38, *57*
Schlaffer, W. G., 102(22), 122(22), 126(22), 127(22), *176*
Schlatter, J. C., 325(42), 331(42), 340(42), *363*
Schmiedel, H., 154(190), 161(216), 162(216), *180, 181*
Schmitz, R. A., 60, 61, 74, *95*
Schomaker, V., 156(202), 161(214), 163(202, 214), *181*
Schoonheydt, R., 152(174), 153(174), *180*
Schoonheydt, R. A., 153, *180*
Schrauzer, G. N., 256(79), *263*
Schreiber, L. B., 121(66), *177*
Schrey, F., 358(105), *365*
Schuit, G. C. A., 191(36), 200(73), 208(106), 209(106), 211, 212, 213(106), *223, 224, 225*, 267, 269(50, 79), 270(37, 59, 100, 114), 271(17), 273(17), 277(37), 278(37, 46, 50), 279(53), 283, 284, 285(79), 286(79), 286, 287, 287(46), 291(117), 294(100), 297(17), 297, 298, 299, 300, 301, 303(7, 17, 100), 304(114), 305, 305(100), *306, 307, 308, 309*
Schultes, H., 3(5), *22*
Schultze, D., 6(11), *22*
Schuman, S. C., 267, *306*
Schwab, G.-M., 2(1, 3, 4, 5, 8), 3(1, 3, 4, 5, 8), 4(13, 14, 15, 16, 17, 18), 7(13, 14, 15, 16, 17, 18), 8(13, 14, 15, 16, 17, 18), 10(13,

14, 15, 16, 17, 18), 13(23, 24, 25, 26, 27).
  18(29, 30, 31, 32, 33), 20(29, 30, 31, 32,
  33), *22, 47, 57*
Schwarz, J. A., 121(75a), 136, *178*
Schwartz, J., 254(75), 255(75), *263*
Scott, K. W., 247(57, 58), *262*
Seaman, G. V. F., 172(241), *182*
Searles, S., 141(152), *179*
Sebalsky, R. T., 123(87), 127(87), *178*
Segal, E., 270(125), *310*
Sekella, T. C., 330, *363*
Selke, E., 254(74), *263*
Selwood, P. W., 23(1, 2), 24(1, 2), 26(1, 2,
  7), 27(10, 11), 27, 28(10, 11), 29, 31,
  33(1, 2, 16, 17), 34(16, 17), 35(16, 17),
  37, 37(16, 17), 39(21, 22, 23, 24), 39,
  40(21, 22, 23, 24), 41(21, 22, 23, 24),
  43(7, 21, 22, 23, 24, 25, 26, 27), 44(21,
  22, 23, 24), 46, 46(25, 26, 27), 47(30),
  48(25, 26, 27), 50(16, 17), 51(16, 17),
  53(16, 17, 21, 22, 23, 24), 54(16, 17), *56,
  57*
Sempels, R. E., 121(62), 135, *177*
Serebryakov, B. R., 203(85), *224*
Seshadri, K. S., 270(33, 39, 40), 276(33),
  277(39, 40), 278(33), 286(33), 287(33),
  294(39, 40), 299(33), *307*
Seyler, J. K., 234(26), *262*
Shalit, H., 267, *306*
Sharma, D. K., 130(123), *179*, 270(81),
  271(27), 275(27), 283(27), 285(81),
  286(27), 291(27), 296(27), 299(27),
  303(27), *307, 308*
Sharma, L. D., 130(123), *179*, 270(81),
  285(81), *308*
Sharp, M. J., 215(125), 215(125), *225*
Shchekochikhin, Y. M., 304(113), *309*
Shelef, M., 317, 322(10), 324(10), 325(10,
  50), 326(10, 50), 330, 341(10, 80), 342,
  344(82), 345(83), 350(82), 354(97),
  355(97), 358(50), *362, 363, 364*
Shepard, J. W., 294(101), *309*
Sherry, H. S., 152(177), *180*
Shiba, T., 165, *181*
Shibata, K., 136(136), 137, *179*
Shiller, J. W., 322(29), 325(29), 328(29),
  337(29), 343(29), *362*
Shimada, Y., 121(70), 135(70), *177*
Shimizu, S., 184(8), *222*
Shimokawa, T., 271(74a), 282(74a),
  297(74a), 303(74a), *308*
Shimomura, K., 271(97), 288(97), *309*

Shirasaki, T., 121(70), 135(70), *177,* 184(8),
  192(44), 193(44), *222, 223*
Shoemaker, D. P., 151(171), 152, *180*
Shooter, D., 184(9), *222*
Short, M. A., 318(12, 13), *362*
Shur, V. B., 233(23), *262*
Shutt, E., 322(22), 324(22), 339(22), 342(22),
  346(22), 356(22), *362*
Siegert, R., 13(23), *22*
Silverman, R. B., 258(83, 84, 87), 260(87),
  263
Simpson, H. D., 350(92), *364*
Simpson, J., 120(64), 121(64), *177*
Sinfelt, J. H., 123(82), 126(82), *178*
Sinkule, J., 89(57), *96*
Sivasanker, S., 271(43), 277(43), 280(43),
  307
Slager, T. L., 270(90), 286(90), *309*
Sleight, A. W., 205(93, 94, 95), 206(96, 97,
  98), 209, *224, 225,* 356(99), *364*
Slinko, M. G., 65(20), 67(20), 69(28), *95,
  96*
Smeets, J. G. M., 298(107), *309*
Smith, C. S., 346(84), 349(84), *364*
Smith, G. V., 303(126), *310*
Smith, R. K., 124(97), *178*
Smith, W. D., 124(103), *178*
Snyder, P. W., 331(53), *363*
Sobolev, B. P., 202(84), *224*
Sochacka, M., 221(140), *225*
Sokolovski, V. D., 211(116), *225*
Solomon, D. H., 170, 171(235, 236), *181*
Solymosi, F., 4(7), 11, *22,* 12(20), 18(20), *22,*
  211, *225*
Sommers, A. L., 102(21), 103(21), 104(21),
  105, 106(21), 118(21), 171, *176*
Sonnemans, J., 270(75), 283(75), 288,
  289(75), *308*
Sontag, P., 269(19), 274, *307*
Spearot, J. A., 347, 348, *364*
Spiridonov, K. N., 197(63), *223*
Spozhakina, A. A., 159(208), 160, *181*
Sprys, J. W., 335(64, 65), 336(64, 65), *363*
Stacey, M., 121(60), *177*
Stamires, D., 115(50), 116(50), 148(50),
  163(50), 167, *177*
Stadtman, T. C., 257(82), *263*
Staudte, B., 154(190), *180*
Stefan, R. J., 341(78), *364*
Steinbach, F., 5, 17(28), *22*
Steinberg, K. H., 132(130), *179*
Steinrücke, D., 250(62), 251(62), *263*

Stepien, H. K., 325(50), 326(50), 330, 345(83), 358(50), *363, 364*
Stern, K. H., 326(45), 352(45), *363*
Stevens, G. C., 271(73), 282(73), 285(73), 286(73), 288, *308*
Stevenson, D. P., 186(16), *222*
Stigter, D., 123, *178*
Stille, J. K., 240(38, 39), *262*
Stone, W. E., 154(189), *180*
Stork, W. H. J., 270(13), 272, 273(14), *307*
Stotter, D. A., 258(85), 259(85), *263*
Street, J. C., 323(34), *363*
Stright, P., 114, 115(48), *177*
Su, E. C., 319, 340(75), *362, 364*
Sugioka, M., 133, *179*
Sulak, R. J., 351, *364*
Sultanov, A. S., 304, *309*
Sumifani, K., 251(66), 252(66), 253(66), *263*
Sumiyoshi, T., 136(136), 137(137), *179*
Summers, J. C., 325(41, 42), 331(41, 42), 340(41, 42), 347(41, 42), 351(93), *363, 364*
Suresh, D. D., 187(20), 192(20), 204(20, 89), *222, 224*
Suzdalev, I. P., 217(134, 135, 136), 218(135, 136), *225*
Suzuki, A., 131(125), *179*
Swift, H. E., 175, *182*, 211, *225*
Swift, J. D., 170(235), 171(235), *181*
Swift, P., 270(63), 281(63), *308*
Szabo, J. J., 205(91), *224*
Szobo, Z. G., 12(20), 18(20), *22*

T

Tabock, J., 324(35), 331(35), 333(35), 334(35), 335(35), 340(35), 354(97), 355, *363, 364*
Takahashi, H., 145(163), 146(163), *180*
Takahashi, M., 121(73a), 135, *178*
Take, J., 107(31), 119(31), 121(69), 133(134), 135, *177, 179*
Takezawa, N., 121(61), *177*
Takifa, Y., 194(50), *223*
Talipov, G. S., 304, 304(115), *309*
Tamagawa, M., 145(164), 146(164), 147(164), *180*
Tamele, M. W., 102(22), 122, 126, 127(22), *176*
Tamele, W. W., 100(11), 131(11), *176*
Tamman, G., 69, 70, *96*
Tamres, M., 141(152), *179*
Tanabe, K., 98(2), 99(2), 100(2), 107(2), 113(2), 130(2), 131, 136, 137(137), 138, 138(143), *176, 179,* 271(98), 288(98), *309*
Tanaka, E., 250(62), 251(62), *263*
Tanaka, M., 121(73), 126, 135(73), *178, 179*
Taniguchi, K., 122, *178*
Tarama, K., 122(80), *178*
Taylor, D., 192(45, 46), 193, 193(45, 46), *223*
Taylor, H. S., 26, *56*
Taylor, K. C., 292(99), 325(37), *309, 363*
Teichner, S. J., 124(102), *178*
Teller, E., 51, *57*
Teranishi, S., 271(74a), 281(66), 282(74a), 297(74a), 303(74a), *308*
Theng, B. K. G., 169(234), *181*
Thiele, E. W., 337, *364*
Thyret, H., 241(43), *262*
Tibbetts, G. G., 358(103), *364*
Timoschenko, V. I., 65(20), 67(20), *95*
Tischer, R. L., 316(9), *362*
Tkhoang, K. S., 147(165, 166), *180*
Tobin, H., 115(49), *177*
Todo, N., 270(47, 95, 106, 121), 271(97), 278(47), 288, 288(95, 97), 295, 295(47), 296(106), *308, 309, 310*
Tomao, K., 251(66), 252(66), 253(66), 254(68), *263*
Tomida, M., 121(73), 135, *178*
Tomlinson, J. R., 304(111), *309*
Topchieva, K. V., 115(51), 116(52), 117(52), 147(165, 166, 167), 148(51, 52), 158, 159(208), 160, 163(222), *177, 180, 181*
Toshida, S., 122(80), *178*
Touillaux, R., 154(187), *180*
Trambouze, Y., 152(178), *180*
Trestianu, D., 270(125), *310*
Trifiro, F., 195, 195(54, 55, 56, 57), 198(65), 213, 213(122), 214, 214(121), 215(126), *223, 224, 225,* 270(57), 279(57), 283, 283(57), 289(57), *308*
Trillo, J. N., 26(8), *56*
Trimble, L. E., 341(76), 358(104), *364, 365*
Truex, T. J., 325(36), 326(36), *363*
Trunov, V. K., 272(12), *307*
Tsuji, J., 261(91, 92), *263*
Tsuruya, T., 121(69), 135(69), *177*
Tsutsui, M., 227(1, 6, 7, 9, 10, 11, 12), 228(9, 10, 11, 12), 229(9, 10, 11, 12), 230(9, 10, 11, 12, 14), 231(15, 16, 17), 232(16, 17), 234(14), 235(14), 236(14), 238(34, 35, 36), 239(14, 34, 35, 36), 242(44, 45, 46, 47, 48), 243(44, 45, 46, 47, 48), 244(45,

46, 47, 48), 245(14), 249(32, 33, 34, 35, 36), 250(64, 65), *261, 262, 263*
Tsutsumi, K., 145(163), 146(163), *180*
Tsyganov, A. D., 217(134), *225*
Turkevich, J., 115(50), 116, 116(50), 124(97), 148, 163, 167, *177, 178, 180*
Turner, I. D. M., 121(60), *177*

U

Uchida, K., 360(112), *365*
Ueda, H., 270(47, 121), 278(47), 295, *308, 309, 310*
Ueno, A., 271(97), 288(97), *309*
Ukihashi, H., 145(162), 146(162), *180*
Ungier, L., 218(138), *225*
Uppal, A., 70(34), 74(42), *96*
Uvarov, A. V., 125(110), *178*
Uytterthoeven, J., 172(238), *182*
Uytterhoeven, J. B., 139(146), 139, 140(146), 141, 141(146), 142, 143(146), 149(169), 150(169), 152, 152(146), 153, 155(199, 200), 155, 158, 158(205), *179, 180, 181*

V

Vaghi, A., 270(25, 51, 83, 102), 278(51), 279(25), 283(51), 285(51, 83), 288(83), 289(83), 295(102), *307, 308, 309*
Van Berge, P. C., 286(85), 288(85), 292(85), *309*
Van Cauwelaert, F. H., 25, 26(5), 29, 30, 37(5), 54(13), *56, 57*
Van Der Aalst, M. J. M., 271(17, 58), 273(17), 279(58), 291(17), 297(17), 303(17), *307, 308*
Van den Elzen, A. F., 200(76), 201, *224*
Van der Meer, C., 271(76), 283(76), *308*
Van de Moesdijk, C. G. M., 208(106), 209(106), 213(106), *225*
Van der Steen, G. H. A. M., 270(100), 294(100), 303(100), 305(100), *309*
Vanderveer, R. T., 319(15), *362*
Van Haandel, A. C., 270(114), 304(114), *309*
Van Hooff, J. H. C., 209(111), *225*
Van Krevelen, D. W., 191, *223*
Van Sint Fiet, T. H. M., 270(100, 114), 271(92), 286(92), 287(92), 288(92), 294(100), 303(100), 304(114), 305(100), *309*
Vaughan, R. W., 121(66), *177*
Veatch, F., 184(3), *222*

Venuto, P. B., 143, 144(160), 161(219), 162, 163, *180, 181*
Verbeek, J., 278(46), 287(46), *307*
Verbeek, J. L., 270(37), 277(37), 278(37), *307*
Verhoeven, W., 275(32), *307*
Vernon, L. W., 269(60), 270(60), 280(60), *308*
Vieth, Z., 47, *57*
Villa, P. L., 198(65), *224*
Vincenzini, J. C., 133(132), *179*
Vishnevskaya, L. M., 147(166), *180*
Voge, H. H., 98(4), 99(4), 120(56), *176, 177,* 184(10), 186, 207, *222, 224*
Vogel, E. M., 358(105), *365*
Volf, J., 121(68, 74), *177, 178*
Vol'pin, M. E., 233(23), *262*
Voltz, S. E., 331(53), *363*
Von Destinon-Forstmann, J., 274(22), *307*
Voorhies, A., Jr., 166(228), *181*
Voorhoeve, R. J. H., 341(76), *364*
Vorob'ev, V. N., 304(113), *309*
Votruba, J., 80(48), 80, 81, 85(54, 55, 56), 89(54, 55, 56, 58), *96*
Vouyanko, I. I., 199(68), *224*

W

Waddey, W. E., 326(49), 347(49), *363*
Wakatsuki, Y., 241(42), 243(42), *262*
Wagner, C. D., 186(16), *222*
Waldrop, M. A., 192(40), *223*
Wallace, D. N., 123(83), 126(83), *178*
Walling, C., 100, 101, 118(12), 169, *176*
Walter, D., 250(62), 251(62), *263*
Walton, R. A., 282, *308*
Walvekar, S. P., 130, 138, *179*
Ward, J. W., 111(42, 43), 113, 114(42, 43), 139, 140, 141, 142, 142(148), 143(148), 143, 144(156), 152, 153(183, 184), 153, 154(193), 155(193), 155, 159, 159(156), 209, 210, 211, 212, 213), 160, 160(156), 175, 176), 161, 161(148, 209, 210, 211, 212, 213), 162, 162(209, 210, 211, 212, 213), 163(209, 210, 211, 212, 213), 164, 164(209, 210, 211, 212, 213), 165, 167, *177, 179, 180, 181*
Watanabe, Y., 159(20), *181*
Waters, R. F., 269(30), 275(30), *307*
Waywell, D. R., 294(101), *309*
Weaver, E. E., 319, 320, 322(19, 20, 29),

AUTHOR INDEX

325(29), 328, 337, 341(19, 20), 342(19, 20), 343, 346(84), 349(84), *362, 364*
Wedding, B., 326(44), 349(44), 349, *363*
Weeks, T. J., Jr., 167(229), *181*
Wei, J., 312(3), 313, 318(3), 338, 339, 345, *361, 364*
Weidenbach, G., 335(56), *363*
Weil-Malherbe, H., 101, *176*
Weise, E. L., 326(45), 352(45), *363*
Weiss, J., 101, *176*
Weisser, O., 266, 267, 304(1), *306*
Weisz, P. B., 60, 64, *95*, 138(144), *179*
Weller, S. W., 31, *57*, 70(35), *96*, 271(69), 282(69), 286(69), *308*
Welling, C. E., 319(14, 16), 323(14), 352(14), *362*
Wentreck, P. R., 192(43), 193(43), 196(43), *223*
Werter, P. G. A. J., 271(92), 286(92), 287(92), 288(92), *309*
West, P. B., 122, *178*
Wheatley, T. F., 68(25), *96*
Wheeler, A., 312, 338, *361*
Wheeler, M. A., 324(35), 331(35), 333(35), 334(35), 335(35), 340(35), *363*
White, H. M., 111(41), 111, 112(41), 126(113), 141, 141(41), 143, 143(41), 145(41), 153, 153(41), 175(41), *177, 178*
White, J. L., 152, *180*
White, R. J., 126(113), *178*, 322(32), 342(32), *362*
Whitehurst, D. D., 245(52), *262*
Whiting, M. C., 227(5), *261*
Whitmore, F. C., 98, *176*
Wicke, E., 65(15, 16), 80, 81, *95, 96*
Wigg, E. E., 322(24, 25), 326(24, 26), 330(24, 25), 341(24, 25), 349(24, 25), 351(24, 25), *362*
Wigner, E., 37, 51, *57*
Wilhelm, F. C., 325(39), 249(39), *363*
Wilke, B., 250(62), 251(62), *263*
Wilke, G., 250(63), *263*
Wilkinson, G., 244(49), 245(49, 50, 51), *262*
Williams, F. L., 352, 353, 355, *364*
Wilson, A. J. C., 201(82), *224*
Wilson, G. R., 335(59, 60), *363*
Wilson, J. N., 100(11), 131(11), *176*
Wilson, J. V. D., 347(88), 347, *364*
Wise, H., 192(43), 193(43), 196(43), 199(67), 214(124), 216(132), *223, 224, 225*, 313, 318(4), 318, *362*
Wise, J. J., 143(160), 144(160), *180*

Wolfs, M. W. J., 209, 218, *225*, 270(114), 304(114), *309*
Wolkenshein, Th., 4(6), *22*
Wong, P. K., 240(39), *262*
Wood, B. J., 199, *224*
Woodward, R. B., 227(5), 248(59), *261, 262*
Wragg, R. D., 192, 192(42), 193(42), 194, 194(42), *223*
Wright, A. C., 174, *182*
Wynblatt, P., 357, *364*

Y

Yamadaya, S., 270(16), 273(16), 274(16), 275(16, 26), 286(16), *307*
Yamagata, N., 271(98), 288, *309*
Yamamoto, K., 254(68), *263*
Yamamura, T., 131, *179*
Yang, C. H., 68, *96*
Yao, H. C., 317(10), 322(10, 30), 324(10), 325(10), 326(10), 341(10), 342(10), 343(30), 354(30), *362*
Yao, Y., 322(21), 323, 329(21), 331(21), *362*
Yao, Y.-F. Y., 325, 344, 349, 349(38, 90), 350(90), 352(81), 358(90), *363, 364*
Yarrington, R. M., 319(18), 322(18), *362*
Yates, D. J. C., 139(149), 159(149), 160(149), 161(149), *179*
Yeddanapalli, L. M., 271(85a), 286(85a), *309*
Yolles, R. S., 199(67), *224*, 313, 318(4), 318, *362*
Yoneda, Y., 107(31), 118(55), 119(31), 121(69), 133(134), 135(69), *177, 179*
Yonemura, M., 271(97), 288(97), *309*
Yoshizumi, H., 121(70), 135, *177*
Yotsuyanagi, T., 121(65), *177*
Young, J. F., 245(50, 51), *262*

Z

Zabala, J. M., 271(18), 273(18), 276(35a), 281(18), 297(18), 302(18), *307*
Zambonini, F., 200, *224*
Zavadil, V., 215(127), *225*
Zazzetta, A., 200(75), 205(75, 92), *224*
Zeiss, H. H., 227(1), 230(1), *261*
Zemann, J., 201, *224*
Zettler, H., 18(29), *22*
Zimmerman, H., 250(62), 251(62), *263*
Zingery, L. W., 121(67), 124(67), *177*
Zwaga, A. C., 270(100), 294(100), 303(100), 305(100), *309*

Subject Index

A

Absorption spectroscopy, 101
Acceptor reaction, 2, 19
Acetaldehyde
 combustion, 189, 190
 formation, 189
Acidity, 284, 285
Acid site
 number of
  from reaction rates, 111–114
  strength of, from catalytic titrations, 114–118
Acrolein
 combustion, 189, 190
 formation, 189
 oxidation of propylene, 184
Acrylic acid, formation, 195
Activation energy, 5, 6, *see also* specific catalyst
 of alloys, 7
 changes in, 19
Active site, 210–221
 dual-site concept, 210
 electrical conductivity, 216, 217
 ESCA, 218, 219
 ESR, 214–216
 infrared spectroscopy, 213, 214
 model, 219–221
 Mössbauer spectroscopy, 217, 218
 molybdena catalyst, 304–306
 for propylene adsorption, 210–213
Acylcarbinol indicators, 104–106
Adsorption, 285, 286
 of gaseous bases, 107–109
Aerogel, colloidal, 100
Air pollution control, 88
Alcohol oxidation, 70
Aldehyde, synthesis of, 70
Alkylbenzene, cracking, 134
Alloy, *see* specific substances
Allyl hydroperoxide, 187–189
Allylic intermediate, 185–187

Alumina, 268, 269
 acid strength, 102, 103
 chlorided, 126–130
  surface acidity, methods for determining, 121
 chlorinated, 98
 copper oxide on, 80–85
 doped, 7
 extrinsic field effects, 26, 30, 31
 fluorided, 125, 126
 interaction with lead, 354, 355
 palladium on, 80–85
 platinum on, 80–85, 87
 poisoning, 124
 propylene oxidation, 188, 196
 surface acidity, 123–131
  methods for determining, 121
Aluminosilicates, crystalline, *see* Zeolite
Aluminum alkyl, 236
Aluminum chloride, 250
Aluminum phosphate, 130
Amine
 chemisorption, IR spectra, 111, 112
 poisoning, 114, 115, 117
Ammonia
 adsorption, 108, 137
 chemisorption, IR, 110
 oxidation, 70
  on Pt wire, 71, 73
Anthraquinone, absorption spectra, 103, 104
Arrhenius relation, 111–113
Ash, 317
Atomic absorption, 317
Auger spectroscopy, 318, 352
Automotive catalyst, *see also* Poisoning
 analysis, 317, 318
 atomic absorption, 317, 318
 base metal, 314
 contaminant distribution, 327–334
 contaminant retention, 321–326
 interaction between poisons and, 352–357

SUBJECT INDEX

mass transfer effects, 337–340
monolithic, 313, 314
noble metal, 314
pelleted, 313, 314
poisoning of, 311–361
  effects, 341–352
poison-resistant, 357, 358
pore plugging, 239
rejuvenation, 358–361
thermal deactivation, 334–337
three-way, 314
X-ray fluorescence, 317, 318

B

Back bonding, 228, 229
Band theory, 2
Barium, in oil, 317
Barium molybdate, 205
Base metal, automotive catalyst, 314
Bentonite, 173
Benzene-d_6, exchange reaction, 245–247
Bismuth iron molybdate, 207–209
  X-ray diffraction, 209
Bismuth molybdate, 184–187, 189, 191–194, 196, 199–204
  active site, 210–213
  α phase, 200
  alumina supported, 203, 204
  β phase, 201
  catalyst reduction, 202–204
  electrical conductivity, 216, 217
  ESCA, 218, 219
  ESR, 214–216
  Fermi energy, 217
  γ phase, 201, 202
  IR, 213, 214
  silica supported, 203
  X-ray diffraction, 200–204
Bismuth molybdovanadate, 205
Bismuth phosphomolybdate, 186
Blow-out phenomenon, 82, 84
Bohr magneton number, 37
Boria, acid strength, 102
Boron, in oil, 317
Brillouin zone, 4
Brønsted acid, 99–101, 105, 110, 111, 113, 114, 118, 120, 122, 125, 127, 130, 131, 132, 133, 136, 137, 148, 166, 174, 304
Butadiene, coupling reactions, 250, 251

Butane, oxidation on Pt wire, 70, 71
Butene
  chemisorption, 285
  dehydrogenation, 191
  isomerization, 124

C

Cadmium molybdate, 206
Calcination, 268, 269
Calcium, in oil, 317
Calcium molybdate, 206
Calorimetry, surface acidity, 121
Carbene complexes, 249
Carbon–hydrogen bond activation, 245, 246
Carbonium ion mechanism, 98
Carbon monoxide oxidation, 14, 15, 65, 67–69
  on copper oxide/alumina, 80
  on Pd/alumina, 80
  periodic activity, 87
  on Pt/alumina, 80, 87
  on Pt wire, 71, 73, 75–77
Carborundum (SiC), silver on, 10–12
Carbon oxide, 189
Catalyst, see also specific substance
  automotive, see Automotive catalyst
  effect of carrier on, 3–6
  monolithic, 313, 314, 327, 330
  multicomponent, 209, 210
  pelleted, 313, 314, 327, 330
  porous, 60–69
  reduction, 202–204
  rejuvenation, 358–361
  solid
    acid strength, 102
    surface acidity, 97–176
  supported
    electronics of, 1–21
    inverse, 4, 12–18
      theory, 12, 13
    normal, 6–12
  surface area, see Surface area
  synergetic promotion, 3–6
Catalyst bed, 327–330
Catalyst-gas, 59, 60
Catalyst mixing, 86, 87
Catalytic reaction, see also specific reactions
  hysteresis and periodic activity, 59–94

SUBJECT INDEX

Catalytic reactor, 60, see also specific types
Catalytic titration, 114–118
  gas-liquid chromatography, 115, 116
  pulse reactors, 115, 116
Cerium, in fuel, 316
Cetane, cracking, 114, 115
Chemisorption, 107, 135, 136, 285, 286
  of oxygen, 192
Chromia
  conversion rates, 38–44
  extrinsic field pattern, 45, 49
Chromium
  poisoning, 317
  sandwich compound, 227
Chromium allyls, 250
Chromium chloride, 237
Chromium complex
  hydrogenation, 254
  $\sigma-\pi$ rearrangements, 230, 231
Chromium dioxide, extrinsic field effects, 46
Clay, see also specific types
  color tests, 101
  structure, 168, 169
  surface acidity, 169–174
    methods for determining, 121
  synthetic, 98
Coal liquefaction, 226, 305, 306
Cobalamin, see Vitamin B_{12}
Cobalt, role of, in molybdena catalysts, 302, 303
Cobalt/alumina
  acidity, 284, 285
  ESCA, 281–283
  ESR, 277
  gravimetric–volumetric determination, 274, 275
  reflectance spectroscopy, 279
  X-ray diffraction, 272, 273
Cobalt chloride, 237
Cobalt complex
  hydroformylation, 244, 245
  $\sigma-\pi$ rearrangements, 234
Cobalt molybdate, 266
  support interactions, 290
Cobalt molybdate/alumina
  acidity, 284, 285
  ESCA, 281–283
  ESR, 277
  gravimetric–volumetric determination, 274, 275
  IR spectra, 283, 284
  magnetic measurements, 280
  oxidized state, 289
  Raman spectra, 284
  reduced state, 291
  reflectance spectroscopy 279
  X-ray diffraction, 272, 273
Cobalt monoxide, field effect, 44, 45
Cobalt sulfide, 275, 276, 286, 287, 294
Coking, 295
Combustion, 189, 190
  reaction scheme, 190, 196
Conductivity
  electrical, 20, 21
    active site, 216, 217
  temperature dependence, 20, 21
Contaminant
  distribution
    along catalyst bed, 327–330
    within porous system, 330–334
  retention, 321–326
Continuous stirred-tank reactor, 74–77
Conversion theory, 50, 51
Copper
  activation energy, 7
  foil, 353
  on MgO, 6, 7
  poisoning, 317
Copper–chromite, 344, 352
Copper–magnesium alloy, 6, 7
Copper oxide, 184–187, 199
  on alumina, 80–85
Copper oxide–manganese oxide, 91, 92
Copper oxide–platinum catalyst, 86–88
Cracking, 98, 105, 107, 134, see also specific compounds
Cracking catalysts, 105, 114, 115
Cresyl diphenylphosphate, 345
Cross-coupling reactions, 251–253
Crystal lattice, 2
Cumene
  cracking, 113–115, 122, 125, 127
  poisoning of, 116, 117, 148–151
Cyclohexene hydrogenation, 93, 94

D

Deamination, 259
Desorption, 135
Desulfurization, 305
Deuterium exchange
  benzene-d_6, 245–247
  on Mo/Al, 275

SUBJECT INDEX

Diffusion coefficient, porous catalyst, 61
Diol dehydrase, 260
Diol dehydration, 258, 260
Donor reaction, 2, 19
Doping, 4–6, 10, 11

E

Electronics
  definition, 1
  of supported catalysts, 1–22
  test reactions, 2, 3
Electron microscopy, 318
Electron spectroscopy for chemical analysis (ESCA), 218, 219
Electron spin resonance (ESR), 276–278
  active sites, 214–216
  of oxides, 197
Electron transition, see Electronics
Eley–Rideal mechanism, 8
Emission-control system, 312
Erbia, conversion rates, 33–35
Ethylene
  hydrogenation, 8, 9, 65
  polymerization, 235–238
Ethylene dibromide, 350, 351
Europium monoxide, extrinsic field effects, 46, 47
Exhaust gas, purification of, 79, 88
Exothermic reaction
  parameters of, 63
  temperature oscillations, 65–67
Extrinsic field effect, 26–48
  catalysts for, 49
  theory, 50, 51

F

Faujasite, 139–166
  acidic sites, 151–154
  alkaline and rare earth forms, 160–165
    amine titration, 163
    infrared studies, 160–163
    surface acidity and catalytic activity, 163–165
  aluminum-deficient, 154–159
  Group Ia form, 159, 160
    catalytic activity, 159, 160
    infrared studies, 159
  hydrogen forms, 139–147
    infrared studies, 139–143
  surface acidity
    amine titration, 145–147
    catalytic activity and, 143–145
  NMR, 154
  poisoning, 147–151
  transition metal forms, 165, 166
  ultrastable, 154–159
    infrared studies, 155, 156
    surface acidity and catalytic activity, 156–159
  X-ray diffraction, 152
Fermi energy, 217
Fermi level, 4, 5
Flow reactor for extrinsic field effects, 24
Fluorescent indicator, 104, 106
Formaldehyde
  combustion, 189, 190
  formation, 189
Formic acid dehydrogenation, 6, 7, 9–11
Friedel–Crafts catalyst, 129
Fuel
  potential poisons in, 315, 316
  unleaded, 312

G

Gadolinia, conversion rates, 35
Gauze, catalytic, 69
Gold
  nickel oxide on, 14, 15
  supported, 187
Gold–palladium alloy, 14, 15

H

Halide, organic, in fuel, 315
Hammett acidity function, 101
Hammett indicator, 100–106, 133
  colors of, 101
  for spectrophotometric study of acid strength, 103
  for visual measurement of acid srength, 102
Heat conduction, 89
Heat convection, 89
Heat recirculation system, 93
Heat transfer
  in continuous stirred-tank reactor, 74–77
  in monolith, 89
  in porous catalyst, 60–63, 68
  in tubular reactor, 79, 82, 87

Hein's complex, 230
Heterogeneous reactions, Sh/Nu ratio, 64
Hydrocarbon
  reactions, 98
  isomerization, 98
  oxidation, 341–345
Hydrodesulfuration, 266, 267, 275
Hydrodesulfurization, 292
Hydroformylation, 244, 245
Hydrogen, oxidation, 65–69
Hydrogenation, 305, 306
  large-scale, 79
Hydrogen cyanide, synthesis of, 69, 70
Hydrogen–deuterium self-conversion, 30
Hydroperoxide intermediate, 187–189
Hysteresis, 59–94
  honeycomb structures, 89
  loop, 65–67, 71, 81–83, 93, 94

I

Ignition–extinction phenomenon, 68, 69, 71, 72, 81
Infrared spectroscopy, 283, 284
  active site, 213, 214
  of allylic intermediate on zinc oxide, 187
  determination of surface acidity, 110, 111
  surface acidity, 121
Iron
  activation energy, 10
  on alumina, 9, 10
  poisoning, 317
Iron complex
  olefin oxidation, 240–244
  $\sigma-\pi$ rearrangement, 231, 232, 234, 235
Iron molybdate, ESR, 215
Iron oxide
  activation energy, 16, 17
  on silver, 14–17
Isotope, natural, 30

K

Kaolinite, 102, 170, 171
Kinetics, porous catalyst, 61

L

Langmuir–Hinshelwood expression, 67
Lanthana, extrinsic field effects, 26, 27, 30

Lattice oxygen, 191
  chemical nature of, 195, 196
  role of, 191–195
Lead
  in fuel, 315
  poisoning, 341–345
  retention, 321–324
Lead scavenger, 315, 350, 351
Lead vanadate, 320, 352
Lewis acid, 99, 110, 111, 117, 124–126, 130, 131, 136, 137, 148, 167, 174, 304
Lewis number, 62, 64, 66
Ligand
  field stabilization energy, 229
  π bonding, 228, 229
  σ bonding, 229
  $\sigma-\pi$ rearrangement, 234, 235
Lithium aluminum silicate as support, 317
Lutetia
  activity, 31, 32
  extrinsic field effects, 27–29, 31, 45, 49

M

Magnesia, acid strength, 102
Magnesia–chromia semiconductors, 18
Magnesium
  activation energy, 7
  in oil, 317
Magnesium oxide
  copper on, 6, 7
  silver on, 6, 7
Magnetic field effects, conditions for, 49
Magnetic moment, 280, 281
Magnetocatalytic effect, 23, 26–48
Manganese
  in fuel, 316
  poisoning, 351, 352
Manganese monoxide, field effect, 45, 46
Mass transfer, 337–340
  in continuous stirred-tank reactor, 74–77
  in monolith, 89
  in porous catalyst, 60–63, 68
  in tubular reactor, 79, 82, 87
Mercaptan, 133
Metal, electronic properties, 2, 4, 5
Metal alkyls, 236
Metallic catalyst, 3, see also specific catalyst
Metal oxide
  adsorbed oxygen on, 196–198

SUBJECT INDEX

binary, surface acidity, 136–138
coordination number, 136
surface acidity, methods for determining, 121
Metal–oxygen bond, 195, 196
Metal–semiconductor interface, 19
Methanol
 decomposition, 18
 oxidation, 18
Methylcyclopentadienyl manganese tricarbonyl, 351, 352
α-Methylstyrene hydrogenation, 93
Mica (synthetic)–montmorillonite, 174–176
Mössbauer spectroscopy
 active site, 217, 218
 of molybdates, 217, 218
Molybdate
 active site models, 219–221
 scheelite structure, 205–207
 acidity, 284, 285
 active sites, 304–307
 activity, 294–298
 adsorption, 285, 286
Molybdena
 calcination, 268, 269
 characterization of, 265–306
  techniques, 267, 269–289
 diffuse reflectance spectroscopy, 278–280
 ESCA, 281–283
 ESR, 276–278
 gravimetric determination, 273–276
 IR spectra, 283, 284
 magnetic measurements, 280, 281
 oxidized state, 289–291
 preparation, 268, 269
 pretreatment, 269
 reduced, 291, 292, 294, 295
  active stite, 304, 305
  role of cobalt, 302, 303
  of nickel, 303, 304
 sulfided, 274, 275, 292–298
  active site, 305, 306
  models, 298–302
   contact synergism, 301
   intercalation, 300, 301
 sulfur analysis, 286, 287
 supports, 268
  interactions, 289
 volumetric determination, 273–276
 X-ray diffraction, 272, 273

Molybdena/alumina (Mo/Al), 130
 acidity, 284, 285
 ESCA, 281–283
 ESR, 276, 277
 gravimetric-volumetric determination, 274, 275
 IR spectra, 283, 284
 oxidized state, 289
 reflectance spectroscopy, 279
 sulfided, 299
 surface area, 288
 X-ray diffraction, 272, 273
Molybdenum complex, σ–π rearrangements, 233, 234
Molybdenum hexacarbonyl, 246
Molybdenum oxide, 266, *see also* Molybdena
Molybdenum oxide–magnesium oxide, 197
Molybdenum sulfide, 275, 276, 286, 287, 293
Monolithic catalyst, 88–93
 multiple steady-state phenomenon, 89–93
Montmorillonite, 100, 102, 170–172
Mordenite, surface acidity, 166–168

N

Néel point, 38–46, 49
Neodymia, conversion rates, 37
Nickel
 activation energy, 7–9
 on alumina, 7, 8
 catalyst for hydrogen oxidation, 67
 extrinsic field effects, 45, 47–49
 poisoning, 317
 on ZnO, 8, 9
Nickel bromide, 237, 238
Nickel complex, cross-coupling reaction, 251–253
Nickel–molybdena catalyst, 303, 304
Nickelocene, 238
Nickel oxide
 activation energy, 14, 15
 on gold, 14
 on silver, 13, 14, 19
Nitric oxide, synthesis, 70
Noble metal
 automotive catalyst, 314
 propylene oxidation, 196

Nuclear magnetic resonance, surface acidity, 121
Nusselt number, 63, 64, 70

O

Oil, poisons in, 316, 317
Olefin
  coupling, 235–238
    double, 238, 239
  isomerization, 125
  metathesis, 246–249
    mechanism, 248
  oxidation, 196, 238–244
    oxide catalysts, 184
  polymerization, 98, 133
*Ortho-para*deuterium, 25, 50
Ortho-para hydrogen conversion, 23 ff.
Oxidation
  large-scale, 79, 82
  propylene, 183–222
  redox mechanism, 191
Oxygen
  activation, 191–198
  adsorbed, 191
    role of, 196–198
  isotopic, 192–194
  lattice, *see* Lattice oxygen

P

Palladium
  on alumina, 80–85
  catalyst for carbon monoxide oxidation, 67
  nickel oxide on, 14, 15
Palladium chloride, 237, 238
  steroid preparation, 255
Palladium chloride–copper chloride, 238–240
Paraffin alkylation, 98
*Para*hydrogen conversion rate
  correlations, 48–50
  effect of magnetic field on, 23–56
  experimental results, 25–48
  measurement of, 24, 25
Peclet number, 80, 81

Periodic activity
  catalytic reactions, 59–94
  experimental observation, 64–73, 75–77, 87, 88
  in tubular fixed-bed reacted, 87
Perovskite, 358
Phosphorus
  in fuel, 316
  in oil, 316, 317
  posioning, 345–349
  retention, 324
Photoelectron spectroscopy (ESCA), 281–283
Platinum
  catalyst for carbon monoxide oxidation, 67
  crystallite, 356
  foil, 352, 353
  wire, 70, 71
Platinum–alumina, 80–85, 87, 128, 129
Platinum complex, olefin oxidation, 241, 242
Platinum–copper oxide, 90, 91
Platinum–copper oxide–manganese oxide, 91, 92
Platinum–palladium catalyst, 348
Poison, 314–317, *see also* specific substance
  analysis of, 317, 318
  distribution, 327–334
  examination of, 317, 318
  in fuel, 315, 316
  interaction between active components and, 352–357
  mass transfer effects, 337–340
  in oil, 316, 317
Poisoning, 114–118
  automotive catalysts, 311–361
  early observations of, 318–321
  effects of, 341–352
  surface, 312
  thermal deactivation vs, 334–337
Polyethylene, 235, 236
Praseodymia
  conversion rates, 33, 34
  extrinsic field pattern, 45, 49
Propylene
  activation, 185–190
  adsorption, active site, 210–213
  allylic intermediate, 185–187
  ammoxidation, 191

^{14}C-labeled, 185
combustion, 189, 190
deuterated, 185, 188
oxidation
  active catalyst systems, 198–210
  active sites, 210–221
  hydroperoxide intermediate, 187–189
  reaction paths, 185, 186
    over noble metals, 188
  selective, 183–222
Propylene oxide, formation, 189
Pyridine
  chemisorption, 107
  IR, 110–112
  poisoning, 133, 134

Q

Quinoline
  chemisorption, 107, 108
  poisoning, 116, 117, 148–151

R

Radiation, 89
Raman spectra, 284
  surface acidity, 121
Rare earths
  conversion rates, 32, 33
  extrinsic field effects, 32–38
  lanthana-supported, 32–34, 38, 45, 49
  paramagnetic, 32–38
    Bohr magneton number, 37
  self-supported, 32, 33
  structure, 32
Reaction rate, 111–114
Reactor, 193, *see also* specific types
Redox reaction, 2
Reduction–sulfidation, 291
Reflectance spectroscopy, diffuse, 278–280
Reynolds number, 82–85, 91
Rhenium oxide, 246
Rhodium, supported, 187, 188
Rhodium complex
  benzene $-d_6$ exchange, 245–247
  hydroformylation, 245
  hydrogenation, 254
Ruthenium, supported, 187, 188

S

Scheelites, 205–207
Semiconductor, 2, 4, 5, 12, 13, 20, 21
  in combustion, 196
  doped, 4
  types of, 3, 20
Semiconductor–semiconductor catalyst system, 18
Sherwood number, 63, 64, 70
$\sigma-\pi$ rearrangements in catalysis, 227–261
  activation of C–H bond, 245, 246
  carbene complexes, 249
  classification of, 230–235
  driving force for, 228, 229
  hydroformylation, 244, 245
  olefin
    coupling, 235–238
    metathesis, 246–249
    oxidation, 238–244
  reactions, 235–260
    of ligand, 234, 235
    on metal, 230–234
    stereospecific 249–255
  vitamin B_{12} catalysis, 256–260
Silica
  acid strength, 102
  propylene oxidation, 196, 203
Silica–alumina, 105, 108, 109, 113–115, 118
  ^{13}C NMR, 136
  hydroxyl groups, 135
  IR spectra, 110
  surface acidity, 131–136
    methods for determining, 121
Silica gel
  acid strength, 102
  surface acidity, 120–122
    methods for determining, 121
Silica–magnesia, 137
Silica–titania, 138
Silver
  activation energy, 7, 11, 18
  on carborundum, 10–12
  iron oxide on, 14–17
  on MgO, 6, 7
  nickel oxide on, 13, 14, 19
  pills, 69
  zinc oxide on, 17, 18
Sodium, poisoning, 132, 133, 135

Stabilizer, 314
Steady-state phenomenon, 60
  multiple, experimental observation, 64–73, 75–77, 81–93
    nonporous catalyst, 69–73
    porous catalyst, 61–63
Stereospecific reaction, 249–255
Steroid, preparation, 255
Strontium molybdate, 207
Structural promotion, 3–6
Sulfur
  in fuel, 316
  in oil, 316, 317
  poisoning, 349, 350
  retention, 324–326
Sulfur dioxide oxidation, 16, 17
Surface, solid, acid strength, 100–104
Surface acidity
  determination of, 99–120
    adsorption of gaseous bases, 107–109
    aqueous methods, 99, 100
    infrared spectroscopy, 110, 111
    methods, 121
      nonaqueous indicator 100–107
    model reactions, 111–118
    measurement recommendations, 118–120
  of solid catalysts, 97–176
  titration of, 104–107
Surface area, supported catalyst, 3–6
Synergetic promotion, 3–6

T

Temperature
  isocatalytic, 6
  oscillations, 65–68
Terbia, conversion rates, 33, 34
Thermal conductivity, porous catalyst, 61
Thermal desorption, surface acidity, 121
Thermal diffusion, 71, 72
Thiophene, hydrogenolysis, 295, 296
Tin molybdate, Mössbauer spectroscopy, 217
Tin oxide–antimony oxide, 184, 187
Tin oxide–molybdenum oxide, 194
Titanium allyls, 250
Titanium chloride, 236
Titanium complex, $\sigma-\pi$ rearrangements, 232, 233

Titanium oxide, coprecipitated gels with metal oxides, 137, 138
Transition metal complex
  energy levels, 228
  $\sigma-\pi$ forms, 228, 229
Transition metal halide, 236
Tributylphosphate, 346
Trickle-bed reactor, 93, 94
Tubular fixed bed reactor, 77–88
  catalyst mixing, 86, 87
  effect of catalyst bed dilution, 85, 86
    of inlet CO concentration, 81, 82
    of inlet gas velocity, 82, 84
    of inlet temperature, 81
    of reactor length, 85
  multiple steady state, 77–87
  periodic activity, 87, 88
  steady state in packed bed, 85

U

Ultraviolet spectra, surface acidity, 121
Uranium antimonate, 184, 187, 204, 205
  X-ray diffraction, 204

V

Vanadia catalyst, 352
Vanadium oxide, 191
Vanadium oxide–magnesium oxide, 197
π-Vinyl alcohol complex, 240–244
  NMR, 243, 244
  X-ray crystallography, 242, 243
π-Vinyl iron complex, 240–244
Vitamin B_{12}
  addition reactions, 259
  catalysis by, 256–260
  deamination, 259
  diol dehydration, 258, 260
Vitamin B_{12a}, 257

W

Wacker process, 238–244
Washcoat, 331–333
Wire, catalytic, 69–73
  multiple steady states, 70, 71

X

X-ray diffraction, 205, 272, 273, 318, see also specific substances
X-ray fluorescence, 132, 317, 318
Xylene isomerization, 108, 109
  model reaction, 113, 114

Y

Yttria
  activity, 31, 32
  extrinsic field effects, 29

Z

Zeolite, see also specific types
  crystalline, 98
  decationated, 100
  faujasitic, 139–166, see also Faujasite
  IR spectra, 111, 112
  mordenite, 166–168, see also Mordenite
  poisoning, 116–118
  structure, 138, 139
  surface acidity, 138–168
    methods for determining, 121
  synthetic, 139
Ziegler–Natta reaction, 235–238
Zinc, in oil, 316, 317
Zinc dialkyldithiophosphate, 347–349
Zinc oxide, 187
  activation energy, 18
  doped, 8, 9
  ESR, 197
  IR, 197
  on silver, 17, 18
Zinc oxide–nickel oxide semiconductor, 18
Zirconia, 332

Contents of Previous Volumes

Volume 1

The Heterogeneity of Catalyst Surfaces for Chemisorption
HUGH S. TAYLOR

Alkylation of Isoparaffins
V. N. IPATIEFF AND LOUIS SCHMERLING

Surface Area Measurements. A New Tool for Studying Contact Catalysts
P. H. EMMETT

The Geometrical Factor in Catalysis
R. H. GRIFFITH

The Fischer-Tropsch and Related Processes for Synthesis of Hydrocarbons by Hydrogenation of Carbon Monoxide
H. H. STORCH

The Catalytic Activation of Hydrogen
D. D. ELEY

Isomerization of Alkanes
HERMAN PINES

The Application of X-Ray Diffraction to the Study of Solid Catalysts
M. H. JELLINEK AND I. FANKUCHEN

Volume 2

The Fundamental Principles of Catalytic Activity
FREDERICK SEITZ

The Mechanism of the Polymerization of Alkenes
LOUIS SCHMERLING AND V. N. IPATIEFF

Early Studies of Multicomponent Catalysts
ALWIN MITTASCH

Catalytic Phenomena Related to Photographic Development
T. H. JAMES

Catalysis and the Adsorption of Hydrogen on Metal Catalysts
OTTO BEECK

Hydrogen Fluoride Catalysis
J. H. SIMONS

Entropy of Adsorption
CHARLES KEMBALL

About the Mechanism of Contact Catalysis
GEORGE-MARIA SCHWAB

Volume 3

Balandin's Contribution to Heterogeneous Catalysis
B. M. W. TRAPNELL

Magnetism and the Structure of Catalytically Active Solids
P. W. SELWOOD

Catalytic Oxidation of Acetylene in Air for Oxygen Manufacture
J. HENRY RUSHTON AND K. A. KRIEGER

The Poisoning of Metallic Catalysts
E. B. MAXTED

Catalytic Cracking of Pure Hydrocarbons
VLADIMIR HAENSEL

Chemical Characteristics and Structure of Cracking Catalysts
A. G. OBLAD, T. H. MILLIKEN, JR., AND G. A. MILLS

Reaction Rates and Selectivity in Catalyst Pores
AHLBORN WHEELER

Nickel Sulfide Catalysts
WILLIAM J. KIRKPATRICK

Volume 4

Chemical Concepts of Catalytic Cracking
R. C. HANSFORD

Decomposition of Hydrogen Peroxide by Catalysts in Homogeneous Aqueous Solution
J. H. BAXENDALE

Structure and Sintering Properties of Cracking Catalysts and Related Materials
HERMAN E. RIES, JR.

Acid-Base Catalysis and Molecular Structure
  R. P. BELL
Theory of Physical Adsorption
  TERRELL L. HILL
The Role of Surface Heterogeneity in Adsorption
  GEORGE D. HALSEY
Twenty-Five Years of Synthesis of Gasoline by Catalytic Conversion of Carbon Monoxide and Hydrogen
  HELMUT PICHLER
The Free Radical Mechanism in the Reactions of Hydrogen Peroxide
  JOSEPH WEISS
The Specific Reactions of Iron in Some Hemoproteins
  PHILIP GEORGE

Some General Aspects of Chemisorption and Catalysis
  TAKAO KWAN
Noble Metal—Synthetic Polymer Catalysts and Studies on the Mechanism of Their Action
  WILLIAM P. DUNWORTH AND F. F. NORD
Interpretation of Measurement in Experimental Catalysis
  P. B. WEISZ AND C. D. PRATER
Commercial Isomerization
  B. L. EVERING
Acidic and Basic Catalysis
  MARTIN KILPATRICK
Industrial Catalytic Cracking
  RODNEY V. SHANKLAND

Volume 5

Latest Developments in Ammonia Synthesis
  ANDERS NIELSEN
Surface Studies with the Vaccum Microbalance: Instrumentation and Low-Temperature Applications
  T. N. RHODIN, JR.
Surface Studies with the Vacuum Microbalance: High-Temperature Reactions
  EARL A. GULBRANSEN
The Hterogeneous Oxidation of Carbon Monoxide
  MORRIS KATZ
Contributions of Russian Scientists to Catalysis
  J. G. TOLPIN, G. S. JOHN, AND E. FIELD
The Elucidation of Reaction Mechanisms by the Method of Intermediates in Quasi-Stationary Concentrations
  J. A. CHRISTIANSEN
Iron Nitrides as Fischer-Tropsch Catalysts
  ROBERT B. ANDERSON
Hydrogenation of Organic Compounds with Synthesis Gas
  MILTON ORCHIN
The Uses of Raney Nickel
  EUGENE LIEBER AND FRED L. MORRITZ

Volume 6

Catalysis and Reaction Kinetics at Liquid Interfaces
  J. T. DAVIES

Volume 7

The Electronic Factor in Heterogeneous Catalysis
  M. MCD. BAKER AND G. I. JENKINS
Chemisorption and Catalysis on Oxide Semiconductors
  G. PARRAVANO AND M. BOUDART
The Compensation Effect in Heterogeneous Catalysis
  E. CREMER
Field Emission Microscopy and Some Applications to Catalysis and Chemisorption
  ROBERT GOMER
Adsorption on Metal Surfaces and Its Bearing on Catalysis
  JOSEPH A. BECKER
The Application of the Theory of Semiconductors to Problems of Heterogeneous Catalysis
  K. HAUFFE
Surface Barrier Effects in Adsorption, Illustrated by Zinc Oxide
  S. ROY MORRISON
Electronic Interaction between Metallic Catalysts and Chemisorbed Molecules
  R. SUHRMANN

Volume 8

Current Problems of Heterogeneous Catalysis
  J. ARVID HEDVALL

Adsorption Phenomena
J. H. DE BOER
Activation of Molecular Hydrogen by Homogeneous Catalysts
S. W. WELLER AND G. A. MILLS
Catalytic Syntheses of Ketones
V. I. KOMAREWSKY AND J. R. COLEY
Polymerization of Olefins from Cracked Gases
EDWIN K. JONES
Coal-Hydrogenation Vapor-Phase Catalysts
E. E. DONATH
The Kinetics of the Cracking of Cumene by Silica-Alumina Catalysts
CHARLES D. PRATER AND RUDOLPH M. LAGO

Surface Potentials and Adsorption Process on Metals
R. V. CULVER AND F. C. TOMPKINS
Gas Reactions of Carbon
P. L. WALKER, JR., FRANK RUSINKO, JR., AND L. G. AUSTIN
The Catalytic Exchange of Hydrocarbons with Deuterium
C. KEMBALL
Immersional Heats and the Nature of Solid Surfaces
J. J. CHESSICK AND A. C. ZETTLEMOYER
The Catalytic Activation of Hydrogen in Homogeneous, Heterogeneous, and Biological Systems
J. HALPERN

Volume 9

Proceedings of the International Congress on Catalysis, Philadelphia, Pennsylvania, 1956

Volume 10

The Infrared Spectra of Adsorbed Molecules
R. P. EISCHENS AND W. A. PLISKIN
The Influence of Crystal Face in Catalysis
ALLAN T. GWATHMEY AND ROBERT E. CUNNINGHAM
The Nature of Active Centres and the Kinetics of Catalytic Dehydrogenation
A. A. BALANDIN
The Structure of the Active Surface of Cholinesterases and the Mechanism of Their Catalytic Action in Ester Hydrolysis
F. BERGMANN
Commercial Alkylation of Paraffins and Aromatics
EDWIN K. JONES
The Reactivity of Oxide Surfaces
E. R. S. WINTER
The Structure and Activity of Metal-on-Silica Catalysts
G. C. A. SCHUIT AND L. L. VAN REIJEN

Volume 11

The Kinetics of the Stereospecific Polymerization of α-Olefins
G. NATTA AND I. PASQUON

Volume 12

The Wave Mechanics of the Surface Bond in Chemisorption
T. B. GRIMLEY
Magnetic Resonance Techniques in Catalytic Research
D. E. O'REILLY
Bare-Catalyzed Reactions of Hydrocarbons
HERMAN PINES AND LUKE A. SCHAAP
The Use of X-Ray and K-Absorption Edges in the Study of Catalytically Active Solids
ROBERT A. VAN NORDSTRAND
The Electron Theory of Catalysis on Semiconductors
TH. WOLKENSTEIN
Molecular Specificity in Physical Adsorption
D. J. C. YATES

Volume 13

Chemisorption and Catalysis on Metallic Oxides
F. S. STONE
Radiation Catalysis
R. COEKELBERGS, A. CRUCQ, AND A. FRENNET
Polyfunctional Heterogeneous Catalysis
PAUL B. WEISZ
A New Electron Diffraction Technique, Potentially Applicable to Research in Catalysis
L. H. GERMER

The Structure and Analysis of Complex Reaction Systems
JAMES WEI AND CHARLES D. PRATER
Catalytic Effect in Isocyanate Reactions
A. FARKAS AND G. A. MILLS

Volume 14

Quantum Conversion in Chloroplasts
MELVIN CALVIN
The Catalytic Decomposition of Formic Acid
P. MARS, J. J. F. SCHOLLEN, AND P. ZWIETERING
Application of Spectrophotometry to the Study of Catalytic Systems
H. P. LEFTIN AND M. C. HOBSON, JR.
Hydrogenation of Pyridines and Quinolines
MORRIS FREIFELDER
Modern Methods in Surface Kinetics: Flash, Desorption, Field Emission Microscopy, and Ultrahigh Vacuum Techniques
GERT EHRLICH
Catalytic Oxidation of Hydrocarbons
L. YA. MARGOLIS

Volume 15

The Atomization of Diatomic Molecules by Metals
D. BRENNAN
The Clean Single-Crystal-Surface Approach to Surface Reactions
N. E. FARNSWORTH
Adsorption Measurements during Surface Catalysis
KENZI TAMARU
The Mechanism of the Hydrogenation of Unsaturated Hydrocarbons on Transition Metal Catalysts
G. C. BOND AND P. B. WELLS
Electronic Spectroscopy of Absorbed Gas Molecules
A. TERENIN
The Catalysis of Isotopic Exchange in Molecular Oxygen
G. K. BORESKOV

Volume 16

The Homogeneous Catalytic Isomerization of Olefins by Transition Metal Complexes
MILTON ORCHIN

The Mechanism of Dehydration of Alcohols over Alumina Catalysts
HERMAN PINES AND JOOST MANASSEN
π Complex Adsorption in Hydrogen Exchange on Group VIII Transition Metal Catalysts
J. L. GARNETT AND W. A. SOLLICH-BAUMGARTNER
Stereochemistry and the Mechanism of Hydrogenation of Unsaturated Hydrocarbons
SAMUEL SIEGEL
Chemical Identification of Surface Groups
H. P. BOEHM

Volume 17

On the Theory of Heterogeneous Catalysis
JURO HORIUTI AND TAKASHI NAKAMURA
Linear Correlations of Substrate Reactivity in Heterogeneous Catalytic Reactions
M. KRAUS
Application of a Temperature-Programmed Desorption Technique to Catalyst Studies
R. J. CVETANOVIC AND Y. AMENOMIYA
Catalytic Oxidation of Olefins
HERVEY H. VOGE AND CHARLES R. ADAMS
The Physical-Chemical Properties of Chromia-Alumina Catalysts
CHARLES P. POOLE, JR. AND D. S. MACIVER
Catalytic Activity and Acidic Property of Solid Metal Sulfates
KOZO TANABE AND TSUNEICHI TAKESHITA
Electrocatalysis
S. SRINIVASEN, H. WROBLOWA, AND J. O'M. BOCKRIS

Volume 18

Stereochemistry and Mechanism of Hydrogenation of Napthalenes in Transition Metal Catalysts and Conformational Analysis of the Products
A. W. WEITKAMP
The Effects of Ionizing Radiation on Solid Catalysts
ELLISON H. TAYLOR
Organic Catalysis over Crystalline Aluminosilicates
P. B. VENUTO AND P. S. LANDIS

On the Transition Metal-Catalyzed Reactions of Norbornadiene and the Concept of π Complex Multicenter Processes
G. N. Schrauzer

Volume 19

Modern State of the Multiplet Theory of Heterogeneous Catalysis
A. A. Balandin

The Polymerization of Olefins by Ziegler Catalysts
M. N. Berger, G. Boocock, and R. N. Hawarr

Dynamic Methods for Characterization of Adsorptive Properties of Solid Catalysts
L. Polinski and L. Naphtali

Enhanced Reactivity at Dislocations in Solids
J. M. Thomas

Volume 20

Chemisorptive and Catalytic Behavior of Chromia
Robert L. Burwell, Jr., Gary L. Haller, Kathleen C. Taylor, and John F. Read

Correlation among Methods of Preparation of Solid Catalysts, Their Structures, and Catalytic Activity
Kiyoshi Morikawa, Takayasu Shirasaki, and Masahide Okada

Catalytic Research on Zeolites
J. Turkevich and Y. Ono

Catalysis by Supported Metals
M. Boudart

Carbon Monoxide Oxidation and Related Reactions on a Highly Divided Nickel Oxide
P. C. Gravelle and S. J. Teichner

Acid-Catalyzed Isomerization of Bicyclic Olefins
Jean Eugene Germain and Michel Blanchard

Molecular Orbital Symmetry Conservation in Transition Metal Catalysis
Frank D. Mango

Catalysis by Electron Donor-Acceptor Complexes
Kenzi Tamaru

Catalysis and Inhibition in Solutions of Synthetic Polymers and in Micellar Solutions
H. Morawetz

Catalytic Activities of Thermal Polyanhydro-α-Amino Acids
Duane L. Rohlfing and Sidney W. Fox

Volume 21

Kinetics of Adsorption and Desorption and the Elovich Equation
C. Aharoni and F. C. Tompkins

Carbon Monoxide Adsorption on the Transition Metals
R. R. Ford

Discovery of Surface Phases by Low Energy Electron Diffraction (LEED)
John W. May

Sorption, Diffusion, and Catalytic Reaction in Zeolites
L. Riekert

Adsorbed Atomic Species as Intermediates in Heterogeneous Catalysis
Carl Wagner

Volume 22

Hydrogenation and Isomerization over Zinc Oxide
R. J. Kokes and A. L. Dent

Chemisorption Complexes and Their Role in Catalytic Reactions on Transition Metals
Z. Knor

Influence of Metal Particle Size in Nickel-on-Aerosil Catalysts on Surface Site Distribution, Catalytic Activity, and Selectivity
R. Van Hardeveld and F. Hartog

Adsorption and Catalysis on Evaporated Alloy Films
R. L. Moss and L. Whalley

Heat-Flow Microcalorimetry and Its Application to Heterogeneous Catalysis
P. C. Gravelle

Electron Spin Resonance in Catalysis
Jack H. Lunsford

Volume 23

Metal Catalyzed Skeletal Reactions of Hydrocarbons
J. R. Anderson

CONTENTS OF PREVIOUS VOLUMES

Specificity in Catalytic Hydrogenolysis by Metals
J. H. SINFELT
The Chemisorption of Benzene
R. B. MOYES AND P. B. WELLS
The Electronic Theory of Photocatalytic Reactions on Semiconductors
TH. WOLKENSTEIN
Cycloamyloses as Catalysts
DAVID W. GRIFFITHS AND MYRON L. BENDER
Pi and Sigma Transition Metal Carbon Compounds as Catalysts for the Polymerization of Vinyl Monomers and Olefins
D. G. H. BALLARD

Volume 24

Kinetics of Coupled Heterogeneous Catalytic Reactions
L. BERÁNEK
Catalysis for Motor Vehicle Emissions
JAMES WEI
The Metathesis of Unsaturated Hydrocarbons Catalyzed by Transition Metal Compounds
J. C. MOL AND J. A. MOULIJN
One-Component Catalysts for Polymerization of Olefins
YU. YERMAKOV AND V. ZAKHAROV
The Economics of Catalytic Processes
J. DEWING AND D. S. DAVIES
Catalytic Reactivity of Hydrogen on Palladium and Nickel Hydride Phases
W. PALCZEWSKA
Laser Raman Spectroscopy and Its Application to the Study of Adsorbed Species
R. P. COONEY, G. CURTHOYS, AND NGUYEN THE TAM
Analysis of Thermal Desorption Data for Adsorption Studies
MILOŠ SMUTEK, SLAVOJ ČERNÝ, AND FRANTIŠEK BUZEK

Volume 25

Application of Molecular Orbital Theory to Catalysis
ROGER C. BAETZOLD
The Stereochemistry of Hydrogenation of α,β-Unsaturated Ketones
ROBERT L. AUGUSTINE
Asymmetric Homogeneous Hydrogenation
J. D. MORRISON, W. F. MASLER, AND M. K. NEUBERG
Stereochemical Approaches to Mechanisms of Hydrocarbon Reactions on Metal Catalysts
J. K. A. CLARKE AND J. J. ROONEY
Specific Poisoning and Characterization of Catalytically Active Oxide Surfaces
HELMUT KNÖZINGER
Metal-Catalyzed Oxidations of Organic Compounds in the Liquid Phase: A Mechanistic Approach
ROGER A. SHELDON AND JAY K. KOCHI

Volume 26

Active Sites in Heterogeneous Catalysis
G. A. SOMORJAI
Surface Composition and Selectivity of Alloy Catalysts
W. M. H. SACHTLER AND R. A. VAN SANTEN
Mössbauer Spectroscopy Applications to Heterogeneous Catalysis
JAMES A. DUMESIC AND HENRIK TOPSØE
Compensation Effect in Heterogeneous Catalysis
A. K. GALWEY
Transition Metal-Catalyzed Reactions of Organic Halides with CO, Olefins, and Acetylenes
R. F. HECK
Manual of Symbols and Terminology for Physicochemical Quantities and Units—Appendix II
Part II: Heterogeneous Catalysis